Rag Paper Manufacture
in the United States,
1801–1900

Rag Paper Manufacture in the United States, 1801–1900

A History, with Directories of Mills and Owners

A J VALENTE

McFarland & Company, Inc., Publishers
Jefferson, North Carolina, and London

LIBRARY OF CONGRESS CATALOGUING-IN-PUBLICATION DATA

Valente, AJ, 1954–
Rag paper manufacture in the United States, 1801–1900 :
a history, with directories of mills and owners / AJ Valente.
    p.   cm.
Includes bibliographical references and index.

ISBN 978-0-7864-5863-9
softcover : 50# alkaline paper ∞

1. Paper industry — United States — History — 19th century. 2. Paper mills — United States — History — 19th century. 3. German Americans — Social life and customs — History — 19th century. 4. Dutch Americans — Social life and customs — History — 19th century. I. Title.
TS1093.U6V35   2010   676'.130973'09034 — dc22   2010037321

British Library cataloguing data are available

© 2010 AJ Valente. All rights reserved

*No part of this book may be reproduced or transmitted in any form
or by any means, electronic or mechanical, including photocopying
or recording, or by any information storage and retrieval system,
without permission in writing from the publisher.*

Front cover: Woodcut engraving of the Gilpin Mill on
Brandywine Creek (circa 1820), site of America's first
paper machine; rag paper © 2010 Shutterstock

Manufactured in the United States of America

*McFarland & Company, Inc., Publishers
Box 611, Jefferson, North Carolina 28640
www.mcfarlandpub.com*

To Frances and Warren

Table of Contents

Acknowledgments	viii
Preface	1

Part One: The Cylinder-Wire Machine

1 — The Rag Market	3
2 — Gilpin vs. Ames	15
3 — Pittsfield Progress	26
4 — The Paper Trade	38
5 — The Growing Empire	60
6 — The Demise of Handmade Paper	76

Part Two: The Moving-Wire Machine

7 — The Machine Manufacturers	89
8 — Ohio and the West	103
9 — North and South	109
10 — Crane vs. Willcox	119
11 — Holyoke, City of Industry	133

Part Three: The Wood Pulp Era

12 — The Curtisville Exponent	153
13 — A Changing Industry	167
14 — West and Northwest	183
Appendix I: Directory of 19th Century Paper Mills	197
Appendix II: Directory of Paper Mill Owners	252
Chapter Notes	295
Bibliography	299
Index	301

Acknowledgments

A great deal of thanks goes to the American Philatelic Society for their unabashed support of the postal history of papermaking exhibit and the U.S. stamp paper project. John Smith, past president of the Oregon Stamp Society, and the Oregon Stamp Society Library have without a doubt been the most precious resource in this instance. Cindy Prescher, past president and founder of the Central Oregon Writers Guild, along with the many volunteers of the group have been most supportive of the effort and really got me on the road to writing. I would also like to thank Cindy Bowden, director of the Robert C. Williams Paper Museum, and staff for all of the help and resources they provided on this project. In addition, Peter Hopkins, curator of the Crane Museum, was very helpful and supplied the bulk of the material found here regarding stamp paper. My sister Frances contributed artwork to the project, and for all her support to little brother over the years, I am greatly appreciative. Lastly, the Oregon Stamp Society, Seattle Collectors Club, American Philatelic Society, U.S. Stamp Society, and the U.S. Philatelic Classics Society have all been excellent resources.

Preface

My interest in the subject goes back about twenty years, beginning with an investigation into sources of paper for classical postage stamps. The work had taken me across the country on several occasions, and as the research grew, so did my interest. I managed to publish a few articles in hobbyist periodicals along the way, but was somewhat stymied by a lack of information on the early machine era. If only someone had written a general history on U.S. paper machines and mills of the nineteenth century, that would make the job so much easier.

The previous paper studies were a purely amateur pursuit before I encountered the Owen & Hurlbut correspondence in 1996. Having close to three hundred documents written in nearly as many hands, it is apparent this investigation was no weekend project. I had been working at the time as a high-tech engineer testing what are called "scalable servers" for database applications used by large banks and insurance companies. Still, I had been a history major before engineering school and somewhere in the back of my mind I recalled a lecture on treatment of source materials and the need to keep an open mind. So, a systematic effort was made to transcribe all the documents before proceeding further. The ensuing research was guided by a recent book on Berkshire County papermaking, and turned out to be all the more valuable, as it coincided exactly with my missing history. Owen & Hurlbut turned out to be industrious paper-makers with a hand in nearly every market, so as I gathered bits and pieces from this and other first-hand accounts, a national perspective slowly emerged.

Conservation was a principal factor in this investigation because understanding what a paper was made of (rag, straw, wood-pulp, etc.) is of paramount importance in preserving our perishable heritage. Long a staple of society, paper changed very little through much of antiquity up until the nineteenth century. Then, in the span of just a hundred years, so many things changed so rapidly it is hard to keep up. Paper history is a tool of conservation because to know your papers is to love them.

Since I have used material from prior articles, permit me to mention the publishers, the *American Philatelist* and the *Postal History Journal.*

The following is a partial list of friends and correspondents to whom I express my thanks and affection, a list that also serves as a record of my contemporaries who have contributed to this discussion. I am most grateful and appreciative to Terry Schaffer, David Snow, Ed Shapiro, Rob Lund, Dave Loving, Rolin Lewis, Lynn Bedsole, John Smith, George Atkins, Keiji Tiara, Bob Hegland, W. Wilson Hulme, Stanley Pillar, Richard Frajola, Diane DeBlois, Ken Lawrence, Harold Rapp, E. Fritz, Roger Rhodes, and Stafford Morse.

Part One

The Cylinder-Wire Machine

1

The Rag Market

The first paper mill in America was erected in 1690 on the Monoshone Creek in Germantown, Pa., by William Rittenhouse. This mill serviced publishing concerns in Philadelphia, but soon after going into production it was wiped out in a spring flood. Publishers and papermakers in Europe had entered a long era of cooperation following the invention of moveable type in 1450, and this tradition transferred over to America; for example, after the flooding of the Rittenhouse Mill, the leading publisher of Philadelphia, William Bradford, offered the papermaker an inexpensive long-term lease on a nearby piece of property situated on higher ground. In this case all the publisher asked for in return was all the mill's output over the next ten years and the right to set prices. While such an agreement seems unthinkable today, these were fairly generous terms considering the uncertainty of the times. A case in point is that Bradford's publication ran afoul of the dominant political party in 1693, and while the publishing house had to move to New York, it footed the cost of freight and continued using paper from the Rittenhouse Mill. Poetry was often used as a form of commentary during the 17th and 18th centuries, and it so happens the following verse was penned about the unusual long-distance relationship between Bradford and the Rittenhouse Mill[1]:

A True Relation of the Flourishing State of Pennsylvania
by John Holme (ca. 1696)

Here dwelt a printer and I find
That he can both print books and bind.

He wants not paper, ink, nor skill
He's owner of a paper mill.

No doubt but he will lay up bags
If he can get good store of rags.

Kind friend, when thy old shift is rent
Let it to the paper mill be sent.

The paper mill is here hard by [near]
And makes good paper frequently.

But the printer, as I do here tell,
Is gone to New York to dwell.

The Rittenhouse mill employed time-honored techniques brought from the old country; in this case the papermaker came from the Ruhr region of Germany. The making of pulp was backbreaking work, as old linen rags were ground by hand using mortar and pestle, a process that took a good worker an entire day to make sufficient pulp for just one-half ream of paper (a ream is 520 sheets of paper). Certain European mills experimented with mechanical hammers or stampers (information on early hand papermaking is available in *Papermaking: The History and Technique of an Ancient Craft* by Dard Hunter).

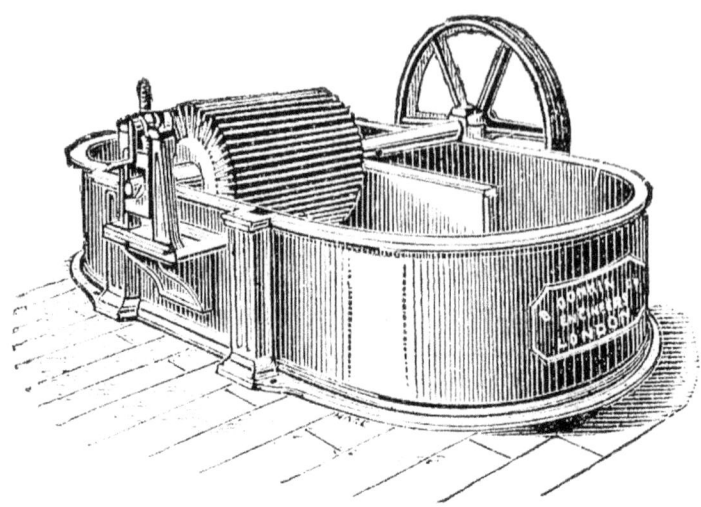

Hollander beater manufactured by Bryan Donkin, London (from advertisement in *Paper Mills of the World*, 1885).

The device that replaced the stamper is called a "hollander" (a.k.a. rag engine or beating engine). This invention is of Dutch origin and uses a set of rotating blades mounted in an oblong wooden tub to beat a solution of ten parts water to one part rags into pulp over a period of about twenty hours. The hollander was originally designed to be powered by windmills; however, German mills using the device around 1710 found the basic waterwheel capable of powering two or more rag engines at a time. The early hollanders did the work of three to four workers, but more importantly, the advancement represented a huge jump in efficiency that allowed mills to put on additional vats.

The hollander first came to America in the mid–18th century, although the exact date of arrival is not known. Sometime around 1755, Jacob Rittenhouse, William Rittenhouse's cousin, made a number of improvements at the mill, including construction of an 18" diameter screw press made of cherrywood. It is possible he also installed a hollander at this time since Jacob reportedly constructed a curved, or slanted, bedplate that helped reduce beating time up to one-third. The rag engine was likely powered by a simple brest wheel that was activated by water falling from a wooden raceway. By the early 1800s a visitor reported the mill employed the more advanced "overshot" waterwheel, which differs from the standard brest wheel in that the blades are folded back against one another, allowing the wheel to turn much faster.[2]

Beach Rag Cutter

Paper had been made by hand for nearly a millennium, and new inventions such as the rag engine came along only rarely. The 19th century, however, would be dramatically different as the industrial revolution took hold. Keeping up with all the changes in the American paper industry would be immensely difficult if not for the work of a contemporary chronicler in Albany, N.Y., named Joel Munsell. Munsell subscribed to a wide variety of publications, and around 1827 he began keeping detailed files on a number of topics including papermaking. He set up a print shop in 1836 and began publishing a series of chronicles, a moneymaking venture. The files on the paper industry first came to light in 1856 under the title of *A Chronology of Paper and Paper-Making*.

Some historians have reason to question the accuracy of certain Munsell's entries, particularly in areas that predate the author. However, Munsell was very accurate when it came to current events, and his records contain details that might otherwise have been lost to history. Confirming such facts, on the other hand, can be challenging. For example, one of Munsell's earliest contemporary entries is the following: "1828. Moses Y. Beach, of Springfield, Mass., afterwards publisher of The Sun newspaper in New York, invented a machine for cutting rags in the manufacture of paper for which he obtained a patent."[3]

As patent office records from this period are unavailable, the above entry is difficult to confirm. Fortunately, Beach sent out a circular commemorating the event, and from this one can begin to explore the depth of changes brought on by industrialization. Unfortunately, not all advancements had the desired effect, and as far as Beach's invention is concerned its impact was marginal at best. Beach's machine prepared the rags for the hollander by cutting them into manageable pieces. Previously, rag preparation had always been done by hand. The basic procedure was for workers to sort the rags according to type and quality into separate baskets. Following this the textiles were cut into strips or squares of 6 to 8 inches, and here's where Beach's invention came into play. There was, however, one hitch, and as Beach himself writes at the bottom of his circular, "I think there can be no machine that takes off hooks and seams."[4] Clasps and stitching were generally thrown away, as they caused lumps or knots in the pulp. So, the cutter was not a panacea. Rags still needed to be prepared by hand.

Moses Y. Beach had something of a colorful history. His career began as an entrepreneur from Springfield who, in 1822, formed a company to build the first stern-wheeled steamship to ply the Connecticut River. There were no engine manufacturers in the United States at this early date, so Beach traveled to Great Britain to bring back the latest marine engine. The steamboat was constructed in a Hartford naval yard, but unfortunately the engine proved too underpowered to overcome the strong river current, and the project was abandoned.

Steamboats were the high-tech enterprise of the day, and Beach's adventures had come to the attention of a paper manufacturer in Springfield by the name of D&J Ames. The brothers David and John Ames were no ordinary papermakers, rather they owned the largest paper mill in the nation as well as being the only paper machine manufacturer in the industry. As knowledge of mechanics and steam plants were highly valuable skills in those days, they enticed Beach to come to work for them. The Ames's South Hadley Mill had twelve rag engines that consumed some 3600 pounds a day and employed some two hundred workers, most engaged in rag preparation. Naturally, any labor-saving device would be welcomed, and to this end Beach was apparently sent to England to investigate the latest mechanical developments.

Beach's rag cutter consisted of a rotating drum with knives strategically mounted around the inside walls. In his circular of June 1828, he reports it was capable of processing 6,000 pounds of rags per day and that "no rights of constructing or vending had yet been given." This seems a rather odd admission given his standing with D&J Ames, so one must assume the purpose of the circular was to drum up support for the invention before selling it.

During his travels in London, Beach must have made the acquaintance of a manufacturer who introduced him to another type of paper machine called the Fourdrinier. This machine was more sophisticated than the one made by D&J Ames, and it is named after the Fourdrinier Brothers of London, who held the patent. Upon returning to the states, Beach formed a new company called Beach, Hommerken & Kerney, and they proceeded to import the first Fourdrinier machine in the country, installing it at a mill in Saugerties, N.Y., formerly owned by Henry Barclay. The Fourdrinier was put into operation in October of 1827 under the supervision of Peter Adams, later of the firm of Adams and Bishop (see Appendix 1 for a compilation of 19th century mills and paper companies). The following year the same installed a second Fourdrinier machine, this built by a firm in Bury, England. About this time Beach acquired the *New York Sun*, and with a two-machine mill at his back, he went on to build the *Sun* into the leading publication of its time. The Saugerties Mill was later sold to J.B. Sheffield & Son, who owned the other two paper mills in town, and the original Fourdriniers continued in operation at the same location until the plant was destroyed by fire in 1872.[5]

Rag Merchants

During the 1700s and early 1800s American manufacturers had something of a dilemma in that before starting a paper mill one had to establish a system of rag collection. The earliest paper mills in America solved the problem by locating close to major population centers as Philadelphia, Boston, and New York. In 1769 a notice in a Boston paper read, "The bell car will go through town about the end of each month to collect rags." Another tactic employed by a Boston-area mill around 1799 was a watermark reading "SAVE YOUR RAGS." It is also said that Long Island, at one point, was home to twenty hand mills. During the Revolutionary War papermaking became a matter of strategic importance, and Committees of Safety in Philadelphia and elsewhere took to organizing rag collection.[6]

The publishing industry was also confined to major towns at first, but eventually newspapers spread to Connecticut, Rhode Island, New Jersey, Delaware, and South Carolina. Here too were opportunities for papermaking, provided enough rags could be found nearby. A pioneering mill in Troy, N.Y., supplied the local farming news, and an editorial comment in support of the collecting effort in 1802 reads: "The ladies in several of the large towns display an elegant work bag as part of the furniture of their parlors into which every rag that is saved for the paper mill is carefully preserved. Were this example imitated more, this state would not be drained of its circulating cash for paper.... The ... saving of rags, will [also help] procure paper and books for schools and farm use."[7]

The far western reaches of Massachusetts had a flourishing agricultural community and a ready availability of waterpower. A young papermaking entrepreneur from Boston came here around 1800, and happened by chance to find a pure artesian spring on a lot with waterpower. Hard water was responsible for much of the brownish tint found in papers of the era, and pure spring water made for a clearer, whiter sheet that commanded nearly twice the price. Still, the entrepreneur had to organize a source of rags before establishing a business. Berkshire County's population was about 33,885 — evidently enough to sustain a venture. The young papermaker subsequently placed an ad in the local farming paper, the *Pittsfield Sun*, reading: "Americans! Encourage your own manufactories and they will improve. Ladies, save your rags! As the subscribers have it in contemplation to erect a paper mill in Dalton the ensuing spring, and the business being very beneficial to the community at large, they flatter themselves that they shall meet with due encouragement. And that every woman who has the good of her country and the interests of her own family at heart, will patronize them by saving their rags, and sending them to their manufactory, or to the nearest storekeeper for which the subscribers will give a generous price."[8]

The rag collection effort in Berkshire County now came to be centered on a depot established in the crossroads community of West Stockbridge. Collecting bins went up in stores and taverns around every small village and hamlet, and rag routes were established. Every fortnight a designated teamster traveled the county stopping in turn at each collecting site. The effort proved so successful that two additional paper mills were established by 1810. At some point the rag route was divided up into franchise areas, and the original mill in Dalton along with a neighboring mill divided Berkshire County into northern and southern districts. A third mill in South Lee held sway over the larger communities of Pittsfield and Great Barrington. Bits and pieces of the collection effort can be found in letters to the South Lee Mill. One letter of 1837 from J&E Peck's sheet iron business in Pittsfield invites the papermakers to come and take away several thousand pounds of white rags. Similarly, in December of 1843 the mill received a letter from B.W. Patterson's store on Main Street in Great Barrington reporting that collections had closed for the season and over three tons of rags were now available for pickup.[9]

Rags came into high demand as the number of paper mills in America grew. To meet their needs mill owners oftentimes accepted in-kind payments, and one such arrangement was made between the South Lee Mill and a merchant in Troy, N.Y., by the name of William Parker. Since this was not a local collecting effort a good deal of the discussion occurred in letter form. From the start Parker kept a single bin where he placed all of his rags both linen and cotton, and when the papermaker later pointed out that white linen rags demanded twice the price, Parker exclaimed, "I had no idea of any difference between the top and bottom of the heap!" There were a lot of rags coming down the Erie Canal on flatboats and with the paper mill dictating prices the shopkeeper later expressed concern about the scheme, writing: "It is highly probable if we continue to buy [at] that [price] our rags will run ... much like that you have received [in the past]. [And,] if you ... lower the price ... here, we run the risk of loosing them (maybe all)."[10]

Papermakers also facilitated rag collection by distributing specially marked burlap sacks. In September of 1840, Warren W. Lyon of Saugerties, N.Y., writes the South Lee Mill to say: "I am now about ready to send you the rags ... would you send me some sacks by your teamster ... I want sacks enough to hold about 3000 pounds ... have them left at Harts Tavern, Hudson [N.Y.] ... your man says he generally stops at Harts on Tuesday and Friday nights." Bronson & Crocker of Oswego, N.Y., reports the collection of thirty-two sacks for the South Lee Mill the year prior. In 1841 Boyston & Whitcomb of Templeton, Mass., sent thirty-two sacks of white rags to Pittsfield to be picked up on the regular route. Nor did the owners themselves bother to bring rags all the way back to the mill, as in the following year one of the partners in the South Lee Mill sent a note informing that he'd left thirty-eight sacks of rags at the West Stockbridge depot. One of the more detailed accounts of this kind of activity comes in an 1838 letter of Dickensen & Curtis, merchants of Troy, who, in this case report not only the number of sacks and types of rags (white rags are linen, brown rags are cotton), but the name of the collector too[11]:

    July 17   5 sacks white rags by Stephen Bradley
                  19 sacks brown rags by Stephen Bradley
                  19 sacks brown rags by E. Starks
                  9 sacks brown rags by E. Starks
    Aug. 9   6 sacks white rags by Nelson Burr
                  3 sacks brown rags by Nelson Burr
    Aug. 17 17 sacks brown rags by T.G. Frisbee
                  6 sacks white rags by T.G. Frisbee
    Aug. 27 40 sacks brown rags by T.G. Frisbee

Over time as the number of mills in Berkshire County continued to grow, papermakers increasingly turned to regional sources of rags. This wide open market is where prices really firmed up. In writing the South Lee Mill in 1836, Don Lovejoy, a merchant in New Canaan, Mass., offered a wagonload of brown rags at a price established by a neighboring mill. Local merchants discovered they could obtain better prices by searching for the mill with the greatest need. For example, in 1842 William Austin of Albany, N.Y., wrote the So. Lee Mill about an accumulation of five tons of rags that he offered on terms of six months at 2.8 cents per pound. Upstate New York was prime territory for rag collection, and in February of 1838, D.S. Wendell writes from Troy, N.Y., with an enormous offer of sixteen tons of rags, likely collected from various boats plying the Erie Canal. For an additional fee some enterprising merchants teamed their goods directly to the mill, such as the offer made by William J. Benedict of Schenectady, N.Y., to the South Lee Mill in 1831. As a last resort, the merchant could always send the goods into New York City, where demand was always high. For example, a merchant of New Canaan writes the South Lee Mill in 1836 offering two and one-half tons of No.1 linen rags at "market" prices lest they go to New York.[12]

Rural paper mills seemed to be better positioned to tap into regional sources following the coming of the railroad. In 1845 Jonathan Bigelow offered the South Lee Mill a windfall of four tons of brown rags and two tons of white rags for pickup at the railroad depot in Springfield, some thirty miles away. Mills thus became inherently less dependent on local rag collection so long as they were situated on or near the rail lines. About this time specialty warehouses in major cities took the lead in the collection and distribution of rags. For example, in 1844 the Philadelphia warehouse of Charles Magarga sent 275 pounds of fine linen rags by railroad to the South Lee Mill. The articles were still in their collection sacks, each sack listed by the number of rags it contained (e.g., #1— 504, #2 — 372, ...), while a further twenty cents deposit was charged for each sack.[13]

The New York Exchange

The U.S. census of 1810 listed 200 mills with a combined annual consumption of 3500 tons of rags (about 18 tons per). It should be noted the number of mills reported in the appendices is fewer than this number, but this is simply the number of physical sites and not the total number of plants as counted in the census. As prices for rags went on the rise, imports from England and the Continent came into high demand. The Jefferson administration foresaw the needs of the industry in 1804 by eliminating the duty on rags. Then, in 1807, the government issued bonds for the construction of the Market Exchange Building in New York. The exchange would become the center of the burgeoning commodities market, and among other things it had a desk for the buying and selling of rags. But, hope for the future was soon dashed by the onset of the Napoleonic Wars and the War of 1812. Following the end of hostilities the U.S. State Department strove to re-establish U.S. trade, including the acquisition of rags. Their efforts proved immensely fruitful, as U.S. merchant marine vessels began conveying large quantities of rags from nearly every port of call. By 1825 U.S. rag imports reached $200,000 a year (about 1450 tons), amounting to forty percent of consumption in pre-war days.

The New York Exchange played an ever-increasing role in the development of a spot market and became the place where mills could always find what they were looking for. The Market Exchange Building was where contracts were bought and sold, and in addition to imported rags, the exchange also dealt in domestic offerings. Merchants simply barged their loads down the Hudson or East River, and the commodities typically sold on the exchange without ever leaving the dockside warehouse. Lots were swiftly transferred by schooner to virtually any port in the northeast, subject to weather conditions of course. In November of 1837 Samuel Vilas of Troy, N.Y., had several tons of white rags collected in Plattsburgh, Clinton County, N.Y., and out of concern that winter navigation on the Hudson was rapidly coming to a close, he offered them on generous terms to the South Lee Mill for immediate pickup.[14]

The price of overseas rags, though, was generally 10 to 30 percent higher. Still, imports remained in demand because shipments often contained a high percentage of linen. Flax is

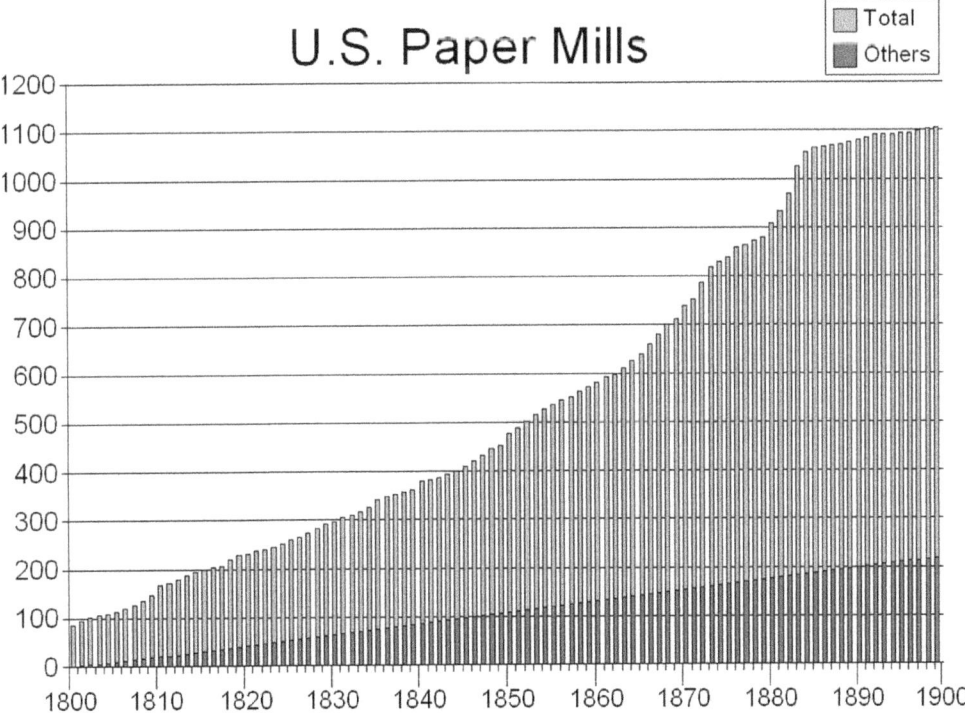

thicker and stronger than cotton, and so was preferred in the manufacture of manuscript, ledger, and fine writing paper where the hardness, or "rattle," of the paper gave it that most enduring quality. As linen grew scarce it was simply diluted with cotton, and the hybrid became ever more popular than the original, especially on price. One of the largest manufacturers in Massachusetts (Tileston & Hollingsworth) was renowned for their high quality papers. Some of their correspondence shows that in 1846 they obtained twenty bags of linen rags for $184.56, or about 9.6 cents per pound not including fifty-seven cents for shipping via railroad from Lowell, Mass. About this time the same bought common rags from a merchant in Providence, R.I., paying just five and three-quarter cents per pound.[15]

The Market Exchange Building was located in the heart of the financial district at the corner of Wall and Front streets. Unfortunately the facility was destroyed in a great fire in December of 1835 that consumed some fifty acres of real estate in the business district of Manhattan. The loss sent traders scrambling to make do any way they could. In 1836 the South Lee Mill must have taken delivery of a shipload of imported rags, and having more material than they could maintain in storage, they sought to sell several lots in New York. As the exchange was closed the lots were offered to the firm of Wilson & Steel. The warehouse, perhaps seeking to profit from the situation, offered no more than the price of the domestic article. Quickly understanding this had become the prevailing attitude, the papermaker turned to a broker of London rags named Able Burrit. Burrit wrote the manufacturer in October to say:

> Gents, In relation to the two lots London rags, I am fearful the estimate by Messrs. Wilson & Steel is too low. Forty dollar[s] a ton is a large discount. I presume they must be full equal in value to the kind I have sent to [the New York warehouse of] Messrs. Ivar Sturge & Co., and Church & Baptiste & Co. ... perhaps even better than those that would promptly fetch $1/4$ or $1/2$ or more. Still, there may be the difference of two cents or less between the sample bale & the remaining two lots. Be this as it may, I want you to gain the worth of them according to the market. We will endeavor to manage the affair right when we next meet. I will thank you to save the sample bale, with two or three bales of the lots. It's probable I shall want them brought to the city for inspection here, as the consignee will like to see them.

After trading on behalf of the client for a year, the broker submitted a ledger copy revealing the extent of the activity. The South Lee account sold $2,000 of imported rags and had six other sales of domestic rags averaging about $200 each. Payment in each case was advanced by six-month notes from named accounts. There was, however, one exception — a special discounted shipment of rags was extended to a neighboring Berkshire manufacturer (John Ingersoll & Co.). On the flip side of the ledger the South Lee Mill generated $2664 in paper sales. The merchant's exchange was re-established the following year at a building near Battery Park, so this brief ledger gives us a rare glimpse of kind of activity handled by the commodities trading desk.[16]

Albert Gallatin was one of the greatest economists of the 19th century, and still holds the record for the longest-serving cabinet member in U.S. history. Gallatin was a free trader who only sought to tax American producers the same as their overseas competition. He reasoned the free flow of goods would give American consumers more choices. In the case of the paper industry, the overseas competition came largely from Great Britain. By 1835 the island nation had an estimated 750 paper mills with annual consumption of 62,500 tons of rags (about 83 tons per). Rag exports to the U.S. were about 5,555 tons, and between

1835 and 1855 rag consumption in England fairly doubled to 125,620 tons, while exports remained about the same. By 1848 British manufacturers were obliged to import some 8,000 tons of rags a year to make up for shortages, and a group of the most influential businesses then appealed to Parliament blaming the current rag shortage on undue foreign influence. The Americans, it was said, were engaged in every foreign port, buying up all the best rags no matter the price. Parliament reached a compromise in 1855 to end the duty on imported rags in exchange for higher taxes on rag exports. The following year British rag imports jumped to 15,000 tons, including some 11,000 tons from Tuscany that had formerly gone to the U.S.A. In the meantime, London shipments to the U.S. dropped to 160 tons of mainly low-grade articles.

In the years before government intervention the London market was a cornucopia of goods and services. Correspondence sent to the aforementioned manufacturer in Boston reveals the New York paper warehouse of Burnap & Babcock provided twenty-one bales of SPFF* rags in 1845 valued at six cents per pound. This lot is said to be of the same quality as thirty-five bales sent the previous year. In 1849 the same discussed an adjustment to one bale that short-weighed 628 pounds. The Boston paper manufacturer evidently traded directly with the London brokerage house of P. Ferguson & Co., but this arrangement was not wholly without risk, as in September of 1850 the broker reported that rags from a certain London warehouse were supposedly "defilth attended," but upon arrival in Boston were deemed totally unfit for use. Nevertheless, the U.S. manufacturer sent a further order for fifty bales to be put on board first ship to Boston. The London broker now reports that Philadelphia was also in great need of rags and so the next two ships from London would sail to that port. It evidently would be a month or more before any new deliveries could be made to Boston, and even those had to go by New York. The London broker did make mention of rags from Hamburg and Slovakia. Continental imports were especially desirable, as they came free of London customhouse duty. Messr. DeBuckemeyer, a Hamburg merchant, was said to be accompanying the cargo. The quote for the shipment was £67 14s for nine tons of rags (£24/6 for SPFF, £23/4 for SPF, and £20/4 for FF). Miscellaneous charges were £3 13s interest on a six-month note, $3.13 insurance, and an eleven pence broker commission.[17]

Once again this is only a brief glimpse of the trading activity in London, but working from numbers provided by Munsell, one can begin to reconstruct the extent of the American trade. In 1838 total rag imports to the U.S. was $465,448, over twice as much as usual. Egypt was one of the more exotic sources, having sent some twenty-three tons of high quality linen wrappings allegedly removed from ancient mummies. The firm of J. Priestly & Co. reported one such shipment of 1215 bales of what were said to be mummy wrappings valued at four cents a pound. By 1846 U.S. rag imports had dropped to a low of 614 tons. Activity then slowly picked up and by 1850 some nineteen different countries exported a total of 1,293 tons to the U.S. The growth of trade around the world at this time is palpable and by 1853 thirty-two different countries sent rags to the U.S. Certain Mediterranean ports such as Alexandria and Smyrna, which normally sent their rags to Constantinople (now Istanbul), agreed to shipments to the U.S. after their obligations to the Ottomans were

---

SPFF stood for special fine first class (mostly white linens), SPF rags were special first class (partly white linen), and FF rags were fine first class (white non-linen).

fulfilled. A similar accommodation was made with Trieste whereby surplus rags collected in Hungary would go to New York or Boston. In the Netherlands papermakers were largely dependent on rags from Belgium, while Belgium, in turn, imported fourteen tons annually from France and the German States. French imports were one hundred tons in 1849 growing to six hundred tons by 1853. Hoping to stem the shortage in their country, one hundred and sixty Dutch papermakers preferred a petition to The Hague in 1854 to place a tax on rag exports. In 1857 France simply banned all rag exports, an example that would be followed shortly thereafter on the Spanish Peninsula. The German States had the largest paper industry in the world with 800 mills in 1850, but composed of mostly hand mills, as annual rag consumption was just ten million kilos (about 13.7 tons per). This may explain how Hamburg and Bremen managed to send 2,000 bales of rags to the U.S. in 1857. These, it seems, would be the last, as both Prussia and German States clapped tight restrictions on rag exports later that year.[18]

As the United States began to encounter significant shortages of linen rags the Italian States became an increasingly important source. The independent provinces and municipalities of Italy had resisted the trend toward cotton textiles in favor of the traditional garb, and here the local economies continued to support the centuries old tradition of growing of flax and weaving their own garments. The most significant port of trade was Livorno (a.k.a. Leghorn), then the capital of the Grand Duchy of Tuscany. Livorno was a suffragan of Pisa, and also called "Little Venice" because it was situated on low, marshy ground interconnected by a network of canals, including a grand canal to Pisa. Livorno had two ports, the old, or Medici, port, and a new facility constructed in 1854 to support increased trade with America. Italian rags were a hot commodity on the New York market. An 1845 correspondence between the New York paper warehouse of Moore & Leggett and the South Lee Mill reveals a consignment of one hundred bales of No. 1 Leghorn rags. The South Lee Mill made a similar offer in 1852 with Babcock, Duly & Hall of New York, this for fifty-eight bales of No. 1 Leghorn rags at eight cents per pound. The protectionist trend eventually caught hold in Rome, which in 1857 banned all rag exports from the Papal states. Other Italian states continued to export rags until 1865.[19]

The difficulty in obtaining sufficient rag stocks caused British paper manufacturers to begin importing large quantities of inexpensive esparto grass from Egypt and North Africa, while Germany largely switched to the manufacture of ground wood paper. As a result the London rag market made something of a comeback in the 1870s, but by this time American producers had largely turned to straw, manila, and wood-pulp production. A number of traditional rag mills in the U.S. such as the one in South Lee continued to resist the trend towards alternative fibers right up until 1893, when paper prices went into freefall as a result of massive overproduction.

2

Gilpin vs. Ames

Paper machines were produced on a small scale in England from 1800 to 1810. Their initial success generated considerable interest on the Continent, and over 1810 to 1830 Great Britain exported machines to Sweden, France, Holland, Belgium, Prussia, and Russia. None, however, were destined for use in British colonies, or former colonies for that matter; to maintain the balance of trade Parliament was against their exportation. This policy put the United States in something of a predicament, as it would be forced to wait on the sidelines while the industrialized nations of the world forged ahead with the latest innovations. Still, a restless few in America would not be denied, and entrepreneurs used whatever degree of stealth and guile was necessary to achieve their goals. The United States was a largely agrarian nation, so its entire manufacturing base had to be built from scratch. The only bright spot was a system of patent law passed down from the English judicial system. So long as the courts gave protection to investments in new technology, Americans would pursue new innovations with ruthless efficiency.

Since Revolutionary War times American publishers had written glowingly of the domestic paper industry, and certain papermaking families rode this wave of patriotism to prominence. During the war Stephen Crane of Boston began his career as an apprentice at a Boston area mill where he attended to shipments of currency paper to the Commonwealth of Massachusetts. In Pennsylvania, the State Assembly and Continental Congress purchased bond and banknote paper from the Willcox Family mill situated on the outskirts of Philadelphia. In the early nineteenth century the most distinguished papermaking family would be the Gilpins of Wilmington, Delaware, who had introduced the nation's first paper machine and became the toast of Philadelphia society.

In 1787 a group of Quakers erected a new paper mill on the Brandywine River two miles north of Wilmington. The Gilpin and Fisher families had been wealthy and influential merchants in Philadelphia, but Quaker sensibilities prevented them from taking sides in war, so the families sold their businesses and moved to Virginia for the duration. Afterward, Mires Fisher returned to Philadelphia to take up law, subsequently becoming successful in both city affairs and the state legislature. The new paper mill was something of a diversion, allowing him to help family friends while rounding out his credentials in the community. For Fisher's brothers-in-law, Joshua and Thomas Gilpin, however, papermaking became a life-long vocation.

For assistance in the venture Mires Fisher approached Benjamin Franklin, who loaned him several volumes on French papermaking. Thereafter the Gilpins began making bond and banknote paper of the finest quality, servicing the needs of the banks of Maryland, North Carolina, Rhode Island, and the United States Treasury. Franklin must have informed Mires Fisher of the latest European innovation called the wove mould. This style was a bit more expensive than the traditional laid mould, but was highly preferred for the manufacture of book paper. The Gilpins had been buying hand moulds from Nathan Sellers, the celebrated wire-worker of Philadelphia, and in 1789 Sellers made them a wove mould, not the first to be used in this country, but certainly the first made in America.[1]

In 1797 the French statesman Brissot de Warville made a tour of America and happened to meet Mires Fisher in New York. The Brandywine Mill still made pulp by hand, and hoping to import a hollander from France, Fisher invited the diplomat to stop at the Brandywine Mill while in Philadelphia. De Warville took up the invitation, writing in his diary: "This town [Wilmington] is famous for its fine mills; the most considerable of which is a mill belonging to Mr. Gilpin and Mires Fisher, that worthy orator and man of science I have often mentioned. Their process in making paper, especially in grinding the rags, is much more simple than ours. I have seen specimens of their paper, both for writing and printing, equal to the finest made in France."[2]

In 1795 Joshua Gilpin went on a grand tour of Europe and studied the latest papermaking techniques. He traveled extensively for five years, setting everything down in a journal that grew to sixty-two volumes. The Brandywine Mill happened to be struck by fire shortly after the turn of the century, so the Gilpins' efforts now turned to rebuilding and modernizing. They hired Thomas Oakes, an English mechanic and engineer, who came over in 1808 to build hollanders and upgrade the mill's waterpower. The Gilpins so admired the rag engines they had him build three additional. Following its reopening the Brandywine Mill employed some forty-four people and sold paper worth $40,000 a year. Things were now going so well that Joshua decided on a second trip to London in 1811, this time taking his family along.[3]

On his earlier tour Gilpin discovered liquid chlorine used for the first time at a textile mill in Edinburgh. The so-called "bleaching liquor" was distilled from out of a large lead retort (kettle) filled with three pails of water, sixty six pounds of common salt, sixty five pounds of manganese, and 119 pounds of vitriol (oil). The retort sat atop a high temperature (cupola) iron furnace and cooked for several hours, all the while releasing deadly chlorine gas. Since the gas also contained a touch of hydrochloric acid, it was drawn off through a pipe and allowed to bubble just beneath the surface of a trough of water. The fumes then collected within a condenser where they ran off in liquid form. As Gilpin continued his journey to Glasgow he encountered a local muslin manufacturer who had invented the process. The inventor told the American that lime, alkalis, or "leaks" would be more cost effective for papermaking; however, Gilpin soon found bleaching liquor already making the rounds at paper mills in Yorkshire and Kent.[4]

Gilpin went to meet with William Stidolph of New Castle, whom he'd learned held the U.S. patent on the chlorine distillation process. Stidolph charged five hundred pounds to establish the process, and when all was agreed he sent an experienced worker to Wilmington to install the system. Laurence Greatrake was a master papermaker who'd worked at the Apsley Mill outside London since 1781. By July of 1804 Greatrake completed construction of a new bleaching room at the Brandywine Mill. The facility contained two bleaching units and three bleaching trays. The retorts were encased in two twelve foot long sand baths, fired from outside the building with hot flues running under the length of each. The bleaching liquor was stored beneath the building in cisterns, and from here pumped into trays filled with rags. The facility could bleach up to six hundred pounds of rags a day at a cost of about $5.25. At less than a penny a pound, the process paid excellent dividends, allowing secondary grade rags (cotton) to be made into fine white paper.[5]

Now back in London, Joshua Gilpin meet with the leading stationers of the city, Henry and Sealy Fourdrinier. The Fourdrinier brothers achieved celebrity status after purchasing rights to the original paper machine from France. This machine employed a circulating wire screen (the endless wire) that imitated the motions of the vat man. The Fourdriniers contracted with a brilliant engineer by the name of Bryan Donkin, who turned the invention into a fine piece of machinery. Two Fourdrinier machines had been erected at a mill near London, and by this time Donkin had already built nearly two dozen other machines. Gilpin would learn the paper machine couldn't be exported without government permission, and to date the only models let out of the country were in Russia, which recently obtained most favored trading status following a state visit by the czar. It seems that Gilpin did manage to obtain a drawing or a plan of the machine, which he immediately forwarded to Wilmington. Thomas Gilpin was the more mechanically inclined of the two, and while he declined to build this kind of machine, he continued to urge his brother on.[6]

Dickinson, Greatrake, Gilpin, and the Cylinder-Wire Machine

An excellent reference for the antique paper machine is *The Paper-making Machine* by Robert H. Clapperton of Oxford University, England. The simplest and most elegant design

Dickinson cylinder-wire machine of 1830. The cylinder is at left. A series of four tandem dryers appears at right. A roll of newly made paper appears in right background (from Evans, *The Endless Web 1804–1954*).

often proved the best solution, and Clapperton nicely annotates the physics behind each advancement. That the basic design of the paper machine persists to this day is testimony to these early successes.

John Dickinson was another London stationer who, among other things, ran on contracts to the British East India Company. In 1804 Dickinson had been using rolls of machine-made paper made by the Fourdriniers on their new machines, but since duty only applied to paper by the sheet, Dickinson contracted with a local mechanic to build a roll paper cutter. It wasn't long after that Dickinson, being mechanically inclined himself, began tinkering with an idea to manufacture paper by the sheet. The concept was fairly well known at the time: a brass barrel turns in a wooden vat filled with rag pulp. The barrel is perforated and covered over in layers of brass screening. A force pump evacuates water from inside through a pipe in the journal bearing. An oval-shaped pickup device inside the barrel literally vacuums the pulp onto the screen. Atop this, a heavy felt roller weighs down on the loose pulp as it emerges from the vat. From there the wet paper traverses onto a waiting conveyor belt, where it is pressed nearly dry. Dickinson's idea was to divide the outside screen into segments so the paper came off in sheets. Dard Hunter calls this kind of cylinder-wire paper "imitation handmade" since it looks exactly like deckled paper except the edges are a bit cleaner and better defined.[7]

Once Dickinson secured a patent on the cylinder-wire machine in 1809 he partnered with a London publishing house to buy an old paper mill in Apsley, a village about 20 miles north of London. Here he constructed his paper machine in a fair amount of secrecy. It seems while the machine made paper of the finest quality, the original version ran a bit slowly, so Dickinson would soon be working on improvements. While residing at Apsley, Dickinson made the acquaintance of Ann Grover, the daughter of a solicitor and banker in the neighboring town of Hemel Hempstead. The two were married in 1810, and as luck would have it, Ann kept a diary that recorded many a historic moment at the mill. Lack of the prescribed waterpower at the mill led to lawsuits, but by 1815 Dickinson was resigned to putting in a steam engine to power all his machinery in times of low flow. Now taking steam from the boiler, Dickinson began experimenting with drying paper over steam-filled rollers or cylinders, a mechanism that he first patented in 1817. In 1818 Dickinson purchased the Nash Mill, situated just a half mile from Apsley, but with considerably more waterpower. Here were installed two new paper machines, and Dickinson began branching off into just about every area of papermaking there was, from musket ball paper to early envelopes and everything in between.[8]

Joshua Gilpin had naturally heard of Dickinson's accomplishments and so desired to pay a visit to Apsley. By this time he was well schooled in the art of persuasion, and in a letter of introduction he gave every assurance he had no intention of acquiring any of Dickinson's inventions, plans, or workers. The normally secretive Dickinson granted the American's request with the understanding he would not be giving anything of any real importance away. And it's true, Gilpin came away with his usual bundle of notes, but his brother could make nothing out of them. Still, after witnessing the machine's performance, Gilpin was in a much better position to assess its potential.

Gilpin left London soon thereafter, and upon arriving home he had a change in mind and so wrote to Laurence Greatrake: "You know that my brother and self are concerned here in the paper manufacture and it has been so admirably improved by him [Dickinson], in my business as to constitute an excellent concern ... to which it is exceedingly important for us to possess, as my brother has begun works which must be suspended till we have them [the plans], and when I assure you that the possession of them, if they do not actually make a fortune for my family ... I am sure you will serve them and me in the business."[9] Greatrake had helped in the rebuilding of the Apsley Mill alongside John Dickinson, and seemed all too willing to return to the United States. Of course everyone knew that Dickinson would never sell the plans, but Joshua Gilpin seems to have found another source. In his travels he befriended J. Wyatt, editor of the massive sixteen volume *London Repertory of Arts, Manufacturers, and Agriculture,* which contained innumerable drawings and specifications obtained through his brother, who was a clerk in the patent office. With the plans in hand things came together rather quickly. Thomas Gilpin built a machine in their workshop while Greatrake supervised construction of the mill. Thomas Gilpin filed the U.S. patent in 1817, and when Dickinson found out he bitterly denounced the Gilpins for, as he saw it, buying Greatrake to gain his invention.[10]

The Gilpins set up their cylinder-wire to make paper by the roll. An article in Wilmington's *American Watchman* in November of 1817 reports: "The [machine] process ... delivers a sheet ... in one continued unbroken succession of fine or coarse materials regulated at pleasure to a greater or lesser thickness. The paper when made is collected from the machine on reels, in succession as they are filled, and these are removed to the further progress of

Gilpins' Mill, 1817, first machine mill in America. A paddle wheel turns in a raceway. A figure stands on a bridge over the tail-race with the Brandywine Creek in foreground.

the manufacture." Essentially the spools were carried up to the loft for cutting and drying in the traditional manner. The machine was soon put to work making newsprint for *Poulson's Daily Advertiser* and book paper for Lavoisne's *Complete Genealogical, Historical, Chronological, and Geographical Atlas*. A further testimonial was given the following year by Matthew Carey and Son of Philadelphia, who say, "We have used the copper-plate paper made on the Brandywine paper machine for our atlas, and cheerfully declare that it is equal to any paper we have ever used, foreign or domestic." Fortune now smiled on the Gilpins, and in celebration they presented a one thousand foot long roll of paper at a meeting of the American Philosophical Society in Philadelphia. By 1820 annual income of the Brandywine Mill had grown to $50,000, a twenty percent increase over prior years.[11]

The Gilpins suffered a serious setback in 1822, as reported in the *Eminent Philadelphian*: "A flood of unprecedented violence in the Brandywine carried away the extensive paper mill of the Messrs. Gilpin, although the building in which their costly machinery was placed, had been erected, it was thought, beyond all possibility of danger from such a cause.... The flood rose to the top of the building. For two days the whirling torrent swept along with fearful turbulence, and when the water at length subsided, the edifice itself was a mass of ruins." What was not reported is that with help from Philadelphia banks rebuilding began almost immediately and the mill was back in operation by 1824.

The South Hadley Mill

All of the talk and testimonial over the Gilpins' new machine created a degree of excitement, but the Gilpins never sought to profit from the patent, and appeared satisfied with keeping the invention bottled up for as long as possible. To this end certain measures were taken to secure the mill and discourage onlookers, and such efforts proved largely successful, as it was nearly five years before another machine appeared on the scene.

Springfield was a small town that straddled two important military roads in central Massachusetts connecting Boston to the western reaches of the state. Shortly after the Revolutionary War the federal arsenal established here had become the object of Shay's Rebellion. In 1794 President Washington assigned Colonel David Ames to the post of superintendent. After eight years Ames retired from the post and bought a local paper mill on the Connecticut River. Erected in 1800, the Springfield Mill was a two vat and two rag engine affair powered by a basic waterwheel.[12]

Colonel Ames specialized in ledger and writing paper for the county court. He also had two sons, David Jr. and John, who worked at the mill from early ages. After Ames retired in 1820 his sons took over operations at the mill, and increased the capacity by doubling the number of rag engines and installing the more powerful overshot waterwheel. John Ames elected to oversee the improvements, and the project evidently put him in touch with a machine manufacturer in Great Britain. The Fourdrinier patent had run its course by 1822, and a number of machine manufacturers now grew up in the trade. Ames was naturally drawn to the cylinder-wire machine, and somewhere along the way obtained plans for the Dickinson machine.

John Ames set up a working model of the machine in his Springfield workshop and recorded a new U.S. patent. The brothers then began searching for a site to erect a new mill, and following recommendations from successful papermakers in Windsor Locks, Conn., they chose a tiny settlement on the Connecticut River called Canal Village. The village was located less than ten miles from Springfield and had grown up around a system of locks and canals that assisted navigation around Hadley Falls. This location possessed ready transportation to Hartford and New York, and would prove the ideal location for a machine mill. Construction was completed in 1824, and sporting twelve rag engines and two paper machines, the South Hadley Mill was by far the largest mill in the county. The plant quickly lived up to expectations, handling the first big order of several tons of book paper for the first edition of *Webster's New Quarto Dictionary* with ease. As business grew the number of rag engines was expanded to sixteen, capable of processing up to three tons of pulp daily. To supplement the demand for rags, D&J Ames also solicited paper shavings from book binderies in which it did business. Clean shavings were valued at two and one-half cents a pound, almost as much as rags themselves. The daily capacity of the South Hadley Mill came to 180 reams of book paper and 80 reams of imperial newsprint. Ames sold book paper for nine cents a pound and ruled writing paper (plain or straw colored) for eighteen cents a pound.[13]

Dryers and Calenders

Given the success of the South Hadley Mill, D&J Ames were soon inundated with requests from like-minded papermakers. To meet this growing demand the firm purchased another mill in 1827 on a choice piece of real estate on the Chicopee River just north of Springfield. Here they built an extensive tool and die shop and began making paper machines at the rate of about 12 to 18 per year. The Ames brand earned a reputation for reliability and longevity, and in the years to come Ames cylinder-wire machines went into mills all around Massachusetts, Connecticut, Vermont, Pennsylvania, New Jersey, and New York.

Alongside the workshops the Ameses also set up a model paper mill called the Water Shops Mill. Here they experimented with various new inventions and the latest developments from Great Britain. It was about this time the firm employed Moses Y. Beach, who brought them the rag cutting machine. The Ameses loft dried their paper as the Gilpins had done, but soon found the amount of loft space needed to support several paper machines was growing out of proportion to the rest of the plant. John Dickinson developed the steam-heated drying cylinder in 1817, but this invention remained a closely held secret. The American version of the same was the fire dryer, which came out of Boston around 1826. The fire dryer was a wood-heated assembly consisting of a large iron cylinder, about ten feet in diameter, with a wood stove mounted inside. As the cylinder slowly turned the stove remained accessible through a door mounted on the side. As the web slowly emerged from the wet end of the paper machine it would pass around the wood-heated cylinder. The only drawback was heat regulation, and uneven drying made the paper prone to wrinkling. Still, this advancement was more than sufficient for the manufacture of inexpensive papers such as wrapping or newsprint.

A former apprentice of D&J Ames later recalled times at the Water Shops Mill when they experimented with the fire dryer. Several years before, a manufacturer in Great Britain developed a process of loading paper pulp with twelve percent gypsum or sulfate of lime, to harden the paper for use in the building trades. According to the mill worker: "One way they had of adding weight was from an old gypsum mine near the Water Shops Mill in Springfield. This they mined and crushed with a crude grinder, and after screening a little, wheeled it to the side of the beaters and shoveled in all they thought the stuff would carry. One of the effects of this kind of pulp was to make the paper quite gritty, like very fine sand paper. The old cylinder machine with one large fire dryer was run about twelve hours per day and during this time the gypsum would accumulate on the dryer so thick that very little heat could get through it. A good strong scraper was then employed to clean it, and the machine was ready to go ahead again."[14]

Meanwhile, the Gilpins of Wilmington seemed to have rebounded from the previous flood, but several years later the Brandywine Mill was plagued by fire once again. Given yet another opportunity to rebuild and modernize, the Gilpins again turned their attention to the latest advancements from Great Britain. This time they found a new piece of equipment called a sheet-calender. The calender was designed to give the paper a smooth finish, replacing the centuries-old tradition of smoothing the paper in a screw press, a slow and labor-intensive process. This device required just two operators, one to feed in the sheets in from the front, and another to collect them on the other side. It was driven by an auxiliary gear off the water wheel, and as the Gilpins were already engaged in rebuilding, a facility for calendering would be easily accommodated.

After securing the U.S. patent to the sheet-calender, the Gilpins leased the rights to a machinist from Philadelphia by the name of Coleman Sellers. Sellers was the mechanically-inclined son of Nathan Sellers, a talented mould-maker from Philadelphia. In 1828 Coleman established a machine shop on Market Street to manufacture cylinder-moulds (the wire-wrapped portion of the cylinder-wire machine), and also during this time he purchased rights to a ream trimmer (used to create uniform edges on book paper). Sellers did not seem entirely satisfied with just auxiliary equipment, and so began designing a cylinder-wire machine of his own using a combination of existing patents along with helpful hints from customers. In starting up the machinery business, he erected a three-story plant just north of the city, which held a tool shop and iron foundry. Unfortunately the venture floundered after only

A drawing of the first sheet calender in Coleman Sellers' letter-book. The iron and paper cylinders are labeled. A sheet of paper to be calendered is shown entering the stack at left.

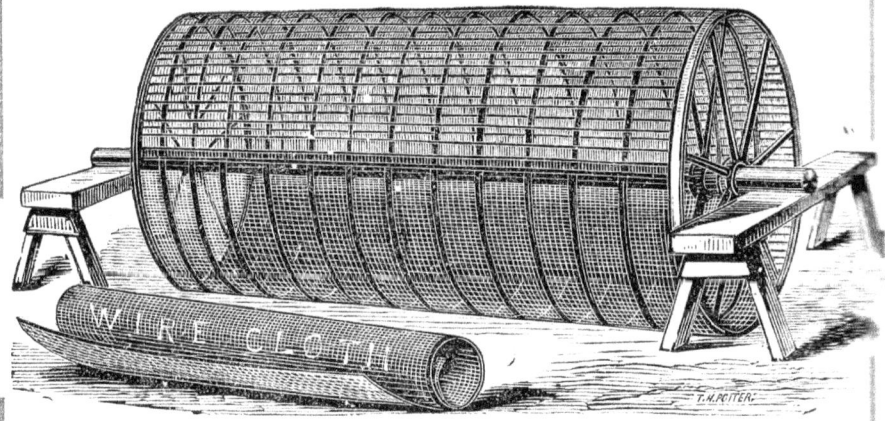

An advertisement for cylinder moulds, wire cloth, and dandy rolls (from *Paper Mills of the World*, 1885).

three paper machines were sold, but in the process he discovered how to build the new steam-heated drying cylinders that had just reached the United States at this time. According to Sellers' daybook, in December of 1832 the Gilpins placed an order for a steam-heated dryer, and this likely prompted a visit to the Brandywine Mill and his introduction to the sheet-calender.[15]

In an 1832 letter to Krepps & Carter of Brownsville, Pa., Sellers describes the new machine as: "Two paper rollers held by two iron ones, thus ... the paper rollers are kept continually polished ... by the finish of the iron rolls. The paper to be pressed is passed between the middle rollers [paper rollers] that produce above 500 tons of pressure. This makes the [finish] as hard as iron."[16] The outer rollers could also be constructed of heavy brass or copper-clad wood, while the inner pair were made of paper, that is, a compressed roll of paper wrapped around a spool. Sellers began marketing the calender to domestic papermakers early the next year, and as chance had it one of his first customers would be the South Lee Mill.[17]

The Great Paper Machine Trial

With competing patents on the cylinder-wire it seemed only a matter of time before events came to a head. The fight over the paper machine came to be the first major patent infringement suit brought in United States courts, and it would be Ames, not Gilpin, who sought the court's protection. The Ames machine had been well received by American producers, and was modestly priced between 2,000 and 4,000 dollars. Buyers were required to pay an annual royalty, designed to help recoup the cost of design and development. In a letter of 1831 David Ames Jr. outlined the program to the South Lee Mill, writing in his a characteristic left-handed style: "It will cost you one hundred fifty dollars for a one engine mill and two hundred dollars for a two engine mill. This will make the finest paper ever. This is the finest machine ever."[18]

It all started in Springfield where the firm of Howard & Lathrop ran a dry goods business. In 1828 they decided to go into the paper business for themselves, erecting a plant in Canal Village alongside the South Hadley Mill. Details are sketchy, but somewhere along the line an Ames foreman appeared and offered to build them a paper machine. This apparently is what happened, and as soon as the machine went into operation D&J Ames promptly filed suit.

The defense team for Howard & Lathrop sought a deposition from the other U.S. patent holder, Thomas Gilpin. Their goal was fairly simple — demonstrate that Ames had violated an earlier patent and force them to withdraw the suit. To this end, Thomas Gilpin testified that in March of 1822 John Ames asked for permission to visit the Brandywine Mill and was promptly turned down. Shortly afterwards a Gilpin workman named Hugh McFee received a note from "an unknown friend" inviting him to a meeting in a Wilmington tavern. Upon his arrival McFee was cornered by Ames, who took him to an upstairs room and locked the door. With a liberal application of liquor McFee was then queried about the details of the machine, but refused to give up any information. Then Ames allegedly attempted a bribe McFee with a "hand-full of bank-notes," along with an offer to relocate him and his family to Springfield where he would be paid a weekly salary of fifteen dollars. Gilpin said that despite taking every precaution, Ames still managed to obtain the necessary plans, and further charged that Ames could have only made cosmetic changes to the invention before filing his patent in May of that year.[19]

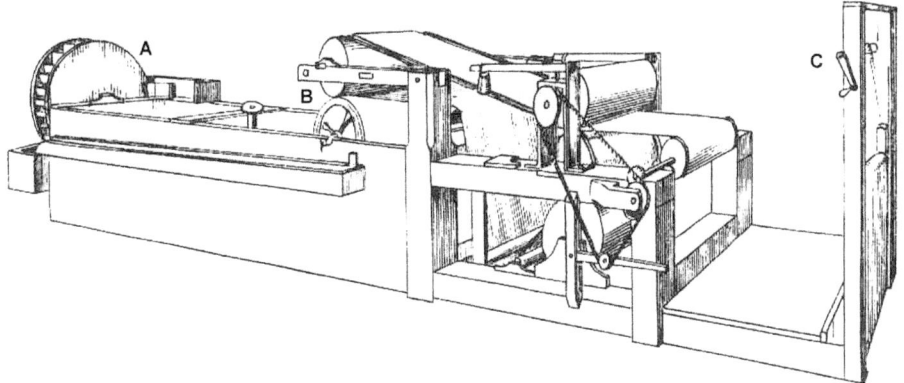

A drawing of a cylinder-wire machine used in 1830 trial of *John Ames vs. Charles Howard and Others*. There are no pulp filters shown; instead a stylized undershot waterwheel appears at far left (A). The vacuum cylinder rides high in the vat (B). There is also no drying section, as the damp paper is simply taken off a reel for further processing (C).

The case was complicated by a fire in the model room of the U.S. Patent Office in 1830, taking with it all the requisite drawings. Therefore, lawyers for both sides sent investigators to England to obtain further evidence. Here they found two of Dickinson's designs in the London patent office, one the original patent from 1810, as well as a subsequent version dating from 1817. It now became apparent to Gilpin that he'd only managed to secure his drawings because Dickinson was in the process of updating his patent. In the newer version the cylinder was mounted lower in the vat, and it had a larger vacuum trough that allowed the machine to run faster. Lawyers had already entered into evidence a recently made drawing of Gilpin's machine, so it rapidly became apparent the Ames machine was based on the newer design and did not infringe on Gilpin's patent.

The case of *John Ames vs. Charles Howard and Others* went before the U.S. Circuit Court of Boston in 1833. Defense lawyers now made the case that the cylinder-wire machine was found in France, Germany, and Italy, and its principal of operation was clearly in the public domain. The facts in the case were that Howard & Lathrop's machine was based on widely available information, and it had only been constructed for their own use. Lawyers for the plaintiff were obliged to agree that the principals of operation were universally known, and that Ames could not lay claim to various parts of the machine such as the vat, rollers, presses, wire-cloth, and felting. Still, they argued, certain features of the Ames design were proprietary and not publicly known. Such improvements were trade secrets and these are what made the Ames brand particularly desirable. Since Howard & Lathrop's machine was shown to be identical in these critical areas, it should be subject to royalties no matter the circumstances under which it was built.[20]

Judge Joseph Story found for the plaintiff, and motions for a new trial were denied. This may have been a costly victory for Ames, but it became the bulwark upon which their future business was built. D&J Ames continued building paper machines for another twenty years, but they lost control of the South Hadley Mill in the wake of the financial panic of 1837. For their part, Howard & Lathrop continued at same location until 1848, when their mill was closed to make way for a new dam across the Connecticut River.

3

Pittsfield Progress

When one thinks of the early centers of papermaking a few places come to mind, such as Chester County, Pa., outside Philadelphia, and Middlesex County, Mass., near Boston. By contrast, Berkshire County, Mass., is far removed from major population centers, and it's amazing to think it would grow to become the largest paper-producing region in the country. The key to this remarkable transformation was New York City. Situated on an island, the city had to look to neighboring communities for its supplies. Upstate New Jersey was the logical choice, but New Jersey had limited resources. Moving up the Hudson River, the town of Hudson sits on the east bank of the river some 35 miles north of New York. Just to the east in Massachusetts lay the Hoosac Mountains, and winding down among its gentle slopes is the mighty Housatonic River. The waterpower available in this area combined with a port on the Hudson River made rural Berkshire County the ideal location for paper mills supplying the New York market. In retrospect, Berkshire County represents a microcosm of events that were unfolding around the country during the 19th century, and the most authoritative work on the subject is Judith McGaw's *Most Wonderful Machine, Mechanization and Social Change in Berkshire Paper Making 1801–1885*. McGaw discovered the Berkshires had so many paper mills at one time that census takers couldn't count them all.

In the 1700s Berkshire County was little more than a series of small farming communities growing up around the crossroads town of Pittsfield. In 1876 the town supplied men and material used in the capture of Fort Ticonderoga. In the aftermath of the siege a large contingent of captured British soldiers were confined at Pittsfield for the duration. Following the war it seems that paper mills were springing up all around the region in Springfield, Mass., Falls Village, Conn., Bennington Vt., and Troy, N.Y., but none ventured into Berkshire County. Rags were seemingly the problem. The mill in Troy was established in 1793 by the firm of Webster, Ensign, & Seymour. The local agricultural newspaper was published by the town postmaster, David Bud. Bud wrote a tribute in 1802 to the local collection effort, saying, "The press contributes more to our knowledge and information than any other medium. Rags are a primary requisite in manufacture of paper, and without paper the newspapers of our country, those cheap, useful, and imaginable companions of the citizen and farmer ... must decline and be extinguished. The paper mills of the state need the poor and the opulent, the farmer and the mechanic, all to be persuaded into the laudable frugality of saving rags."[1]

Pittsfield got its first paper mill in 1801 when a papermaker from Worcester erected a one-vat mill in the village of Dalton about five miles to the east. Called the Berkshire Mill, the enterprise answered the fundamental needs of region, but little more. In 1806 a local landowner and entrepreneur named Samuel Church founded a two-vat mill in the village of South Lee some eleven miles south of Pittsfield. This new enterprise proved so successful that Church expanded the facility to four vats capable of making ten reams a day. As other members of the Church family wanted to get in on the act, Samuel Church erected a second plant in Lee Town Center called the Union Mill. Lee was very accommodating to the business, so in 1822 Church sold the South Lee Mill and built yet another plant in Lee called the Castle Mill.[2]

An Owen & Hurlbut ream label circa 1835. The louvers (resembling shutters) in the drying loft are most prominent. The background to the right shows the old bridge across the Housatonic River and the original South Lee Mill.

Thomas Hurlbut had been managing a paper in Suffield, Conn., and was keen to buy the South Lee Mill. Suffield is a town halfway between Springfield and Hartford, and is where the Eagle Mill was built by brothers Simeon and Asa Butler in 1789. The Butlers engaged in the manufacture of writing paper for the Hartford and New York markets, and had recently been awarded the first writing paper contract issued by the U.S. Senate, so it's easy to see what Hurlbut had in mind for the South Lee Mill. His new partner was Major Charles Owen, who brought leadership and organizational skills from his time in the army. The firm of Owen & Hurlbut (O&H) went in for major expansion by converting a nearby grist mill owned by Billings Brown's into a new manufacturing facility called the Phoenix Mill. Next, to secure the entire water right across the Housatonic, O&H purchased a puddling furnace called "The Forge," which sat on the opposite bank. Aliajah Merrill, the local blacksmith, simply picked up and moved to Pittsfield, where he established a similar ironworks. To serve the growing worker community the papermakers then opened a small country store and converted The Forge into a flour mill. In 1827 Hurlbut also started a small post office at the back of the store so residents no longer had to travel the two miles to Lee for their mail.[3]

Local residents in South Lee must have been more than pleased with the new ownership of the paper mill. The Merrill family, largest in South Lee, proved to be of the greatest utility, as when one or both partners were out of town on business, J.T. Merrill, who ran the local tavern, looked after the both the village store and post office. Another member of the Merrill family, William, apparently managed the mill as he had probably done for Samuel Church. In 1822 William Merrill was tasked with obtaining new mould for the mill. Hurlbut had evidently known a tool-maker in Cornell, Conn., a town just south of Suffield, who could produce a spindle used to create a wove mould. In April, William Merrill received a reply: "Your favr. of the 20th has arrived. We would make a paper mill spindle within a fortnight for you, but after the most diligent search we cannot find the pattern for the one made for Messrs. Butler & Ward [of Suffield]. If you could send us the shape we would try to make it for you immediately."[4]

Pittsfield's Book Fair

By 1809 the total output of Berkshire mills came to two-thirds of a ream for every person in the county, and no new mills were built over the next ten years. The War of 1812 saw Pittsfield once again turned into a prisoner-of-war depot. The commissary established on North Street was commanded by Maj. Thomas Melville, whose nephew would one day be the famous author Herman Melville. At the urging of the War Department a modern all-weather road running through Pittsfield was built to help carry arms and material to the Canadian and western fronts. Henry Fulton's steamboats were also employed to supplement Hudson River sloops transporting war materials to Albany. The steamboats made the trip from New York to Albany in thirty-four hours and charged less than one-fifteenth the cost of overland freight.[5] The military road through Pittsfield made for easier access to Albany, and from there to New York. Gradual improvements to the steam engine eventually cut travel time to New York down to ten hours, and in 1824 the Supreme Court struck down the old Fulton-Livingston monopoly, further lowering steamboat fairs.

The biggest factor in the growth of the paper market was the completion of the Erie Canal in 1825. New York would now be transformed into a major Atlantic port for the entire Great Lakes region. A result of the increase in shipping business, the counting houses and merchants of that city consumed ever-greater volumes of journals, ledgers and fine writing paper. At the same time, demand for news and information drove print media to unprecedented levels. The demand for book and newsprint began to peak at a time when the paper machine was just coming onto the market. The growth spurt of New York's population now called for mass quantities of low-cost wrapping and sack paper, and increasingly affluent homeowners created a ready market for wallpaper. The latter business was so lucrative that artisans relocated from as far away as London and Paris. By 1832 Berkshire County would send fully sixty-five percent of their output, or about 46,000 reams, to New York City, and Berkshire County now consumed less than ten percent of the paper manufactured in the region. A new wave of optimism called a "paper fever" was sweeping the region, and this movement saw ten new mills built in Berkshire County during the 1820s.[6]

In retrospect Owen & Hurlbut were in the vanguard of a movement, and the South Lee Mill soon found itself brimming with business. The search for rags was made easier by the New York Mercantile Exchange, which provided an estimated seventy-five percent of all Berkshire mills' needs. No doubt Hurlbut had experience with book paper, as the firm begun using wove moulds from the start. He established a book-bindery that produced, among other things, fine leather-covered ledgers with geld pages. The bindery was particularly active during winter months, essentially giving workers something to do at times when the river was frozen over and the hollanders idled. The South Lee Mill also seemed to do the bidding of a local bookseller. In December of 1827 George Kellog wrote from nearby Sheffield asking for two dozen Bentley spelling books and six copies of the *Woodbridge Geography & Atlas* (papermaker notation: Answer'd & Ch[arge]d).[7]

Being in the bookbinding business, O&H also sold supplies to a local bookbinder in Pittsfield by the name of Samuel Wardwick. Due to the distances involved, and because Thomas Hurlbut, as postmaster of South Lee, could absorb much of the cost of postage, a good deal of their communications went by the mail. It seems Wardwick's store in Pittsfield had a rag bin out of which he traded for blank book paper and post paper for writing. Wardwick also saved his clean shavings from trimming reams since these were worth nearly as much per pound as rags themselves. The amount of communications between the two underscores a trusting relationship. In 1833 Owen & Hurlbut set up a new sheet-calender obtained from Coleman Sellers. The papermakers then asked Wardwick to test several samples to see how well it held up in binding, and the bookbinder's response was that the stock should not be calendered more than once.[8]

As New York now received all its grain from the Midwest, Berkshire farms began switching towards more profitable specialties for the urban market such as livestock and dairy farming. Ready access to the New York market also attracted a range of industries to the Berkshires, including cotton mills, woolen mills, turning mills, carriage makers, forges, machine shops, tanneries, and clothiers. Given the growing prosperity of the region, and given that Berkshire County was awash in paper, Pittsfield now began holding an annual book fair organized by local bookseller Bangs, Richards, & Platt. Such an event was fairly unprecedented outside of major population centers, but it was apparently well attended and offered a range of blank books, ledgers, and fine writing paper by the quire (26 sheets, $1/20$

of a ream) or ream. In advance of the 1838 fair Samuel Wardwick placed an order with the South Lee Mill for one hundred reams of heavy paper valued at $3 per ream. These he planned to rule and bind, but running short of cash, he offered to sell a number of books on consignment to the mill. Owen & Hurlbut may not have found the terms to their liking, and they countered by offering Wardwick a job preparing a similar number of articles for them. The bookbinder accepted on condition of a cash advance of $25 to defray day labor costs he said he needed to complete the job in time for the fair. This request seems justified, as Wardwick's contribution was significant, it including sixty long folio day books, sixty-six broad day books, thirty-seven journals, forty-five double ruled ledgers, forty-four single ruled ledgers, twenty-four invoice books, and twenty-four cash books of varying sizes. The merchandise was subsequently packed in six wooden boxes costing fifty cents each, and in submitting his bill Wardwick asks the papermakers keep their arrangement confidential lest word get out of his actual costs, which came out as follows[9]:

```
Dec. 4, 1837:
For ruling and binding 46 journals & ledgers      $1.25     $57.50
" binding 37 ledgers ruled and done up in boards    .50     $18.50
" binding 17 ledgers ruled & cut without boards     .62     $10.62
" transporting books to and from S.Lee                       $2.00
                                                (Total)    _____
                                                            $88.62
```

The *Pittsfield Eagle*

With the establishment of the paper industry in Berkshire County rag collecting bins were set up at various shops and taverns, and a centralized rag collection depot was set up in the crossroads town of West Stockbridge. Stockbridge was the original mission established among the Housatonic Indians in 1734, and West Stockbridge (est. 1774) occupied an important junction between Stockbridge and Pittsfield. The rag route eventually merged with the newspaper delivery service.[10] The original Berkshire farming newspaper, *The Western Star*, had been published in Stockbridge since 1788, and with the arrival of the paper mill in Dalton the publisher moved to Pittsfield and changed the flag to the *Pittsfield Eagle*. Subscribers could have the paper delivered by mail for a fee, but the original mail system served only Stockbridge, Pittsfield, Sheffield, Great Barrington, Williamstown, Lenox, and Lanesboro. It seems the same person who delivered newspapers doubled as postal rider, and a post office established at West Stockbridge in 1804 allowed rag collection to be added to the mix. Over time postal delivery was extended to Hinsdale, Richmond, New Marlboro, Lee, Otis, Sandisfield, Cheshire, Adams, Dalton, Becket, and North Adams, which gives an idea of the number of newspapers delivered around the county on a regular basis.[11]

The paper machine came to Berkshire County in 1826, the initial effect being a drop in the price of newsprint. Over the long run, however, prices went up, and by 1840 the *Eagle* found itself shopping around for bargains. In a letter datelined Lenox, 13 January 1841, the publisher of the *Eagle* writes the South Lee Mill to say: "From what encouragement you gave me when at your office ... I have not engaged paper as yet from another mill and am desirous of purchasing some of you and not knowing if alterations may be possible in the size of the paper. In the next few months I should rather not purchase more than 26

The *Eagle* building in Pittsfield from a postcard circa 1900.

reams or so. Can't you furnish me with that amount, if not at the price spoken, at $4.00 per ream? ... I send you a specimen sheet of the lower priced paper ... which is made at Northampton for $2.00."[12]

Over the next three months the South Lee Mill supplied the *Eagle* with fifty reams of 28 pound newsprint measuring 23" × 32" and priced at $3.75 per ream. The following year the publisher appeared eager to wrap up a long-term contract, unfortunately, the papermaker's response regarding price and availability had been misplaced at the post office in Pittsfield, and as a result, the *Eagle*'s order arrived too late. The publisher now had to scramble to find an alternate source. Business did get back on track the following year with the publisher now writing O&H concerning the calendering of the stock, saying: "I am making the *Eagle* paper & I wish you would have it calendered pretty hard. No matter if it does not make it look thin — it works better. I have some paper of T. Sedgwick & Co.'s [Thomas Sedwick & Co., owners of the Pleasant Valley Mill in Lenox] which pulls off on the type so that it fills them full before we get half our edition off. It would work well I think if it had been calendered hard, although the stack is miserable. They offered to furnish me at 10 cents per pound [$2.80 per ream]." The publisher now placed another order for fifty reams of thirty-weight paper to be made one inch longer than before, and this stock proved so satisfactory the *Eagle* placed an identical order the following year.[13]

Over 1844–45 Charles Montague, publisher of the *Eagle*, had been unable to travel to South Lee because of poor health and so elected to transact business by mail. Montague sent O&H a note for $270, of which $170 was to pay down debts. In August of 1845 Montague asked the papermaker for fifty reams of newsprint, adding: "I have met with so many large losses this spring that I find that I shall be unable to meet the note that comes due next month.... I had hoped that I should have an increase of circulation, and collect my

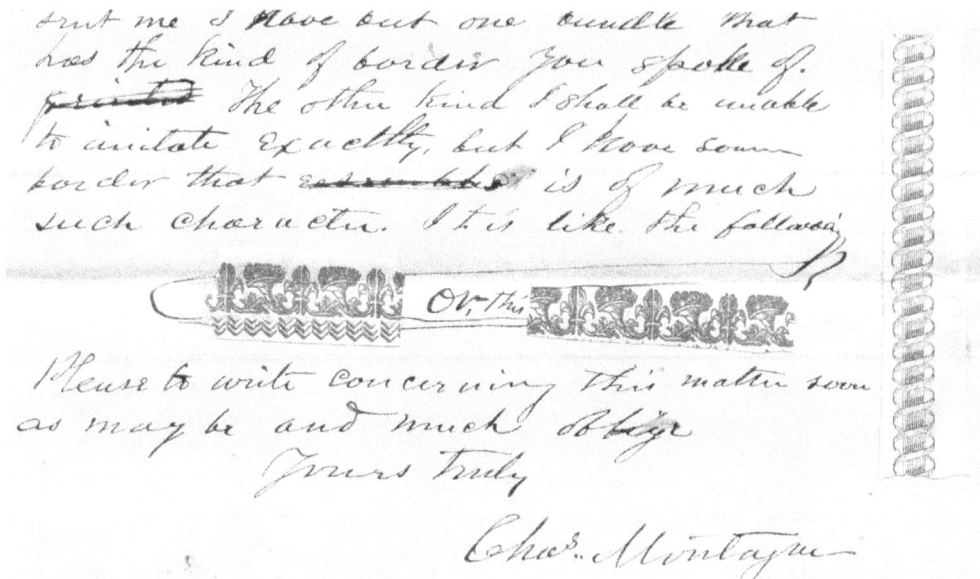

Details from a Charles Montague letter of 1845 with samples of artwork pasted on.

Details of a Norman Rockwell mural depicting Pittsfield in 1850. The post office is second from the right.

debts so that I should have means to clear off the whole of the old score." But things did not work out as planned and in May of 1846 Montague writes again to say: "Sedwick & Sabin [Thomas Sedwick and George Sabin of the Pleasant Valley Mill, Lenox] have made my paper since your misfortune [?] & they charged me but 11 cents per pound delivered at 4 months credit." So, it seems both papermaker and printer were having their difficulties at this time, and this would be the last mention of newsprint between the two. In retrospect, the South Lee Mill's "misfortune" may have been construction related, as letters from their insurance agent reveal they were building an addition at the time.[14]

Ream Labels

If price increases hit the *Pittsfield Eagle* hard, the problems posed to publishers in New York were magnified a hundred fold. Back in 1833 Hoe & Co. of New York introduced the Napier press, the first commercial steam-powered printing press. This development in combination with inexpensive machine-made paper so reduced the cost of printing that newspapers in New York sold for a mere penny. By the 1840s, however, publishers were again feeling the increased cost of raw materials, so Hoe developed the "lightning press," an innovation that printed upon four typeset plates at a time. While it cost $20,000, the lightning press helped to stabilize newsstand prices in New York at around a nickel.

In 1844, Charles Montague, the publisher of the *Eagle*, ordered twelve reams of straw colored envelope paper from the South Lee Mill. Paper mills at the time sold envelopes in their unfolded and ungummed state, to which printers applied customer advertising to the backflap (the address side still being strictly regulated by the postal service). In the course of conversation it seems the papermaker discovered the printer was strapped for cash and looking for odd jobs. As the South Lee Mill was in need of new ream labels, Hurlbut offered the printing job to Montague, and upon receiving the offer the printer wrote back: "[I] should be glad of any assistance you could give me in way of wrapper printing & shall feel grateful for the same." The wrappers were printed in basic offset type and returned to the mill in December of 1844. Something of an improved business relationship developed between the firms, and in May of 1845 Montague wrote: "I wish ... to purchase all my print of you ... on account of your befriending and helping me, & giving me your work." Montague was also making handbills for the local cattle show and the Pittsfield Jubilee, and at one point placed a small order for two and a half reams of 19" × 21.5" handbill paper. The print shop was subsequently billed eighteen cents per pound (about $4 a ream), to which Montague expressed his dismay. The quality of the paper was said to be no better than common newsprint at ten cents a pound. The comment must have missed the mark because sometime later the same ordered 19.5" × 24" handbill paper, offering no more than thirteen cents a pound. One can only imagine Montague's surprise to learn the added cost was for cutting paper to specification.[15]

Owen & Hurlbut's next order of ream labels was completed in March of 1846. Charles Montague said he put them on a railroad car to South Lee (a very early reference to the Stockton & Pittsfield Railroad). This would be the final hand printing, as in May of that year the *Eagle* installed a Napier press. Montague now created a copperplate engraving for O&H's new ream labels, and experimenting with the steam-powered press, the printer asked the papermakers to supply two samples of paper; the usual brown wrapper and a white foolscap. The fine white paper obviously did a better job, but Montague could not talk Owen & Hurlbut out of using the inexpensive wrappers and here's were the trouble began. The wrapping paper sent by O&H was sticking together at the edges, so Montague wrote to say: "Can you not trim the edges so as to remedy this evil — or calender your paper lightly? ... the girl who calenders [has] both hands at liberty to separate the sheets much better than we.... The press when on its regular motion gives us no time to separate sheets ... because we have to use one hand to place the sheet in the press & the other to take it off.... [The press] cannot be stopped to do this without a great waste of time and labor as you will readily perceive.... We printed 6500 wrappers for Platner & Smith in less time than 4500 of yours — and it was all owing to sheets sticking together on the edges."[16]

In 1850 Owen & Hurlbut placed another order for ream labels, these to be printed in gold ink on brown wrapper paper. Evidently to avoid the previous episode the beater engineer added starch to the pulp, giving the paper a slightly oily film. Not long after the printer began the work the papermaker received an urgent letter from Montague, who wrote: "I had a call from the ink manufacturer who furnished me with gold ink & ... says the reason why the ink would not dry is because there is something ... in the size of our paper which prevents ink from drying. He showed me other work done with the same ink which was perfect. He says that ... the paper is either so fully sized as to prevent the ink from being fully absorbed, or there is oil in the size.... I write in hopes that you can avoid this source of trouble if the difficulty rests here. I have had no complaint from other manufacturers for whom I have printed on dry paper with the same ink as I have used on your wrappers."[17]

Papermakers used two kinds of sizing at this time. The first was traditional sizing (a.k.a. tub sizing), where a gelatinous glue made from ground-up hides, horns, and hooves of farm animals was applied to the surface of the paper. But external sizing interfered with typeset printing, so book paper and newsprint had for centuries been made from unsized paper called "waterleaf." In 1806, H. Hig of Erbgach, Germany, found that rosin soap added to the hollander produced a smoother sheet ideal for typeset printing. Hig precipitated the soap with alum, a colorless, non-magnetic metal powder. This method of sizing is commonly called "engine sizing" because the ingredients are added in the rag engine. Still, this was an era when chemistry was not well understood, and while rosin was shown to be the only true sizing agent, beater engineers continued to add things like starch or caustic soda (soda ash) as they saw fit.

The time had come for Owen & Hurlbut to begin using white ream labels; still, with the price of rags on the rise the owners were reluctant to waste so much good paper on throw-away wrappers. The papermakers eventually compromised by printing labels on hand-bill paper and gluing them to the wrappers.

Touring a Berkshire Mill

The Western Railroad from Springfield and the Housatonic Railroad from Connecticut arrived in Berkshire County in 1842. This was quite a remarkable development for rural Berkshire County, as one could now take the train to Boston or New York. That same year Anson H. Clark of Stockbridge invented the daguerreotype, and soon thereafter Pittsfield became the most photographed small town in America. Pittsfield's postmaster, Phineas Allen, finally moved the post office out of the family store to a new location on North Street, where it shared a building with the Berkshire Life Insurance Co. A century later the latter engaged Norman Rockwell of Stockbridge to paint a mural dedicating their new office building, and guided by photographs of Pittsfield from the 1840s, Rockwell created an exotic parade on North Street in front of the business block. Their world became even smaller in 1856 when Hiram Silby established the Western Union Telegraph Co. in Pittsfield.

Henry Wadsworth Longfellow was a native of Pittsfield. In 1843 he returned home on the train and much to his delight he found the town so well stocked with paper and blank books that he came back to spend two wonderful summers over 1847-48; there he wrote such classic poems as "The Old Clock on the Stairs." Other writers, such as Oliver Wendell Holmes, had also discovered the region's charm. When he retired in 1849 Holmes built a home in Pittsfield and informed all his friends, "The best of all tonics is the Housatonic." Nathaniel Hawthorne and Herman Melville also came to the area in 1850 and moved into adjoining houses in Tanglewood. Hawthorne, Melville, and Holmes then held a famous literary outing on Monument Mountain. Melville evidently had settled in for the long term, beginning work on a new novel he would call *Moby Dick*.

This was Pittsfield's golden age, so to speak, as the small town now seemed to be at the center of the literary world. Naturally the veteran writers toured the Berkshire countryside and used the experience to guide their works. As luck would have it Melville visited a nearby paper mill and recorded his experiences in a short story, "The Paradise of Bachelors and the Tartarus of Maids." Judging by descriptions in the text (e.g., the kinds of paper made at the mill), the most likely possibility would be the Berkshire Mill in Dalton, which was run on a single cylinder-wire machine at the time. Melville's tour of the mill must have gone something like the story he wrote:

[Melville's guide is a young boy called Cupid, who begins the tour in the rag room]:
He took me up a wet and rickety stair to a great light room, furnished with no visible thing but rude, manger-like receptacles running all round its sides; and up to these mangers, like so many mares haltered to the rack, stood rows of girls. Before each was vertically thrust up a long, glittering scythe, immovably fixed at bottom to the manger-edge. The curve of the scythe made it look exactly like a sword. To and fro, across the sharp edge, the girls forever dragged long strips of rags, washed white, picked from baskets at one side; thus ripping asunder every seam, and converting the tatters almost into lint. The air swam with the fine, poisonous particles, which from all sides darted, subtlety, as motes in sunbeams, into the lungs.
"This is the rag-room," coughed the boy....
"Where do you get such hosts of rags?" picking up a handful from a basket.
"Some from the country round about; some from far over sea—Leghorn and London."
"The edges of those swords, they are turned outward from the girls, if I see right; but their rags and fingers fly so, I can not distinctly see."
"Turned outward." Yes, I murmured to myself; I see it now; turned outward; and each erected sword is so borne, edge-outward, before each girl.
"Those scythes look very sharp...."
"Yes; they have to keep them so. Look!" That moment two of the girls, dropping their rags, plied each a whetstone up and down the sword blade.
[Leaving the rag room, the tour proceeds to the river's edge]:
"Come ... and see the water-wheel," said this lively lad, with the air of boyishly-brisk importance. ... we crossed some damp, cold boards, and stood beneath a great wet shed, incessantly showering with foam, like the green barnacled bow of some East Indiaman in a

gale. Round and round here went the enormous revolutions of the dark colossal water wheel, grim with its one immutable purpose.

"This sets our whole machinery a-going, sir; in every part of all these buildings...."

[Next to the wheel house is the engine room where the hollanders, or rag engines, are at work]:

I crossed a large, bespattered place, with two great round vats in it, full of a white, wet, woolly-looking stuff, not unlike the albuminous part of an egg, soft-boiled.

"There," said Cupid, tapping the vats carelessly, "these are the first beginnings of the paper, this white pulp you see. Look how it swims bubbling round and round, moved by the paddle here. From hence it pours from both vats into that one common channel yonder, and so goes, mixed up and leisurely, to the great machine."

[The tour continues to the factory floor where the cylinder-wire machine is in operation]:

"And now," said he, cheerily, "I suppose you want to see our great machine, which cost us twelve thousand dollars only last autumn. That's the machine that makes the paper, too. This way, sir."

He led me into a room, stifling with a strange, blood-like, abdominal heat, as if here, true enough, were being finally developed the germinous [sic] particles lately seen.

Before me, rolled out like some long Eastern manuscript, lay stretched one continuous length of iron framework — multitudinous and mystical, with all sorts of rollers, wheels, and cylinders, in slowly measured and unceasing motion.

"Here first comes the pulp now," said Cupid, pointing to the highest end of the machine. "See; first it pours out and spreads itself upon this wide, sloping board; and then — look — slides, thin and quivering, beneath the first roller there. Follow on now, and see it as it slides from under that to the next cylinder."

"There; see how it has become just a very little less pulpy now. One step more, and it grows still more to some slight consistence. Still another cylinder, and it is so knitted — though as yet mere dragonfly wing — that it forms an air-bridge here, like a suspended cobweb, between two more separated rollers [press section]; and flowing over the last one, and under again, and doubling about there out of sight for a minute among all those mixed cylinders you indistinctly see, it reappears here, looking now at last a little less like pulp and more like paper, but still quite delicate and defective yet awhile. But — a little further onward, sir, if you please — here now, at this further point, it puts on something of a real look, as if it might turn out to be something you might possibly handle in the end. But it's not yet done, sir. Good way to travel yet, and plenty more of cylinders must roll it."

"Bless my soul!" said I, amazed at the elongation, interminable convolutions, and deliberate slowness of the machine; "It must take a long time for the pulp to pass from end to end and come out paper."

"Oh! not so long," smiled the precocious lad, with a superior and patronizing air; "only nine minutes."

Pacing slowly to and fro along the involved machine, still humming with its play, I was struck as well by the inevitability as the evolvement-power in all its motions.

[The web of paper now exits the press section and feeds into the steam-heated dryer section]:

"Does that thin cobweb there," said I, pointing to the sheet in its more imperfect stage, "does that never tear or break? It is marvelous fragile, and yet this machine it passes through is so mighty."

"It never is known to tear a hair's point."

"Does it never stop — get clogged?"

"No. It must go. The machinery makes it go just so; just that very way, and at that very pace you there plainly see it go. The web can't help going."

Something of awe now stole over me, as I gazed upon this inflexible iron animal....

"Halloa! The heat of the room is too much for you," cried Cupid, staring at me.

[After passing around the drying cylinders the web passes to the end of the line were automated cutters cut the paper into sheets]:

Previously absorbed by the wheels and cylinders, my attention was now directed to a sad-looking woman standing by. "That is rather an elderly person so silently tending the machine-end here. She would not seem wholly used to it either."

"Oh," knowingly whispered Cupid, through the din, "she only came last week. She was a nurse formerly. But the business is poor in these parts, and she's left it. But look at the paper she is piling there."

"Aye, foolscap," handling the piles of moist, warm sheets, which continually were being delivered into the woman's waiting hands.

"Don't you turn out anything but foolscap at this machine?"

"Oh, sometimes, but not often, we turn out finer work — cream-laid and royal sheets, we call them. But foolscap being in chief demand, we turn out foolscap most."

It was very curious. Looking at that blank paper continually dropping, dropping, dropping, my mind ran on in wondering of those strange uses to which those thousand sheets eventually would be put....

[Finishing operations, such as ruling and calendering, were conducted within a separate facility]:

Immediately I found myself standing in a spacious place intolerably lighted by long rows of windows.... Seated before a long apparatus, strung with long, slender strings like any harp, another girl was feeding it with foolscap sheets which, so soon as they curiously traveled from her on the cords, were withdrawn at the opposite end of the machine by a second girl [ruling machine]. They came to the first girl blank; they went to the second girl ruled.

Perched high upon a narrow platform, and still higher upon a high stool crowning it, sat another figure serving some other iron animal; while below the platform sat her mate in some sort of reciprocal attendance [sheet-calender]. Not a syllable was breathed. Nothing was heard but the low, steady overruling hum of the iron animals.

At rows of blank-looking counters sat rows of blank-looking girls, with blank, white folders in their blank hands, all blankly folding blank paper.

In one corner stood some huge frame of ponderous iron, with a vertical thing like a piston periodically rising and falling upon a heavy wooden block [stamping machine]. Before it — its tame minister — stood a tall girl, feeding the iron animal with half-quires of rose-hued note-paper which, at every downward dab of the piston-like machine, received in the corner the impress of a wreath of roses. I looked from the rosy paper to the pallid cheek, but said nothing.

[A separate facility was used for sorting, counting, and packaging]:

In a few moments, feeling revived a little, I went into the counting-room ... and where the desk for transacting business stood, surrounded by the blank counters and blank girls engaged at them.

"Cupid here has led me a strange tour," said I to the dark-complexioned man ... the principal proprietor."

"[This mill operates] twelve hours to the day, day after day, through the three hundred and sixty-five days, excepting Sundays, Thanksgiving, and Fast-days."

"Yours is a most wonderful factory. Your great machine is a miracle of inscrutable intricacy."

"Yes, all our visitors think it so."[18]

4

The Paper Trade

The 17th and 18th centuries saw successive waves of Dutch and German immigrants arrive in Pennsylvania, bringing the papermaking tradition along with them. By 1779 there were some 53 mills in eastern Pennsylvania, most situated along the Schuylkill River and tributaries thereof in Chester and Montgomery counties.

On the river road from Conshohocken to Roxborough, which runs along the northern bank of the Schuylkill, the waterway is largely inaccessible except for a narrow, winding cart path running along Trout Creek. This track led to perhaps the most famous paper mill in the state, established in 1746 on a fifty-four acre parcel by Anthony Newhouse and Benjamin Franklin. Publishing and papermaking had been well established in Philadelphia by the time Benjamin Franklin arrived on the scene. In his autobiography, Franklin writes about maintaining a rag bin at his general store and carting paper through town in a wheelbarrow, but in fact he was much more heavily invested in the paper industry than he ever admitted. Franklin established perhaps the first rag warehouse in Philadelphia, and his ledgers from 1741 to 1749 show he sold twenty-five tons of rags to the aforementioned Anthony Newhouse, as well as nearly twenty-three tons to Thomas Willcox of the Ivy Mill. By his own recollection Franklin helped establish or helped to supply eighteen different paper mills in the Philadelphia region; for instance, he sent a large quantity of rags to a new mill being started in Williamsburg, Pa., by William Parks.[1]

The Trout Creek Mill also made a special paper watermarked "Pro Patria" for the 1748 edition of *Poor Richard's Almanac*, and Newhouse and Franklin went on to make a small fortune selling currency paper to the Virginia Assembly. Newhouse retired to Germantown and sold the mill to Jacob Hagy, who continued to work in cooperation with Franklin. Franklin proceeded to establish a chain of printing franchises, and Hagy found so much business he erected a second plant in 1764.

Franklin moved to London for several years as the representative of Pennsylvania's legislature to oversee matters of trade and postal communications, and about this time wove paper made its initial appearance in Birmington, England. The mould of woven wire was designed to make book paper where the sheets are twice folded to create four pages (quatro). Traditional laid paper looked irregular in this role, and furthermore was difficult to print on against the grain of the watermark. By contrast, wove paper could be folded and still maintain a smoothness and consistency no matter which direction the print ran. Franklin greatly admired his 1757 copy of John Baskerville's *Virgil*, the first book printed on wove paper. Years later when he went to Paris as the U.S. plenipotentiary, he introduced the wove mould to Louis 18th, gaining immediate recognition for the *papier velin* brand, and cementing his reputation as a friend of the court.[2]

During the Revolutionary War, British forces destroyed every mill in New York, New

Jersey, and Pennsylvania they came in contact with. The situation became so dire that important leaders such as John Adams and George Washington reported they could scarcely find scraps of paper on which to write. Meanwhile, newspapers nearly ceased to function. One publication, the *Albany Register*, reports it had been reduced at times to pasting together broken sheets just to complete a printing.

Following the capture of Philadelphia in 1777 the Continental Congress removed to a county courthouse in York, Pa. With the urgent need to print money, they sought out the most skilled mould-maker in the land. Nathan Sellers, a talented wire-worker of Chester County, Pa., was then serving in the Continental army in New Jersey, and within ten days he was brought to York, where he commenced designing watermarks for the new Continental currency. Upon his return to Philadelphia, Sellers worked out of a small shop with his brother, James. The Sellers brothers created moulds of every size and description, even importing mahogany and other exotic hardwoods on occasion, and the pages of their day books would become something of a who's who of the American industry. Sellers' son, Coleman, was admitted to the firm in 1811, and he would carry on the business after his father's retirement in 1817. Freed from further business responsibilities, Nathan Sellers began devoting his time to a new invention called the chain mould. This device accepted the frames of hand moulds, dipping them into a vat and imitating the motions of the vat man. It's never been reported how many machines were ever built or where they were employed, but whatever success they may have enjoyed would soon overshadowed by the arrival of the cylinder-wire.[3]

Long after the war Phillip J. King erected a small paper mill on the grounds of the old courthouse in York where Congress once met. Business proved so good the Congress Mill, as it was called, was expanded in 1812. Following Phillip King's death in 1829 the mill passed to his eldest son, George. The younger King secured a contract to supply paper to the U.S. Congress and began embossing their stationery in the upper left corner with a "Congress" trademark. Congress paper became widely popular with paper dealers in New York and elsewhere, and other papermakers who supplied the government would subsequently adopt the brand.

At the turn of the century, Penn state mills numbered sixty-four with annual production of 165,000 reams, growing to eighty-seven mills by 1840, the most of any state. One of the most celebrated papermakers in the Pennsylvania tradition was Thomas Amies of Philadelphia. Dr. Dard Hunter relates the story of a vat man at the Willcox mill who set himself up in business in Lower Merton in Montgomery County. This region was popular for papermaking and had up to fifteen mills by 1810. Amies made book paper and newsprint that sold for nine cents a pound, and ruled writing for eighteen cents per pound. The papermaker came to prominence after contracting to make the paper for the official 1819 reproduction of the Declaration of Independence. As reported in the April 30, 1817, edition of the *Democratic Press*, Amies had special moulds measuring 36 by 26 inches made just for the occasion, and the 100 percent white linen stock would be valued at an unprecedented $25 per ream. Following this publicity the Amies brand became especially popular, a Washington paper merchant wrote the South Lee Mill in 1829 about his laid quatro post, saying that while others sold for $4.25, Amies charged $5.50. The merchant lamented: "The only reason I can give for it is that he sorts his paper better, and is more uniform in sizing. I am compelled to keep it notwithstanding his extravagant price!" The Amies' Lower Merton Mill succumbed to fire in 1830, but was rebuilt and continued in operation through the early 1840s.[4]

In 1788, General John Steele built an addition to his Octoraro Valley grist mill and set up a paper mill to supply newsprint to the *Lancaster Intelligencer and Weekly Advertiser*, published by his brothers-in-law, Jacob and Francis Bailey. According to Nathan Sellers' day book the Steelville Mill purchased paper moulds up through the 1820s, although Gen. Steele retired in 1808, handing over operations to his son, James. James Steel was an active entrepreneur who owned a gristmill, sawmill, and two cotton factories on 400 acres of land in nearby Lancaster County. He started up a paper machine at the Steelville Mill in the mid–1830s, but his firm failed during financial panic of 1837. Local farmers James A. Love and Newton B. Love bought the mill at a sheriff's auction in 1844 and leased it out. It continued in operation until 1855 before being demolished to reduce tax assessments.[5]

Delaware County broke off from Chester County in 1789 in a dispute over the location of the county courthouse. Chester County remained a popular place for papermaking, and tax records reveal that up to thirty-four vats still operated within the district, with six more built before 1800. Between 1800 and 1810 Joseph Dickey, a real estate developer in Chester County, established what he hoped would become an industrial village on Leach Run, a tributary of the Octoraro Creek. The village was called Mt. Vernon in honor of the late president, and here was established a cotton mill, two paper mills, a blacksmith, post office, and numerous shops and tenements. The cotton factory sat between the two paper mills, named the Stone Mill and Mt. Vernon Mill. Papermaker David Lefever managed the two single-vat mills during the 1820s, and these came under his ownership after he married Dickey's daughter, Margaret. Following Lefever's death in 1834, controlling interest returned to the Dickey family, who operated the mills in partnership with William Lysle and his brother James. The firm of Dickey & Lysle installed a cylinder wire machine in the lower mill (Mt. Vernon Mill), which thereafter ran on wrapping and newsprint for the Philadelphia market.[6]

Traditional papermaking was more deeply rooted in the cultures of England, France, and Germany, where it would continue for another generation or two after introduction of the machine. The decline of hand papermaking in the U.S., however, was more abrupt.

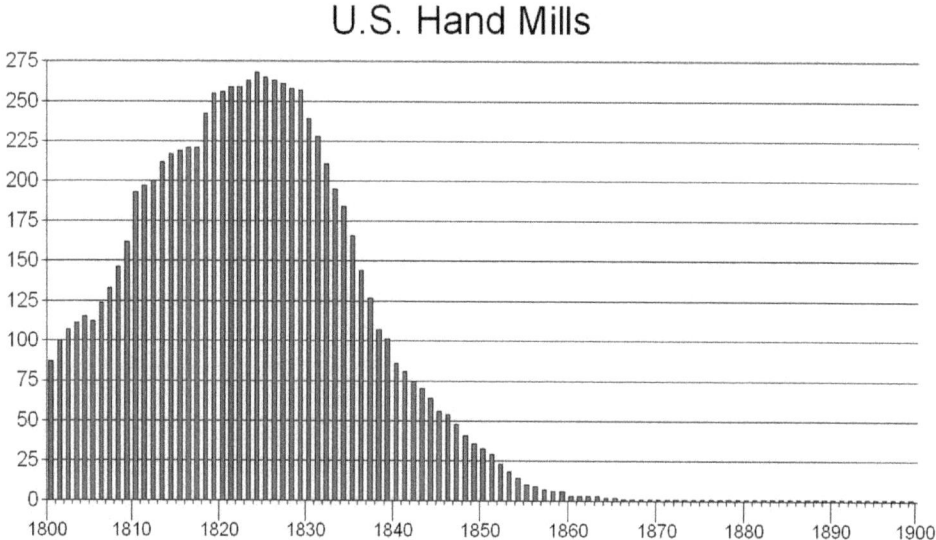

After 1830, paper machines begin making serious inroads, and mills in Pennsylvania were quick to adapt. Over 1833-34 the first cylinder-wires in Chester County went up at the Dorlan and Eagle mills. In Philadelphia County, paper machines replaced hand operations at the Manayunk, Schuylkill, City, and Flat Rock mills. Some papermakers embraced the new technology while choosing to keep their long-standing mills. For example, James M. Willcox, whose family had been making paper in the former Chester County since 1729, built a new plant called the Glen Mill in 1835 and started up the first Fourdrinier machine in the state.

Philadelphia's Market Decline

Philadelphia had been the center of the paper trade since the founding of the country and before. The city's rag trade developed fairly early on, and merchants also organized collection efforts in Virginia, where paper mills were scarce. Naturally the city was also a center of publishing, and in the revised census of 1810 Isaiah Thomas reported that Philadelphia had the nation's largest bookbinding industry. For example, *Rees Encyclopedia* of 1808 and 1822 was said to have consumed 30,000 reams of paper, or nearly twenty percent of the state's annual production. It was about this time that paper trimmings from book manufacture came to be used as another source of raw material. The hollander easily digested clean trimmings, and pound for pound paper made of linen was nearly as fiber rich as rags themselves.

The hand mills of eastern Pennsylvania remained relatively insulated so long as the Philadelphia market was strong, but the 1830s saw the beginning of tough times, as much of the national business went to New York. Things got worse in the 1840s with the completion of the Pennsylvania Railroad, allowing the city to become inundated with machine-made paper from New England and New York. Most of the local hand mills went under at this time; for example, James Fulton, a papermaker of Fountain Green, Chester Co., Pa., wrote his congressman in 1841 to say: "I have been engaged in the manufacture of paper all my life and by misfortune and other pressure of the times I will soon be out of the business." While it has been said that many traditional papermakers simply refused to yield to progress, it's also possible that many operations either lacked sufficient water rights or waited too long and were unable to secure financing. Whatever the reason, the decline of hand manufacturing during the 1840s was palpable, and by 1850 the total number of paper mills in Pennsylvania had fallen to sixty-one. But, with failure came opportunity, and the Philadelphia market was down but not out. The remote location of Newhouse's old Trout Creek Mill in Montgomery County made it an unlikely site for redevelopment; however, in 1856, Edwin Cope, a Philadelphia businessman, purchased the property and started up a machine mill. By 1858 the Cope Mill employed more than 30 people and was making 3,000 pounds a day.

As cash transactions were scarce, Philadelphia warehouses generally offered credit of three to six months' duration. Staying in the paper business was largely a matter of great sales and careful bookkeeping, as the survival of most houses depended on interest made from customer accounts. To further improve their margins, warehouses obtained paper on a commission basis, which also gave the papermaker maximum leverage on price. Owen &

Hurlbut, for one, imposed strict price controls on commission sales. R.M. Feeters of Philadelphia wrote from Philadelphia in 1852 saying:

> I have noted the instructions and am perfectly willing you should keep debit and credit information of my sales.... My sales ... to date ... are over six thousand dollars, and besides opening for you many customers, which you could not have had before, I'm willing to represent yours as Magasgu & Co. I know they are on much better terms if they only sell the amount of your paper in the year that was suggested to me when in your office in South Lee. I consider it a paltry amount compared to the large commission they must possess in their

Circular for Philadelphia trade sale, 1853. This was the forty-first annual event, suggesting it had been going on since 1812.

long-standing business.... If at the end of my arrangement, my exertions (also the credit side of the ledger) have failed to give you satisfaction, you will consider yourself at liberty to make any other arrangements you deem necessary.

The aforementioned Magasgu & Co. was the largest commission warehouse in Philadelphia, and their affiliation with Owen & Hurlbut began with the arrival of the railroad in the early 1840s. Political events temporarily overshadowed business in 1844 when the Know Nothing Party took to the streets of Philadelphia in what were billed as anti–Catholic riots. As Owen avoided travel to the city, Magasgu began sending statements by mail. The first ledger (written on a sheet watermarked "Amies, Phila.") describes how, after a month of martial law under a force of 500 militia, everything was finally quiet. In January of 1845 business began picking up and Magasgu reported fifty-seven individual sales ranging from forty reams down to as little as one-half ream. Most sales were for cap and post, the rest divided between Phoenix cap, printing paper (28" × 32"), and envelopes. Monthly sales averaged $585.72, and for their part Magasgu & Co. earned a $28.38 commission. Paper sales in February topped $1100, and the warehouse now asked for a change in the monthly quota from 280 reams to 480 reams, saying that super cap (plain and ruled), and super post (plain and ruled) were in high demand. This business arrangement continued until 1852, when the firm's principal partner, J. Magasgu, retired. The long-term owner wrote O&H to say: "Having met with an accident in fact caused the fracture of lower part of my knee and joint 3 months since. It may be a while before I can return.... It is the intention of our new friend to confine ourselves to a commission business if we can make such an agreement.... Mr. Aesyeng is a very popular salesman, and Mr. Bark is a good businessman."[7]

During the 1850s the Southworth Manufacturing Co. of Mittineague, Mass., set up a manufacturer's warehouse at 3 Minor St., where they could offer their paper directly, as well as gain access to Philadelphia's immense rag market. J.M. Willcox & Co., a local manufacturer of both hand and machine paper, had earlier established their warehouse at 7 Minor St., which was then at the core of the business district and an ideal location for such an enterprise.

Shyrock Straw

In 1790, John Shyrock founded the Hollywell Mill in Chambersburg, Franklin County, Pa. Here was made printing and wrapping paper as well as fine quality banknote paper for the Treasury Department. The mill had become something of a fixture in the region when in 1827 a farmer named William Magaw, who was manufacturing potash in nearby Meadville, came up with what he believed to be a viable straw pulp. The farmer had lined his hoppers with long straw that became crushed in the process and noticed the macerated pulp looked promising enough for paper manufacture. Magaw carted a small load over to the Hollywell Mill, where it garnered great interest. Straw has only a 35 percent cellulose content, and Shyrock found the fibers too short and brittle for most applications, but when mixed with rags the paper gained a hardness and a rattle equal to fine papers.

Shyrock patented the straw pulping process in 1828. When used in conjunction with the cylinder-wire machine, it produced a yellowish paper that was strong and durable enough for good wrapping or sack paper. Following this success the process was tested on a number of other materials, including wheat, rye, barley, oats, buckwheat, corn, white pine, and willow. In the course of these trials Shyrock ordered cylinder moulds of varying gauge from Coleman Sellers in Philadelphia, and the papermaker found that a mesh of 30 to 36 wires per inch was best for straw, while a finer mesh of 50–54 wires per inch was needed for wheat and other grains. In the final analysis, rye, wheat, and straw made the best paper, while oat was the least desirable.[8]

The straw process rapidly made the rounds in the industry and began to pay big dividends in the form of royalties. The papermaker now spent $35,000 for improvements to the mill, including a steam plant, drying house, four new beaters, a cylinder-wire machine, a new mill dam, and an overshot waterwheel. The mill was soon boiling up to 1000 tons of straw pulp a season and Shyrock now set his sights on the manufacture of newsprint. Imperial stock was then selling at $2 a ream and Shyrock's machine was capable of making up to 300 reams a day. The tipping point seemed to be about forty percent rags, and the *Niles Weekly Register* and the *Philadelphia Bulletin* successfully used straw newsprint in 1829. With the rising cost of rags, however, the papermaker faced diminishing returns, so the manufacture of newsprint was abandoned.

When John Shyrock retired in 1830, ownership of the Hollywell Mill along with rights to the patent went to his son, George. The Hollywell Mill was proving inadequate for the company's needs, so George Shyrock took on private investment and built what would be called the Mammoth Mill. Coleman Sellers assisted in the design of the three-story, 150 by 50 foot facility built on Conococheague Creek near Chambersburg, Pa., and he recognized that while the cylinder-machine ran rather slowly, putting on more machines could offset

Gunpowder Mill, Baltimore County, Maryland. A double-wide overshot waterwheel appears just above ground level at left. The boiler chimney is at far right.

this disadvantage. To this end the mill had eight paper machines with combined output of 100 pounds per hour supplied by eight rag engines with a capacity of 150 pounds per hour. The mill required an immense drying loft with an estimated two miles of drying poles, making the Mammoth Mill truly a mammoth operation. The plant ran on wrapper and book binder board for some thirty years until July of 1864, when it was destroyed by Confederate troops under General J.A. McCausland. Following this loss, Shyrock & Co. built the Papyrus Mill in Shippensburgh along similar lines.

The success of the straw process lay primarily in the simplicity and versatility of the cylinder-wire machine. Straw mills rapidly grew up in agricultural regions of the country from eastern Pennsylvania to Indiana, and from Maryland to New York. Taking advantage of the abundance of local grain in Shamburg, Baltimore County, Md., the old Gunpowder Mill (est.1780) and the nearby Eagle Mill (est.1800) were modernized with cylinder-wire machines for the manufacture of straw paper. Other straw mills in Baltimore County were the Glenmount Mill in Freeland built by Abraham Shaver, and the old Excelsior Mill in White Hall converted by A.J. Burke. The firm of William H. Hoffman & Sons bucked the trend and built two rag mills in the vicinity that ran on book and news for the Baltimore market. The firm later purchased the Gunpowder Mill, and their cluster of plants in central Baltimore County came to be known as "Hoffmanville."

The Milton Mills

The third mill in North America and the first in Massachusetts was chartered in 1728. Daniel Henchman, a bookseller and publisher with a shop in Cornhill, Boston, converted an old grist mill near the bridge on the Neponset River in Milton. Henchman hired a papermaker from England to run the plant. In 1741 Jeremiah Smith bought the mill along with seven acres, and hired two English papermakers in succession to run the facility. In 1765 a former Smith employee partnered with James Boies to erect a second facility. The Neponset Mills were then sold to Daniel Voss, who continued in the businesses over 1769–1789.[9]

Upon selling his mill, James Boise founded a new plant on the other side of the Neponset, and in 1793 he partnered with a son-in-law to establish another mill in nearby Dorchester. Mark Hollingsworth, a sixteen-year-old native, began his career at the Dorchester Mill, eventually rising to supervisor. Following Bose's retirement in 1801, Hollingsworth formed a partnership with Edmund Tileston to buy the Dorchester plant. These plans fell through when the mill was destroyed by fire in 1805, but Tileston & Hollingsworth (T&H) continued in the business, buying old Gillespie Mill in East Walpole. Continuing to expand the business, T&H converted a chocolate factory in Dorchester to paper making, then bought the aforementioned Neponset Mills when it became available in 1828.[10]

Mark Hollingsworth would have been fifty-one years of age when the sons of the owners, Edmund P. Tileston and Amour Hollingsworth, were admitted to the firm. In 1839 the next generation of Tileston & Hollingsworth purchased the Eagle Mill in Milton. Here the firm went in for mechanical improvements, installing a 58" Fourdrinier machine, and hiring John L. Seaverns of Worcester to manage the facility. T&H put up a warehouse in Boston selling book, news, wrapping, and specialty papers such as lithograph, etching and chart. Another specialty of T&H was a blue wrapping paper made for a local chocolate manufac-

Letter of 1845 addressed to Tileston & Hollingsworth, "Paper Manufacturers," Boston, Mass.

turer. In 1846 the New York warehouse of R.S. & S.A. Stuart bought fifteen bundles of 10" × 15" and forty-five bundles of 18" × 24" book paper. Burnap & Babcock of New York requested T&H's 37 pound royal (24" × 28") with ruled columns and "as smooth a finish as may be had without calendering." In 1847 the same wrote T&H asking for music paper to be sent immediately by the new propeller packet *Virginia*, which by traveling along inland passages was swifter than the train.[11]

Tileston & Hollingsworth also specialized in the manufacture of "plate" paper that was used primarily in flat plate printing. Plate printing, as opposed to moveable type, was used to render complex engravings for things like banknotes, bonds, and stock certificates. It required a very smooth and flat surface in order to create the perfect image, and this kind of paper cost five times that of ordinary printing stock. The paper itself was plated, that is to say, each sheet was pressed between boards in a screw press (or hydraulic press when they became available in the 1840s) to apply the final finish. Plating the paper was time consuming and labor intensive, but printers simply couldn't do without it. During the 1850s a new invention came along whereby the sheets were slid between thin metal plates and passed between heavy metal rollers. The plating machine saved time and money and greatly speeded up the process.

During the mid–1840s, Tileston & Hollingsworth sold a considerable amount of plate paper in New York through the firm of Burnap & Babcock (B&B). In November of 1844, B&B ordered thirty reams of 19" × 24" 50 pound plate, which they sold at 20 cents apiece (about $100 a ream). Standard plate paper sold for about seventeen cents a sheet, and in April of the following year B&B ask for additional 19" × 24" plate, along with one hundred reams of 40 pound 21" × 28" plate, which was in considerable demand. Business was booming. In 1847 the same ordered even more of the large format plate, and despite raising the price to 25 cents a sheet (about $125 a ream) it was still selling briskly. Burnap & Babcock later asked for twenty reams of the same to be sent immediately by train, with a further 30–50 reams by packet (steamboat). In closing, the dealer wrote, "If you have any remnant of the fine 25" × 35" 80lb. please send it to us as ... quires will be wanted."[12]

The weight of the above is also worth noting, as machine-made paper sold by the thousand-fold for ease of accounting. A handmade paper of 80 pounds to the ream was unheard of, and not even Thomas Amies could pull such large and heavy sheets in any meaningful quantities. The conversion from hand to machine is about two reams per thousand-weight, so an 80 pound machinemade paper is roughly equal to a handmade article of 40 pounds to the ream.

The Falls at Newton

The town of Needham, Massachusetts, sits adjacent to Newton Falls about ten miles up the Charles River from Boston. Three dams built here created a twenty-two foot drop sufficient to power a small industrial community. In 1801 Ephraim Jackson converted an old grist mill on Newton's upper dam to papermaking. In 1822 Amos Lyon purchased the Jackson Mill along with its water rights; he rebuilt the mill and installed a paper machine. After twenty years in the business Lyon sold the mill to the firm of Nathaniel Wales and William Mills. Wales & Mills went on to manufacture fine book and writing paper, and in 1854 began making the newsprint for the *Boston Congregationalist*.[13]

After retiring from papermaking Amos Lyon remained in Needham as leader of the local Masons guild. A man of high moral conviction, Lyon was deacon at St. Mary's Church and member of the Knight's Templar. Politically he was firmly antislavery and an outspoken member of the Temperance Society. The Lyons were a family of papermakers who first worked at the old Neponset Mills in Milton. Peter Lyon, Amos' elder brother, purchased the land adjoining the Jackson Mill in 1810. Here he erected a paper mill powered by a small dam below the main dam. Peter divested of a smaller adjoining piece of land with water rights to William Clark and Reuben Ware, who built a machine shop, and then in 1831 sold his main mill to William Hurd and Lemuel Crehore. Fire destroyed both mill and machine shop in 1834 with combined losses of $50,000. Hurd & Crehore quickly rebuilt, installing four rag engines and a 62" Fourdrinier for the manufacture of book and newsprint.

The burned out workshop of Ware & Clark was also rebuilt, and the firm went on to become one of the most successful millwrights in the Boston area. Ruben Ware had grown up in the paper business in Newton. His father, John Ware, came here in 1790 and converted an old sawmill on Washington Street to papermaking. The elder Ware had been a soldier in the Revolutionary War who saw action at Bunker Hill. He went on to become a surveyor, trader, and magistrate, and was a trusted adjutant to General Benjamin Lincoln during the suppression of Shay's Rebellion in 1786. Ware sold his mill in 1815 to a group of investors headed by the aforementioned William Hurd. Hurd was joined by Lemuel Crehore in 1825, and together they started up a 36" Ames cylinder-wire at what would be called the Crehore Mill. Having gained some measure of success with the paper machine, Charles Crehore expanded his holdings in 1834 by acquiring the tiny Grant Mill next door. The Grant mill had been established in 1809 by the firm of Moses Grant & Son of Boston, who had previously sold the adjoining piece of land to John Ware in 1790. Crehore now tore down both mills and built a new facility around a cylinder-wire machine reconfigured for the manufacture of pressboard. Pressboard was invented by the mill superintendent, Surner Shattuck. The essence of the technique was to affix the wet web to a spool, and as the paper slowly wraps around a press roller compacts the layers into a solid mass. Once the desired thickness is achieved a little bell summonses the machine tender, who cuts the board from the web and from the spool with a single widthwise cut. The process begins again when there's enough slack to wrap the web around the spool. Surner Shattuck also used his skills as a cabinetmaker to design mortise and tenon joints to assemble pressboard boxes. At first the boxes were made for local woolen mills that needed inexpensive and lightweight packaging to ship their goods around the country. The mill could also turn out boxes with a glazed finish, another innovation that was well received. Hurd & Crehore went on to contract with the U.S. government to make specialized pressboard boxes for a variety of official uses.

The falls at Newton had attracted a number of papermakers from Milton such as Solomon Curtis and Thomas Annis. In 1789 Curtis & Annis formed a partnership with a local leather maker named Edward Jackson, and hired Hezekiah R. Miller, a millwright from Dorchester, to erect a new mill on the upper dam. The leathermaker then partnered with Simon Elliott to erect another mill in the immediate vicinity. Annis later relinquished his rights to Jackson, who sold out Simon Elliott, and Elliott and Curtis then merged their operations and ran the two mills as one. In 1814 Elliott & Curtis went on to buy a neighboring grist mill and converted that to papermaking as well. In time Elliott sold out, and

Sketch of Thomas Rice Mill, Newton, Mass., about 1864. The falls over the mill dam are visible at left foreground (from Wiswall, *One Hundred Years of Paper Making: A History of the Industry on the Charles River at Newton Lower Falls, Massachusetts*).

Solomon Curtis died suddenly in 1818. The Curtis Mills were inherited by Allen and William Curtis, who formed the partnership of A.C. & W. Curtis. This company turned to the manufacture of bond and banknote paper. The mill's operations were supplemented by a pulp mill built on an artesian spring in the nearby town of Natick. In 1820 they made currency paper for the bank in Dedham with random threads of mohair as a security precaution. Sometime in the mid–1820s, A.C. & W. Curtis made extensive waterpower improvements, doubling the number of rag engines, and added a cylinder-wire machine — the first in Newton Lower Falls. The number of hollanders would double again to eight in 1828 after the firm imported two 60" Fourdrinier machines from England. The ramshackle facilities soon proved inadequate, so in 1834 a new mill of stone construction was built in its place. For a time the Curtis Mill was the most advanced facility in the state.

The waterpower of Newton Lower Falls was largely exploited through the use of raceways. One particularly clever location was an island near Pratt's bridge that connected Needham to Newton. The raceway had been dug about 1793 to power a snuff mill and trip-hammer grist mill. Both were eventually converted to papermaking and the raceway now supported three rag engines. The Island Mills came into possession of Joseph Foster in 1845, and Foster rebuilt and combined the waterpower to set up a wet machine for the manufacture of bookbinder boards. Surner Shattuck's wet machine had by now become a standard in the industry.

In 1788 Edward Jackson and William Hoggs, an investor from Boston, constructed a third dam south of Pratt's bridge, then proceeded build a paper mill on the northern side

Sketch of Rice and Garfield Mill, Newton, Mass. The building at left is believed to be part of the Ware & Clark machine shop. Both facilities were rebuilt after a fire in 1834 (from Wiswall, *One Hundred Years of Paper Making: A History of the Industry on the Charles River at Newton Lower Falls Massachusetts*).

of Pratt's bridge. The Nehoiden Mill, as it came to be known, would be managed by Stephen Crane, who learned the trade in the employ of Daniel Voss in Milton. Crane went on to make fine vellum cap and standard white foolscap, but his stake in the operation was rather small, and eventually fellow shareholders sold out to William Hoggs & Son, who took over operations for themselves. Hoggs now opened a general store in Newton, then a sleepy village of just a half-dozen homes. The store stocked various sundries; boots, handkerchiefs, etc., and to boost sales Hoggs began producing wrapping paper, a rare commodity in those days. The smart packaging helped bring in customers from as far off as Sherborn and Hopkinton. The Nehoiden Mill changed hands in 1816 and 1835. By 1847 the plant came into possession of Joseph Greenwood and Paul Dewing, who completely rebuilt the plant and installed a 62" Fourdrinier machine for the manufacture of book paper and newsprint. In 1851, the Nehoiden Mill was sold to A.C. Curtis and Son, with George B. Curtis as superintendent.

Stephen Crane's younger brother, Luthor, came to Newton and formed a partnership with John L. Rice in 1831 to buy the old Peter Lyon Mill established in 1810 on the Needham side of the river. Rice & Crane immediately went in for waterpower improvements, and when the plant burned in 1834, the firm rebuilt with a 62" cylinder-wire machine for the manufacture of book and news. In 1836 Luthor Crane sold his interest to Thomas Rice, Sr., who managed the facility as the Thomas Rice Mill until 1852 when his son, Thomas Rice, Jr., took the helm.

Boston's Market

Following its success at South Hadley, the paper machine would quickly spread to other areas of Massachusetts. One of the first was erected in 1824, a machine at a Worcester-area mill owned by Isaac Burbank. The Burbank Mill (est. 1776) was built on the outlet of Crooked Pond, now called Singletary Lake, near the village of Sutton. The original owner, Abijah Burbank, retired in 1788 and left the business to sons Caleb, Elijah, and Abijah Jr. The brothers added another vat, two beaters rated at 200 pounds a day, and a new twelve-foot tall brest wheel. The plant was capable of ten reams a day and ran on book paper, in particular, supplying the stock used for *Webster's American Spelling Books*. The owners had been keen to obtain a paper machine ever since Gilpin's success in 1817. The Ames machine at this time was said to be just 30" wide by 26" tall. Caleb Burbank and his brother Elijah owned another paper mill in the nearby town of Sutton, and in 1828 they installed a cylinder machine, several new hollanders, and a (Beach) rag cutter for the manufacture of newsprint. Caleb Burbank was a respected Boston publisher and financier; unfortunately after the financial panic of 1837 he ended up selling both family mills to creditors.

The paper machine was readily adopted by New England papermakers, spreading much along the lines of the industry in Great Britain. In 1838 the Massachusetts secretary of state compiled a census and found 89 paper mills, the largest number of any state. In the Boston area alone there were six paper mills in Milton; five each in Newton and Leominster; three each in Dedham, Pepperell, and Harvard; two each in Newton, Braintree, Dorchester, Walpole, Swansea, Methuen, Framingham, Shirley, Watertown, Fitchburg, Hardwick, Worcester, and Amherst; and one each in Middleton, Groton, Sudbury, Waltham, Athol, Auburn,

Millbury, Blandford, New Marlborough, Tyringham, Fairhaven, Taunton, Bridgewater, and Wareham.

By 1860 Boston had four papermaker warehouses and over three-dozen commission warehouses. The city also led the country in the manufacture of builders' papers, the most notable being the New England Roofing and Manufacturing Co., which made gypsum-loaded roofing shingles and tarred sheeting paper. The largest conventional concern was Grant, Dennis, & Co., later Grant, Warren, & Co., who had a paper warehouse on Milk Street and rag warehouse on Congress Street. Dealers in printing, writing, and colored papers, they also imported rags, wires and felts, and kept a stock of chemicals such as ultramarine, alum, bleaching powder, soda ash, aluminous cake (alum), sal-soda, sizing, and rosin soap. In 1854 Grant, Warren & Co. bought the Copesecook Mill in Gardiner, Maine, the first mill in that state with a Fourdrinier machine.

While the Massachusetts industry as a whole had gone over to the machine, one firm went the opposite way. Hand mills had virtually disappeared from the Boston landscape, yet the city continued to import fine handmade paper from Britain. In 1850 a Berkshire papermaker named Levi L. Brown took the train to Boston to show his wares to various dealers. He and his two uncles, Daniel and William Jenks, had started a machine mill in the village of Adams and were anxious to make new sales. In checking out the market, Levi Brown realized the strong demand for handmade paper went unchallenged, and upon returning home set up a single vat in an out building. The handmade effort proved so successful, L.L. Brown & Co. built a new hand mill in 1857. The Stone Mill went on to make ledger, bond, exchange cap (writing), manuscript, and parchment paper, sending about 200 pounds a day to the Boston and New York markets.

Paper mills in the center and west of the state were fairly cut off from the Boston market until the arrival of the railroad. After the Western Railroad reached Dalton in 1841, local mills began to record twice as many sales of paper in Boston than in New York. Other Berkshire mills followed suit; for example, once the railroad reached Lee, Mass., fully a third of production would go to Boston.

Owen & Hurlbut began working the Boston market in the 1850s. Here they managed to set up commission sales with a variety of dealers, as they had done in other cities. In 1850 they were doing business with Wilkins, Rice, & Kendall at 16 Water St. However, the warehouse fell under hard times, and after reorganizing and reopening in 1852 as Rice & Kendall, Charles S. Kendall wrote the South Lee Mill for cases of blue cap and letter, saying, "And if they strike us favorably we might do considerable with them as possible." Owen & Hurlbut also worked with J.M. March & Co., providing them an extensive stock of laid watermarked papers, these generally in letter or note size, ruled, and in blue or white.[14]

The Politics of Papermaking

The lifting of duty on imported rags had given the American industry a real boost; however, during the War of 1812 the government placed an annual tax on paper vats that assessed a sum of either $200 or $250 depending on the type of paper made. Naturally, the measure increased the price of paper, and after the war the American public, being acclimated to higher prices, developed a preference for British and French papers. Domestic stocks were hard hit, and politically minded publishers became very concerned over the high price of paper. A group of Boston-area publishers petitioned Congress in 1818 for relief from the vat tax, to no avail. The next year a convention of papermakers from Pennsylvania and Delaware met in Baltimore to discuss the issue, and a draft report made it into the *Niles Weekly Register* for January 22, 1820, reading: "In the districts represented, it appears there were 70 paper mills, with 95 vats, in full operation until the importation after the late war. These establishments cost about 500,000 dollars and employed 950 workers whose annual wages amounted to 217,000 dollars, consuming 2,600 tons of rags a year, and producing paper worth 800,000 dollars — but now only 17 vats are at work, the wages amount to only 45,000 dollars; the production no more than 136,000 — and 775 persons out of the employ." The actual impact to the industry is difficult to assess, but other than a flat spot around 1817-1818, growth of the hand industry in the early 1820s was very strong.[15]

The Baltimore memorial was written by none other than Thomas Gilpin, who in retrospect may have been among the least affected by the vat tax. Gilpin proposed a tariff by weight since it would benefit papermakers the most. The Baltimore memorial reads, in part: "Pray that Congress will lay a duty of 25 cents per pound on all writing, printing, and copperplate papers, and 15 cents on all others." The proposal, in effect, represented a 70 percent increase over existing tariffs established by Congress in 1816. The Massachusetts publishers sent a second delegation to Washington in 1819 to oppose exacting tariffs by weight because that would grossly inflate the cost of common papers. An 1821 pamphlet published in New York by Gould & Banks argues: "Mr. Henry Baldwin [of Pennsylvania] proposes a new mode of collecting the duty on paper, because, had he directly proposed a duty of 50 percent ad valorem [by value], the House would probably have revolted."

Congress still couldn't seem to avoid the issue, as controversy now erupted over writing paper watermarked "Napoleon Empereur Et Roi" used by certain of its members (Jeffersonians?). Then on August 5, 1820, the editor of the patriotic *Niles Weekly Register* wrote in a column: "Last winter we indignantly noticed the receipt of a letter from the clerk of the House of Representatives of the United States, written on paper stamped and marked with the royal crown of England.... The paper was of a very fine quality, better, perhaps, than four-fifths of the members of Congress ever used, perhaps ever saw before their arrival in Washington.... A friend in the senate sent us a sheet of the paper usually laid on the desks of its members, dignified with the same emblem of royalty, at which we were again mortified.... [Please] let it take any shape but that, — a codfish or a hoe-cake — a yoke of oxen or a race horse — anything but the regal crown of England."

Addressing the political issue first, Congress agreed to put all their paper contracts out to bid. The Senate's first writing paper contract went to Simeon and Asa Butler of Suffield, Conn. Congress then deliberated the tax issue, and in 1824 passed a revenue neutral bill replacing the vat tax with higher tariffs. Under this new measure wallpaper continued to be taxed ad valorem, while all other papers would be taxed by the pound. What publishers had long fought against now became law, and common papers became more expensive to import, leading directly to higher prices of all domestic papers.

In 1829 the son of a Capitol Hill paper dealer held a meeting in New York with Charles Owen of Owen & Hurlbut. The Berkshire papers evidently made the proper impression, and soon thereafter William A. Davis wrote the South Lee Mill to say: "It is a very considerable length of time since I was at New York. Any arrangement ... made by my son or nephew will undoubtedly be fulfilled. I get the finest blue wove and laid quatro post for $4.50, and white for $4.25, with the exception of Amies of Pennsylvania, [who] charges $5.50.... I am mortified that our Northern papermakers will not do as well! I want the finest of papers, and if you have superior cap or quatro post, full size, you are at liberty to send $100 worth. But, I do not want it at all unless it is superior!" Fortunately this was not the only dealer in Washington, as the following year the firm of Farnham and Blanchard purchased two cases of O&H's fine letter and cap, paying a total of $340.

Congress pressed home another tariff increase in 1828 that further jolted the publishing industry to action. An anti-tariff convention was held in Philadelphia in the fall of 1831 with over two hundred delegates from fifteen states along with representatives of the iron, textile, and paper industries. The chairman and organizer of the convention was none other than Albert Gallatin. Gallatin had been treasury secretary under Jefferson and Madison, and had been instrumental in removing duties on imported rags in the first place. However, during the War of 1812 Gallatin had gone to St. Petersburg and then to Ghent to negotiate peace with Great Britain. Following this success he remained in Europe to establish further treaties, not returning to the U.S. until 1823. Having been unable to further effect a change in trade policy, Gallatin established the anti-tariff convention, which sent a summary report to the Senate detailing the industry's position on a range of issues including the duty on paper.

Here it is said the current tax of ten cents per pound on common paper and seventeen cents per pound on writing acted as a prohibition on papers from France and Italy. The current tariff increased the price of the domestic article from five to seven cents a pound, and effectively handed American producers a monopoly that acted as a break on production. In other words, domestic producers simply priced their goods the same or slightly less than the imported article. Thus the payment of the tariff lay squarely on the shoulders of the publishing industry, and Gallatin reasoned the rates could be reduced without injury to domestic papermakers. He argued the price of paper could not go down by the entire amount of the duty, and that domestic producers still enjoyed the advantage of duty free rags. The market, in turn, would welcome lower paper prices, and manufacturers would ultimately be compensated by an increase in demand.

Gallatin made an elegant argument, but none too timely as far as the paper industry was concerned. The tariff designed to protect the nation's industry had spurred investment in the paper machine, and by 1832 the machine had come of age, causing the price of newsprint to drop by twenty-five percent. Trade policy would now taken up by President Andrew Jackson, who put a number of high-profile industrialists in his cabinet, including the former paper manufacturer Amos Kendall of Kentucky. On the tariff issue Jackson sought a compromise with Congress, and a new bill — passed on July 14, 1832 — returned tariffs to their former 1824 levels. Still there were aspects of the bill that chaffed the South Carolina legislature, which passed an ordinance of nullification drafted by John C. Calhoun. Under the umbrella of states' rights, the South Carolina resolution declared the federal tariff null and void, and not binding on the state, its officers, or citizens. South Carolina forbade collection of duties within the state and threatened secession if the president should attempt enforcement. Congress retaliated and passed a bill in March of 1833 authorizing the use of force, but mollified the act the same day with the passage of the so-called Compromise Tariff offered by Henry Clay.

Again, within the realm of the paper industry, the time tables for reduction of tariffs seemed impressive, but the compromise really only applied to ad valorem duties, or in other words, to wall paper. Whatever the actions taken on iron and textiles, the bill must have had the desired effect. South Carolina repealed the nullification ordinance. Years later, in 1842, Congress returned to the tariff levels of 1832. Then in 1846 Congress passed the Walker Tariff, a bill named for the free-trading secretary of the treasury from Mississippi, Robert Walker. Walker's bill essentially returned to a system of ad valorem rates for every kind of paper, offset by a five percent tariff (ad valorem) on imported rags. A general reduction of rates occurred in 1857, at which time all rags except for wool were exempted from duty. After South Carolina and rest of the southern states seceded in 1861, Congress raised tariffs by thirty percent across the board.

Across the Atlantic, a similar series of events were happening in London. The paper machine was the engine of production, and manufacturers were eager to export to the U.S. The only problem was that British papers were twice taxed before reaching American consumers. The landed aristocracy of Parliament had used tariffs to subsidize their losses in agriculture, and paper manufacturers often lobbied for a reduction in duties, but to no avail. Finally, upon the ascension of Queen Victoria in 1837, the tariff was significantly reduced. Unfortunately the action came too late, as the combination of tariffs and duties had already given American producers the edge they needed to modernize and upgrade. By 1845 total paper production in the United States exceeded that of Britain and France combined.

The federal government's new procurement system would turn Washington, D.C., into one of the most lucrative markets in the country. Even Owen and Hurlbut got into the act; in a letter of 1842 a Washington paper dealer (W. Naga & Sons) writes them: "[The firm of] Scales & Scallion have taken the balance of your paper of 111 reams, and given a draft on the Clerk of the House payable 15th March." Another Washington merchant, Frank Taylor, later writes: "I saw one of your firm [Owen?] while at Sheffield's, New York ... and he said that he should be able to procure for me 20 reams of blue laid letter for which I had previously written for.... I must [obtain] a substitute in some other quarter if it is not going to come from you." Paper dealer R. Farnham writes O&H from Washington in 1855: "You will have the goodness to write soon in reference to the writing papers for the proposals we are in on the 30th (November). If you intend to put it through to me ... send samples by Adams Express." Owen & Hurlbut had so much success in this market they also took on the "Congress Paper" brand, and Farnham asks of them a few months later: "You will please send me one case white laid Congress cap ruled ... also the same article in blue."[16]

The Panic of 1837 and Manila Paper

Thus far the financial panic of 1837 has been mentioned in conjunction with General Caleb Burbank of Boston, the Gilpins of Wilmington, D&J Ames, and James Steel of Steelville, Pa. The Panic of 1837 was a major financial crisis leading to the failure of over 600 banks and perhaps scores of papermakers. This was a time when banks issued their own paper currency backed by gold. The banks were technically required to have 200 percent reserves in their vaults for every dollar in circulation, but they often failed in the timely replacement of deposits or pulled notes from circulation when the situation called for it. As a result, by 1837 most banks were overexposed in real estate and under invested in specie. The crisis began after the failure of Herman Briggs & Co., a major New Orleans cotton broker. Following a series of margin calls, real estate loans began to be called in and depositors rushed the banks to withdraw their savings before the predictable collapse. Unable to meet depositor demands, the New York banks suspended specie payments on May 10, 1837, and when all was said and done some $741 million was lost and 39,000 firms were bankrupt.

The paper industry was largely insulated from market fluctuations; however, many mills in 1837 were in the midst of building or expanding. So, depending on their current state of affairs, the financial panic hit some harder than others. The Panic of 1837 did have one unintended consequence, that being the invention of manila paper. Lyman Hollingsworth (son of Mark Hollingsworth) and James Whitney of South Braintree established a machine mill in West Groton, Conn., in 1835. In the aftermath of the panic the firm was short of cash and unable to buy rags, so they sought raw material in the refuse of the New England fishing fleet. They gathered up old manila rope, rag-bale ropes, hemp sails, canvas sheets, and the like, then Lyman Hollingsworth cut up some rope bolts and tossed them into the rag engine. Things such as rope, yarn, and burlap were made of bast fibers of jute, flax, and hemp because these have higher tensile strength. Therefore, the beating time must have been considerable, but in the end the cylinder-wire machine lapped up the pulp and made a smooth manila-colored paper. The fiber content of hemp and jute is similar to flax, about 80 percent cellulose, except it could not be bleached as white. Even so, the manila-colored paper had immediate applications as sack or wrapping paper, and from this point forward the West Groton Mill made only manila paper. The one drawback to hemp and jute was that it came only from southeastern states, but over time the New York Mercantile Exchange found additional supplies overseas.[17]

In 1843 the Hollingsworth brothers (John, Mark, and Lyman) imported a rag boiler from England. The rag boiler had been developed by Bryan Donkin to remove dirt and grease from secondary grade rags. The spherical boiler was filled with water and rags and then pressurized with steam to about 100 psi. After rotating within the iron sphere about four hours, the rags came out fairly tenderized. The addition of caustic soda or soda ash created a bleaching action that significantly whitened the fibers. The Hollingsworth Brothers patented the bleaching process and soon made a fortune, as numerous mills around the country now turned to manila manufacture.[18]

The fallout in the hand industry combined with the invention of manila paper brought temporary relief to the rag market. However, as economic conditions improved the number of machine mills grew out of proportion to demand. Given the built-in advantage over imported papers, many investors were lulled into a false sense of security. New ownership often lacked essential papermaking experience, and in many cases sold paper at market prices lower than the cost of production. Inefficiencies put additional demand on the rag market, which by 1851 reached of eight cents a pound for common rags.

William Hoffman of Hoffmanville reported in a letter to a machine manufacturer in 1851 that sales in Baltimore were scarce because it was difficult to get a price above 9 or 10 cents per pound. Newsprint had long sold for 9–10 cents per pound, but those days rapidly came to an end, and many newspapers had to reduce the size of their offerings in order to stay in business. The public also began cutting back on consumption, and commission warehouses were especially hard hit. Many large accounts saved money by going directly to the manufacturer; for example, in 1851 the Philadelphia firm of Hogan, Perkins & Co. paid cash for three cases (120 reams) of blue and white writing paper from the South Lee Mill. Papermakers with built in reserves of cash or rags could afford to wait out the market, while many others were faced with the prospect of either converting to alternative materials or simply going out of business. As mills came into receivership, banks tended to keep the facility intact while seeking new ownership. This environment created something of a revolving door of investors who were most susceptible to repeating the pattern of failure, until eventually the mill either came into possession of a capable papermaker or was torn down.[19]

Long-term customers were often dismayed at the higher prices. For instance, when O&H raised the price of letter paper to $2 per ream, C.W. Rose of New York wrote in May of 1852: "I use considerable paper similar to enclosed half sheet which is of your make.... The last cast price of what I use is $1.50 (this I write upon also cost me that). If you choose to send me a case of the perfect like the enclosed sample @ $1.50 on the usual terms you will please do so immediately as I can scarcely wait for it." A note in the margin indicates that O&H answered, "We have no paper on hand as ordered." In September of that year O&H received a standing order for two cases fine white ruled letter from the New York warehouse of Duncan, Lewis & Bartow, but when the buyer balked at the higher price the answer was again the same. The South Lee Mill had recently filled an order for thirty-two reams of the same to Bunting & Foot of New York, so it's not clear if there was a real paper shortage or if the manufacturer was simply playing possum.[20]

Envelope Evolution

Baltimore was the major port of entry for Southern states, and Owen & Hurlbut had explored this market fairly early on. The papermakers established a business relationship with a paper dealer named Otis Spears, but in 1831 that relationship was on the brink after Spears' note for $74.45 was protested (bounced) by the Fulton bank. Recognizing the severity of the situation, Spears wrote a long letter explaining that he was in Washington on business and his son managed the business in his absence. Business was hopping in Baltimore and other warehouses were clamoring for business. Only that year the Baltimore firm of Houghton & Johnson had written the South Lee Mill to say that blue laid and white wove were in high demand, and they could easily sell a case (40 reams) of fine post every month. Houghton & Johnson had only been in business for a year, but gave William Parker of Troy as a reference. Still, O&H stayed with Otis Spears, who proved himself over the next twenty years to be a most resourceful in the Baltimore market.[21]

An ocean away in Paris, France, an emerging trend was the enclosure of notes and letters in a paper packet sealed by a wax lozenge. The fashion eventually made its way across the English Channel where John Dickinson developed a machine to produce the "pockets" en mass. Victorian society readily adopted the trend and by 1841, with the introduction of penny-postage, fully half of all letters posted in Great Britain used an envelope. The envelope machine reached the U.S. during the late 1830s, and according to McGaw, Berkshire manufacturers obtained die cutting machines from Springfield, Mass. Owen & Hurlbut began registering sales of envelopes in New York and Philadelphia around 1839; the earliest mention comes in an 1840 letter from Otis Spears: "I have just closed sales of your envelope paper calendered and I have now got it introduced into the post office here. [This] can no doubt affect sales to some amount therefore have the goodness to send me 50 to 100 reams of it as soon as you can." This is a very early date for envelopes in the U.S. The U.S. Postal Service resisted the trend by charging double the rate of postage if a letter was enclosed in an envelope. Under the mantle of reform, the postal service introduced lower rates in 1845 and quietly dropped the penalty for envelopes. The post office in Baltimore came to be among the first in the nation to issue postage-paid envelopes.[22]

Owen & Hurlbut registered numerous envelope sales before 1845 in Baltimore, Philadelphia, Pittsfield, and New York. The trend in New York was rapidly sweeping the country, as on December 24, 1845, Owen & Hurlbut received a request from the New York warehouse of Moore & Legget, saying: "If the envelope you offer is of the size wanted by one of our out-of-town customers ... we shall want 100 reams of it ... upon the terms you name.... But, till we hear from him we care to order it at a higher price ... $3 per ream less 7½% at which price you are at liberty to send us 50 reams."[23]

Increasingly use of envelopes prompted the manufacture of $8^{3}/_{4}$" × 11" letter paper. When encased in a standard envelope the packet weighed less than one-half ounce, commensurate with the minimum post office rate. For the longest time post and cap were the most popular brand of letter sheet, but Otis Spears informed O&H in 1847: "I am selling off your papers and think I will get through them in a very short time, except some of those caps [foolscap] that have been on second so long. I write to ask if you cannot give me some latitude on them or allow me to sell them some less than your invoice prices. If you can allow me to sell them for about $2.00, and some of them a little less, I can sell them off. I wish to get through them so that I can solicit fresh goods. I have lost sales ... daily ... by not selling your more modern style of papers and so keep along with the times. I enclose a sample of several cases for your consideration and reply. I should ... want ... more fine white ruled cap $2.00, super white ruled letter $2.25, super blue ruled letter $2.25.... And, if you make any lower priced letter or cap paper a few cases of them. I have a constant call for low priced papers & the stock in market appears to be light."[24]

The envelope die cutting machine was first patented in the U.S. by Nesbitt & Co. of New York. The city soon spawned a cottage industry in hand-folded envelopes, and stationers who formerly engaged in printing blank forms for institutional use now created envelopes to match. This became big business and as manufacturing stationers such as Park and Watson of New York, Ezra Coleman of Philadelphia, and R.L. Hawes of Boston gained the upper hand in envelope manufacture, papermakers such as Owen & Hurlbut were driven from the market. Manufacturing stationers went on to develop ever more fanciful die cuts, embossing, borders, and watermarks, cumulating in the 1860s with the offering of boxed sets of matching stationery in the European style. Over time stationers joined forces with papermakers or acquired paper mills of their own.

The evolution of the envelope combined with stringent market conditions of the 1850s created punishing conditions for writing paper manufacturers. The public was cutting back in the face of higher prices, and now the postal service recognized it needed to slash prices in order to maintain demand. In 1851 the ten-cent rate between contiguous states was reduced to three cents, and this seemingly had the desired effect, as the number of letters

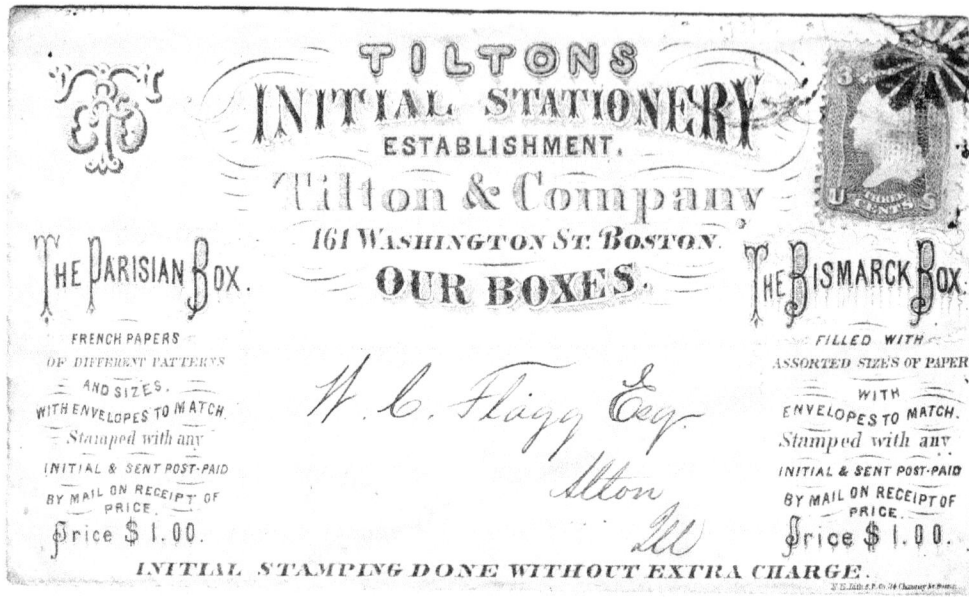

Advertising for Tilton's Stationery in 1867. Boxed stationery in the European style first became popular at the end of the Civil War period.

quickly went on the rise. First class mail volume fairly doubled over the next three years and doubled again by 1857. Still, the use of the folded letter in the form of post paper fell off significantly as inexpensive envelopes made of straw or manila flooded the market. Baltimore paper merchant R. McFeeters wrote to O&H in 1853 to cancel a prior order from September, saying, "Business is very dull here now."[25]

Papermakers such as Owen & Hurlbut managed to survive the crunch by charging nearly as much for letter-sheet as they had for post. In January of 1854 the Philadelphia paper merchant A. Diamond wrote Owen & Hurlbut: "I have this day received your September prices for 1854.... Let mine be the old price and quality as good as you can afford (the fine white laid cap at $1.75 and fine white wove letter at $1.50)." Another cost-saving trend in the marketplace was the note envelope (a.k.a. ladies envelope), which was essentially a half-sized version of the standard envelope. Papermakers were now asked to create even smaller note paper to fit the diminutive pockets. Owen & Hurlbut were quick to fill this need, evidently making their note paper from half sheets of post, and in October of 1852 they received an order from the New York paper warehouse of Wm. H. Hulbert & Co., with the comment: "Our stock of blue laid note now large."[26]

The application of gum to the back flaps of the envelope was an innovation that began around 1850. Gumming was done by hand until an automated process was patented by Thomas Weymouth in 1856. Further improvements to die cutting machines were made in the 1850s by M.G. Puffer, White & Corbin, Duff & Keating, and S.E. Pettee. Pettee later perfected a method of arranging the narrow folds at angles to one another, allowing more blanks to be cut from the same sheet. A folding machine emerged from England in the early 1850s that allowed stationers to automate to a high degree. Competition for government contracts was a major force in the industry. Crane & Co., a Berkshire papermaker, was solicited to make paper in conjunction with a Philadelphia stationer bidding on the first prestamped post office stationery envelopes in 1853, but that contract ultimately went to Nesbitt & Co. of New York, which with advanced equipment was able to underbid by a significant amount. George H. Reay of Brooklyn, N.Y., went on to develop the most advanced envelope-processing machine of the period, one that combined all of the elements of cutting, folding, and gumming. Not surprisingly, Reay went on to capture the lucrative post office stationery contract in 1871.

5

The Growing Empire

The first paper mill in New York was built by John Keating on Manhattan Island between Fly and Burling Slip in 1768. By 1774 the Keeting Mill was removed to Peekskill, a town on the Hudson River that during the Revolutionary War was successfully defended by Maj. General Israel Putnam against the relentless advance of the British Army under Henry Clinton. This location also served as Washington's headquarters and seat of the Continental Army's high command. Following the war in 1790, President George Washington stopped for the night on Long Island in the village of Roslyn as the guest of Hendrick Onderonk, whose paper mill supplied the *New York Gazette*. As Washington toured the plant, the papermaker offered to let him try his hand at the vat, which he did. Onderonk carefully couched the sheet formed by the president and kept it as a lasting memento.

By 1800 the U.S. census reported only twelve mills in all of New York State, most on the Brooklyn River in Queens County. Paper in New York then sold for $3.50 a ream for writing, $3 per ream for book, and 83 cents a ream for wrapping. In 1793 John Walsh started a one-vat mill in Newburgh, Orange County, where he made fine letter and ledger paper for the New York Market. Walsh & Co. went go on to become one of the most respected papermakers of the nineteenth century. New York State saw a flurry of activity in the opening decade of the new century. In 1807 Gen. Walter Martin erected a mill in Martinsburg, Lewis County, and to maintain supplies rags were collected from all over the county and into Upper Canada. The following year S & A Hawley & Co. constructed a mill near Fort Edward at Moreau, Saratoga County, for the production of writing, printing, and book paper. In February of '09 the firm of Wood & Reddington erected a new mill on the Great Western turnpike near the Schoharie Bridge, and in 1810 Henry W. Starin built a mill in Esperance, Schoharie County. By this time New York had no less than twenty-eight paper mills, and certainly would have had many more if not for the War of 1812 that fairly emptied the northern part of the state of its inhabitants. Growth of the New York industry now seemed largely confined to the Catskills, as in 1813 the Hope Mill was erected by Nathan Benjamin in Catskill, Greene County, and Elisah Pitkin established the first paper mill in Columbia County by converting an old flour mill at Stuyvesant Falls.

The War of 1812 underscored a serious weakness in the country's ability defend its territory in the Midwest. Military supplies had to be packed over long mountainous routes at great cost, severely limiting the number of troops that could operate in the theater. A canal project had long been planned for the dense wilderness of New York, but its cost was seemingly prohibitive. Land values in upstate New York became severely depressed after the burning of Buffalo, creating an ideal opportunity for the public project. Following the war the long delayed enterprise was finally approved, with work beginning in 1817. When completed, the Erie Canal extended some 280 miles to connect Lake Erie with the Hudson

River. The waterway officially opened in 1825, and the trade it created far exceeded even the most optimistic projections.

With growing prosperity for all of New York the number of paper mills went on the rise. During the 1820s the firm of Gilman & Silbley started a mill in Rochester, and about the same time a new mill was built in Annsville, Westchester County, by General Pierre Van Cortland, while R.L Underhill & Co. started a new plant in Urbana. A paper mill at Martinsburg reported making writing, wrapping, and wall paper without the benefit of a machine, but the machine was not long in coming. J.B. Sheffield & Son erected the first cylinder-wire machine in 1828 at their plant in Saugerties, Ulster County. More machines went up during the 1830s at the Silver Lake Mill in Gibsonville, Columbia County, and at the Scandaga Mill in Brodlbin, Fulton County. During the 1840s, machine mills were started in Woodville, Jefferson County, at Malden Bridge in Columbia County, and in Lewis County at Lyons Falls and Lyonsdale. By 1850 New York had 106 paper mills, the most of any state. Still the pace of construction never wavered, as more machine mills went up in the towns of Livingston, Ithaca, Sand Lake, Philmont, Hulls Mills, Gargoa, Coeymans, Sandy Hill, Walesvile, Shelby, Accord, Ft. Edward, Rock City Falls, Angelica, Fayetteville, Leona, Milton, Canaan Four Corners, Moriches, and Little Falls.

John Dickinson had demonstrated the ease at which workers could be trained to attend the cylinder machine. Still, the industry in New York was growing out of proportion to all others, and investors increasingly turned to recruiting from mills in New England. A mill worker at Paper Mill Village, N.H., wrote the firm of Kilmer & Ashman at Ballston Spa on March 31, 1848, to say: "Yours of the 18th [March] I did not receive until Saturday.... You say Mr. Gibbons informed you that I wished a situation on the machine. I wrote him of my situation here, and what wages I receive, which is $7.00 per week.... My price [will be] $1.25 per day.... Please inform me how many engines you run, also whether your machine is single or double [cylinder].... I have worked for six years steady.... I will furnish a recommendation from paper manufacturers if necessary. Yours respectfully, J. Bingley Blake."

The double-cylinder was invented in 1830 by John Dickinson, who configured two machines to run simultaneously and arranged the webs to be bonded together using a common press felt and rollers. A good number of mills had double cylinders by the 1850s, but at the time of the above letter it was still something new. Before the end of the century about seventy-seven mills in the U.S. had double or triple cylinder machines; over fifty of these ran on manila, while the rest were straw mills. The kinds of products made at these mills varied, but generally they were tissue, glazed, building and roofing, cord, hardware, box, and carpet lining. It is difficult to tell if the writer had actual experience on a double-cylinder or was simply name dropping since the Cold River Mill in Paper Mill Village is not reported to have had one.[1]

As New York papermakers gained in experience they began to develop equipment of their own. For example in 1852 Norman White of Saugerties patented an improvement in drying sized paper. Horace W. Peaslee of Malden Bridge developed a new rag washer in 1855, and J.S. Blake of Claremont built an automated paper trimmer in 1857. Many of these advancements represented homespun techniques, and may or may not have been ploys to avoid the payment of royalties. Another such example is Addison H. Laflin and his brother Bryon Laflin of the Herkimer Mill. In 1849 Laflin developed a laid wire for the paper machine. This method of watermarking had been tried before in Great Britain; however,

Flooding at Jefferson Paper Mill on the Black River circa 1900. Smoke belches from an unidentified mill in the background.

the industry had long-since moved on to the "dandy" roller. The dandy was a modified version of the suction roller, whose purpose it was to lift the web from the belt as it exited the wet end of the machine. When covered in a layer of wire mesh, the suction roller created watermarks of laid, quadrille, or other repeating patterns such as lilies, etc. By the mid–1830s mass-produced dandy rollers were commercially available, and as a rule the watermark in "imitation handmade" paper is generally clearer and stronger that that of the mould-made variety. As stock dandys wore out after about a year of constant use, they would be replaced with the latest off-the-shelf model. Successive batches of dandys had slightly different dimensions (number of laid lines and chain lines per inch) and through careful observation it may be possible to date the period of use. This method can be very useful in assessing or confirming the date of documents or even envelopes. As one might predict, Laflin's wire belt never really caught on, and in 1857 the Herkimer Mill, along with the extensive wire works, was sold to the Kent Paper Co. for the impressive sum of $70,000.[2]

Fire had always been a threat to paper mills, and investors needed considerable insurance before buying or building a mill property. In 1838, E. Camp erected a mill in Jefferson County that burned down in less than a year at a loss of $8,000 with no insurance. That same year a mill at Urbana burned twice in a twelve-month period. Investors also needed to make sure their stock and rags were insured. In 1853 John Satterly's mill at Little Falls was partially burnt with a loss of $10,000 in stock, and a $20,000 mill in Dansville was lost to fire with insurance of only $10,000. The entire paper industry was so fire and flood prone it's remarkable the insurance companies made any money at all. Other New York mills lost to fire during the 1850s were in Woodville, Nassau, New Baltimore, Balliston Spa, Bath Island, Chatham Four Corners, Manlius, Ogdetisburgh, and Middleburgh.[3]

Black River Straw

In 1830 a papermaking firm from Connecticut named Hamilton & Wright came to Chatham Four Corners (now Chatham) in Columbia County to start a new paper mill. The firm had obtained the straw process from John Shyrock, who had just patented it the year before. The straw mill on the banks of the Steinkill River began the manufacture of wrapping paper, and following this success Chatham Four Corners became something of the center of straw production in the region. Another mill built by Backus and Thomas Wheeler reportedly used a fire dryer, as did the mill of William Davis and Plato Moore built in 1837.[4]

Straw paper came to Rensselaer County in 1846 when John B. Davis and Peter C. Tompkins erected a mill on the Kinderhook Creek in Nassau. Here a 36" cylinder-wire machine with two beating engines and a large bleaching vat produced excellent quality white wrapping paper. A few years later, a 41" cylinder-wire machine was added along with new steam-heated drying cylinders to alleviate the need for loft drying. Peter Tompkins later left the firm, and with his brother, Staats, built another mill in East Chatham where they made paper entirely of straw construction with little or no rag stock.[5]

The Wheeler Mill in Chatham Four Corners later diverged from the traditional straw-rag mix and began making a new variety of straw-manila based on the bleaching process patented by the Hollingsworth brothers in 1843. This new variety, called "bogus manila," was made of 20 percent straw and 80 percent hemp rope or burlap bags, with a bit of cotton rag thrown in just for texture. The popularity of bogus manila caught on at other manila mills, and as a result the rag boiler became widely adopted by 1850. At the Crystal Palace Exhibition in 1853 a new bleaching process was unveiled for straw paper by a French manufacturer. In 1855 the Saratoga *Flag* was printed on paper made at a mill in Rock City Falls, where the manufacturer claimed to use a new French bleaching liquor. It is said the stock contained three-quarters straw and made good printing and writing paper. A few years later

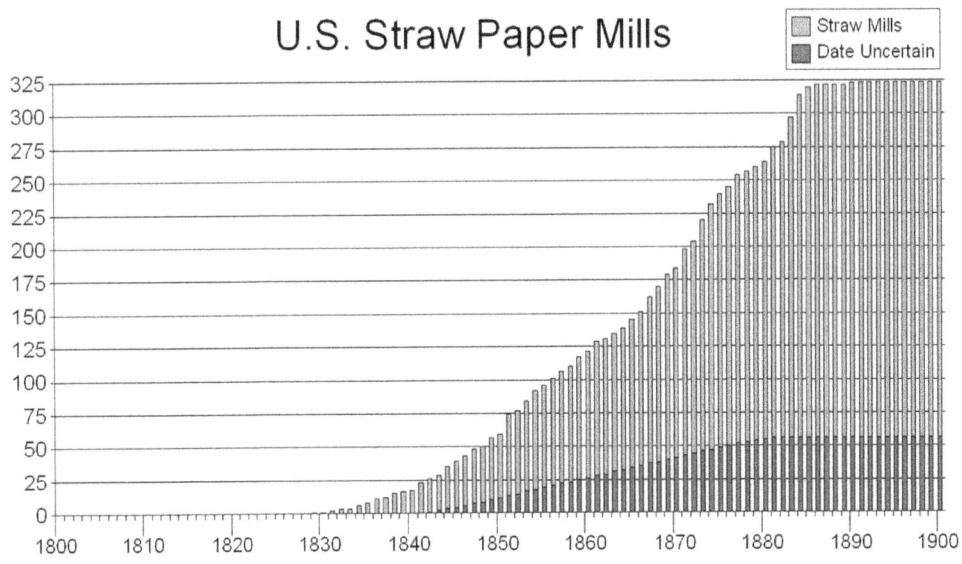

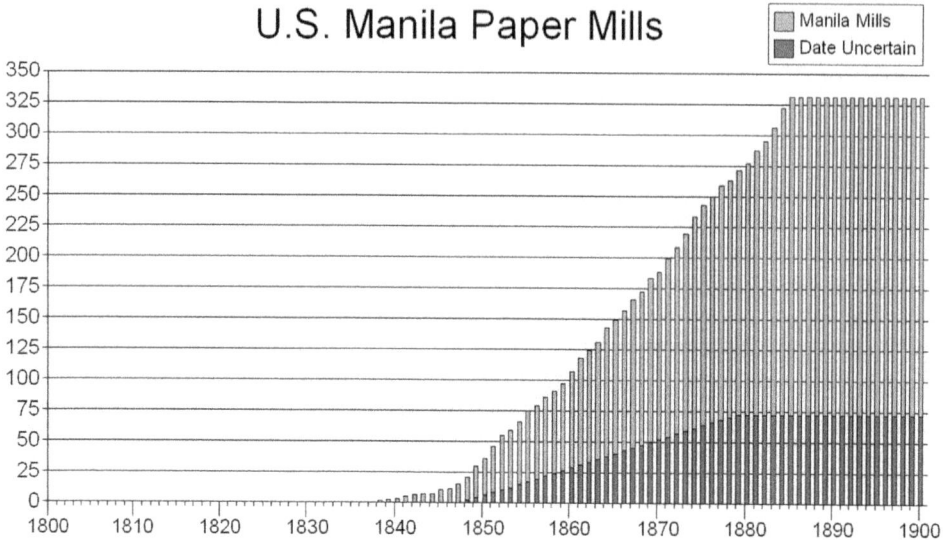

C.S. Buchanan of Ballston Spa, N.Y., transferred the bleaching process to the rotary boiler, which thereafter became known as a "bleach boiler." Few straw mills experimented with printing or writing paper, but all came to employ the bleach boiler to whiten up the stock.[6]

One of the pioneering paper mills of Columbia County in Stuyvesant Falls had earlier been converted to straw by John Hoes. Stephen Rossman, foreman in 1853, patented a device called a "lifting-roller" to prevent the web from tearing as it passed between the wet and dry ends of the machine. One of the principal problems of the paper machine was the synchronization of the two sections, and any jolt to the system could easily tear the web at this point. The lifting-roller won wide acclaim and was featured in an article in the magazine *Scientific American*.

In 1860 Howland & Palmer of Fort Edwards in upstate New York patented the stable fiber method for rye, wheat, and other stocks. As field crops, these fibers invariably contained a mixture of dirt and weeds that could not be economically filtered out, resulting in a paper with noticeable impurities. The stable fiber process employed a line of machinery to first winnow out any impurities, then crush the fibers between heavy iron rollers. The combination of stable fiber and bleach boiling eventually produced a brand of paper called "whisky rye" that became popular with the New York dailies during the Civil War period.

On the Erie Canal

Albany, the capital of New York State, seemed a fitting terminus for the great canal, and Owen & Hurlbut had maintained a number of contacts here over the years. In 1852, E.H. Pease & Co. wrote the South Lee Mill and named the kinds of paper most favored in the city: packet post, demy, single and double flat cap, laid commercial, and royal. The papermakers seemed willing to make sales both great and small, as in 1848 a comparatively minor order was filled for James Henry, another Albany merchant, who ordered just a dozen reams of white ruled cap and six reams of blue flat cap.[7]

The first stop along the canal was Troy, where Owen & Hurlbut had conducted business with William Parker since 1825. This had been a strategic location to intercept rags before they made it to New York, and Parker was very attuned to the trade. Naturally, Parker also sold paper, and in 1830 placed an order for thick demy writing paper for blank work, saying: "If in New York ... send an order for it as we shall be at New York in a few days ... will take it there if you cannot send it from the mill." Following the Panic of 1837 merchants in Troy experienced a serious decline in business, and in May of that year J. Hasford gave up their inventory of letter, demy, and cap formerly held on commission, adding: "Above you have a statement of my account having closed my business as I leave for the West tomorrow. I left the 2 reams of demy with Belcher & Burton bookbinders subject to your order." In May of 1837, Wm. J. Smith & Co. report to O&H on a past due note: "Since Jan. 8th our business has been most wretched. This month, instead of receiving, as we should in ordinary times, at least $3000, we have not received $300.... Our business is completely at a stand for ordinary sales & we do not like to loan our stock at a sacrifice to meet our engagements."[8]

One of the original canal commissioners from 1816 was Stephen Van Rensselaer, and following this experience he became so committed to the advancement of science and engineering that he founded Troy's Rensselaer Polytechnic Institute in 1824. The institute helped Troy become a center of innovation, and in the realm of papermaking, the firm of Thomas & Woodcock began making paper machines here around 1830. It is also reported that the Troy firm of Jarvis & French had been experimenting with a new type of calender with steam-heated rollers, the first of these installed at the Falls Creek Mill at Ithaca. The mill burned in 1842 with a loss of $8000, and there is no mention of what became of the invention. In later years the industry would increasingly turn to "chilled" rolls—a contemporary reference to hardened steel. There were also a surprising number of automated paper cutters to come out of Troy, one patented by J.C. Kneeland & George Phelps, another by Edward Pine, and yet another by Alonzo Gillman. Kneeland would go on to achieve notoriety with an advanced version of the automated layboy, a device that stacked the paper at the end of the line.

The sheer drop in elevation when the canal reached the gorge at Little Falls required twenty-four locks for canal boats to negotiate, and for this reason most passengers took the conveyance from Albany to Troy to link up with their boat in the nearby village of Remington Corners. The canal had become something of a tourist attraction, and so in 1832 the Mohawk & Albany railroad was established on a thirteen-mile line between Albany and Schenectady to take passengers to meet their boats. The railroad, being the first in the nation, had become something of an attraction in itself. Schenectady, situated on the Mohawk River, took everything in stride, as it had grown up as a major staging point for pioneers heading west. The first paper mill was erected here in 1808 by the firm of Van Veghten & Son, which published a local newspaper under the banner of the *Western Budget*.

Up the long plains north of Schenectady is Utica. Here the canal ran straight through town, later giving rise to a considerable light manufacturing industry. Here a sales agent working on behalf of mills in Lee, South Lee, and East Lee passed through in 1830 and spoke with a local merchant named Charles Walbridge. Following this exchange Walbridge wrote the South Lee Mill and placed a large order for Navarino bonnets. These were paper bonnets, or bonnet liners, and evidently quite the rage in hot summer weather. These are said to have been the invention of a papermaker in East Lee, and demand was such that mills in South Lee and Lee Town Center were enlisted to assist.[9]

One of the biggest problems for the canal builders was breaching the Genesee River at Rochester, with the ultimate solution being construction of a stone aqueduct more than three city blocks long. Now supplied with ample waterpower from Genesee Falls, the tiny community of Rochester would be turned into a city of industry. Here were erected numerous iron foundries, textile plants, flour mills, carriage factories, and paper mills. In 1843 O&H got a request from a merchant in Rochester identified as D. Hoyt for ten reams of letter paper to be combined with an order from William Aling, a local bookseller. Aling asked for an assorted case of laid paper consisting of thirty-two reams of cap (plain and ruled), and the rest of letter (blue, white and white vellum). Aling directed the shipment to go by way of the six-day line from Albany, adding that another Berkshire papermaker (Platner & Smith) shipped free of charge. Evidently the freight charges were minimal for goods going up the canal, so another merchant taking advantage of free shipping was Saul Hamilton, who placed a similar sized order for letter and foolscap, saying: "I wish you would ship the above immediately so that it may be at Albany by the 13th of next month (April) to come up by the first boat on the opening of navigation. Mark it by the canal from Albany. I expect you will pay the freight on this (and all other paper that I may order from you) as other manufactories do."[10]

Continuing along the canal, instead of traversing the short distance from Rochester to Lake Ontario, it takes a strategic turn to the east towards the mighty headwaters of the Niagara Falls. Buffalo was then a small tourist town for visitors to the falls that during the War of 1812 had been burned by the Americans to keep it from becoming a base of operations for the enemy. Now with the coming of the canal, the sleepy little village would be transformed into one of the most important ports on the Great Lakes. Also, on the other side of the border in Canada (then known as British North America) sat the frontier town of Law, later Toronto. Paper goods were allowed to travel into Canada free of tax, but would again be subject to British tariff should they wish to re-enter the U.S. This situation is made fairly plain in an 1831 letter from the aforementioned sales agent, who wrote Owen & Hurlbut: "Since I came here ... I have labored under every ... disadvantage.... The amount of paper sold [6 reams of cap and letter] is ... $11.25. You [also] requested [to] know how many black bonnets I received — 20 doz.... Beside the cost of getting them here, when I sold I left the both (paper and bonnets) on the other side [in Canada], not wishing to run the risk of smuggling it across knowing I could not pay the duties without great sacrifice. I sold it to a friend here in Buffalo and he found ... he could not get as much as I gave for it."[11]

Early Watertown

Watertown is located in the foothills of the Adirondack Mountains about 10 miles from Saskets Harbor on Lake Ontario, and the first mill here was erected in 1808 on the Black River by Gurdon Caswell. Caswell learned the trade while working for his father at small mill near Utica. In 1816 the Black River Mill would be sold to the firm of Holbrook & Fessenden of Brattleborough, Vt. Evidently the Vermonters had some difficulty managing the mill from a distance and so leased it to George Knowlton and Clark Rice around 1824.

Knowlton was a former printer and bookbinder, thus a book bindery with a ream cutter and a water-powered press would be added to the facility, and the firm of Knowlton & Rice came to be known for their fine book and printing paper. In December of 1829 Knowlton & Rice took out an ad in the *Watertown Register* announcing, "By having in employ none but good workmen, and by industry and attention to business they hope to deserve an amount of public patronage as shall enable them to do a permanent and successful business."[12]

In 1831 Knowlton & Rice wrote to Holbrook & Fessenden about their plans to erect a paper machine, saying: "We have been making preparations to put in a [paper] machine next spring, for we begin to feel the influence of these things, both in regard to the quality of the paper and the quantity. We cannot make enough with our one vat for our own use, to say nothing of sending to New York markets to raise cash. On an average, machine paper is better than handmade, too, for it does not so much depend on the skill of the workman. It has been a matter of some consideration whether we should get the machine made here, or by Thomas & Woodcock of Troy, but have concluded, as our machinists are inexperienced, to employ T&W provided they can give us credit…. We have your and their instructions as to the necessary designs for putting in a machine." As these plans took shape, K&R wrote again to say: "You can tell T&W that there is a machine in operation here made by one of our town's mechanics, and that the only reason for employing them [Thomas & Woodcock] is, that from their experience, we look for a superior article. Hardness of terms would still give the preference here. They must go the full extent in the nicety of the machine, and yet have the price clever, for us."

Apparently, after finding streaks in K&R's new paper, the landlord suggested a pulp dresser made by Thomas & Woodcock. The pulp dresser was an agitator to insure the fibers picked up by the machine were randomly mixed so that the fabric of the paper would be strong and "tight," as papermakers like to say. In the hand mill a "hog" was a simple two bladed fan that stirred the pulp and kept it from settling to the bottom of the vat. In the evolution of the paper machine the "hog box" replaced the vat, and from here the pulp would be conveyed to the machine. A series of strainers prevented any lumps or knots from reaching the machine, the only problem being upon exiting the strainers the fibers tended to run in parallel to one another, and this weakened the fabric of the paper and caused the characteristic streaking. The solution was a pulp dresser. This device was a simple rotating cog that agitated the fibers, sending them in random directions just before reaching the machine. A pulp dresser developed by Coleman Sellers entirely did away with board strainers. His self-cleaning pulp dresser consisted of a set of star-shaped strainers mounted around a hollow center shaft that allowed pulp to pass directly to the machine. The device was rotated at 40 revolutions per minute, so pulp passing through was thoroughly mixed before reaching the machine, while any lumps or knots stopped at the through-holes were simply flipped back into the hog box where they eventually settled to the bottom.

It seems Clark Rice had been penny-pinching, with the result that their machine produced an inferior product. Now looking at the situation along purely financial lines, Rice writes: "On further reflection about the pulp dresser, we have a mind to propose that we will take one on trial for one or two years, and pay an interest of 10 per cent on its price, for the use, and then if we conclude, 5 percent shall be refunded, or go into the purchase money [purchase price]. We could then judge its utility by comparing our business with the other machine in town. All this is on the supposition that the pulp dresser can be put in and taken out without injury to the machine. I hear the 'shake' or 'agitation' that we have been told of as an improvement to machines. We wish to know if that [can be] appended to T&W's machine at $425 price." One can occasionally find early letter-sheet with such streaking, and so Clark Rice was not alone in this regard.

Knowlton & Rice were very proud of their accomplishment, and the *Watertown Freeman* reports in April of 1832: "[The mill is] now rolling off in a style, and a quality, superior ... to any ever made in the county." However, disaster struck a year later when a fire in the tannery next door spread to the mill and ruined it completely. The loss was reported at $9,000 with insurance of $5,000. In the immediate aftermath the papermakers quickly revived their hand operations as newspaper later reported: "Knowlton and Rice beg leave to state to merchants, in this and neighboring counties, that, not withstanding the severe loss they have sustained on the destruction of their paper mill by fire, they will endeavor to keep up supplies as usual of books, paper, etc., and would invite a continuance of patronage."

The machine mill was closed for nearly three years while the owners struggled to rebuild. The cylinder-wire could be rebuilt, but a lot of the rollers and other small parts had to be fabricated in a local machine shop and this became a real learning experience. Holbrook & Fessenden now sold the property to Knowlton & Rice, who renamed it the Auburn Mill. Business later picked up rather well, and in 1836 the *Freeman* reports: "The proprietors of the Auburn Paper Mill, with a view to accommodate the citizens of Jefferson County with a little better article of paper than they have heretofore been furnished with, have made a depository [warehouse] where they can at all times to obtain writing, letter, and wrapping paper."

Clark Rice learned a good deal about the inner workings of the machine and in 1840 patented what he called a "plunge washer," which was a device to keep the cylinder-wire free of pulp in the washing stage. Rice ran into some disputes with competing patents, so

Knowlton Brothers Mill in Waterton about 1900. The mill dam is at the right.

he never realized much if any profit from it. Then, fire struck the mill once again in 1848, and this time Rice departed the firm. The plant was then completely rebuilt, and the Knowlton Brothers' Mill, as it was known thereafter, went on to produce up to 2400 pounds a day of fine writing paper.

On Beekman Street

New York had the third largest port in the colonies after Philadelphia and Boston. Following General Washington's retreat in 1777, nearly half the town was destroyed by fire, while the rest remained under military occupation until 1783. The British came back to trade in the post-war period, but this era of prosperity would soon be interrupted by the Napoleonic Wars and the War of 1812. Hemmed in by the British, French, Spanish, the United States was viewed in Europe as little more than a string of ports on the Atlantic until Napoleon ceded the Louisiana Purchase in 1804. Still, America lacked the military capacity to protect those interests, as witnessed by the burning of Detroit during the War of 1812. A grand canal running across the wilderness of New York would solve this problem, but despite the obvious benefit both the Congress and the president rejected all appeals for federal support. Finally, at a meeting in New York's City Hotel in 1816, patriotism ran high and the usual political opposition was swept aside, as New Yorkers threw their unequivocal support behind the canal. At the very least the huge project was viewed as a jobs program, so with seed money from the State of New York and individual investors footing the bill, the project broke ground in 1817. After its completion in 1825 the Erie Canal exceeded even the most optimistic predictions, and in connecting the Great Lakes to the Atlantic, the port of New York went on to become the largest shipping facility in the country, if not the world.

In New York the publishing industry was growing by leaps and bounds. By 1828 it consumed some 15,000 reams of book and news paper valued at $75,000. Reporting on these events was the *New York Journal of Commerce*, which began as a small financial paper run out of a basement office of the Mercantile Exchange. In 1831 publishers David Hale and James Buchanan reported that as a result of the paper machine, the size of the offering grew by twenty-five percent, yet the cost of newsprint over the same period had declined by a similar percentage. The center of the paper trade was on Beekman Street, where Johanthan Seymour established the first paper warehouse in the city in 1818. Seymour was a printer from Hartford, who in 1793 was contracted by Noah Webster, then practicing law in the city, to publish his book, *Minerva*. Seymour partnered with George F. Hopkins and set up a bookselling shop at 45 John St. As added income Seymour began stockpiling paper and supplies to sell to other printers, and this enterprise became so successful that in 1818 he sold the book shop and established a commission paper warehouse at 27 Beekman St. By 1840 Seymour admitted his son to the firm, making it Seymour & Son, and in 1850, Hamilton Smith, who had a mill at Whippany, N.J., joined the firm, making it Seymour & Co. The firm did a great deal of business with paper dealer Cyrus Field, William Sergeant of Sergeant Brothers Rag Dealers on 78 William St., J.E. Preble's blank book factory, and envelope manufacturer Samuel Raynor. Charles O'Hara, a former clerk for the paper house of Hann & Beebe on Fulton St., joined the firm as a supervisor of shipping and receiving, then rose to become president in 1859. Seymour & Co. then bought the failed Persee &

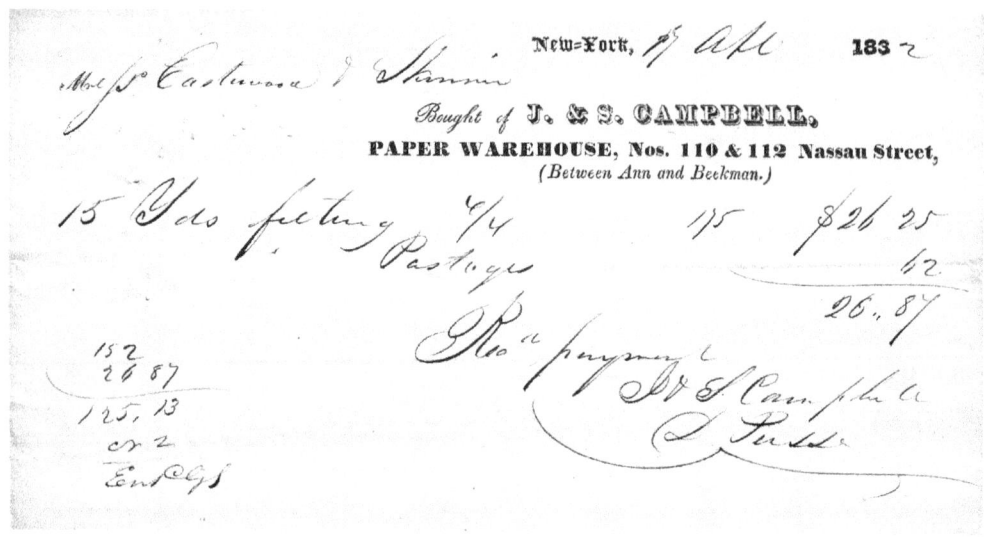

J. & S. Campbell paper warehouse receipt for sheet felting, 1832.

Brooks Mill in Windsor Locks, Conn., and after extensive refurbishment it was reopened as the Pacific Mills. At this point the Seymour Paper Co. became the largest supplier of book paper in the city.[13]

The second paper warehouse in the city was started by Samuel Campbell, a papermaker from Scotland who settled in Millburn, N.J., around 1789. Here he bought an old forge with its waterpower and converted it to papermaking, calling it the Thistle Mill. The mill was destroyed by fire in 1805, but quickly rebuilt, as it was one of the few quality paper mills in the vicinity of New York. Campbell reportedly made banknote paper for a Philadelphia bank, which sent an agent to account for every sheet. In 1820 Campbell opened a paper warehouse at 110-112 Nassau St. between Ann and Beekman. Campbell would be succeeded in business by his son, John, who focused his energy on the warehouse while leasing the mill to James Clark and Oliver E. Bailey. In later years, the mill was equipped with a cylinder-wire machine to make newsprint for the New York market. The Thistle Mill succumbed to fire in 1857 and was not rebuilt. John Campbell's son-in-law, Richard Augustine Smith, joined the firm in 1843, and the business would be reorganized once again in 1854 when John H. Hall was admitted. Hall had been a member of Babcock, Dubisson, and Hall, which broke up after Babcock drowned while on a European vacation.[14]

New York's paper warehouses would take on even greater importance following the completion of the canal. For example, the Red Mill in Berkshire County registered no sales whatsoever in New York, though by 1826 the papermaker had named his newborn son Seymour in appreciation of all the sales made through Seymour's warehouse. According to Dr. Hunter, the first machine mill north of the border had been erected by the firm of John Eastwood and Colin Skinner on the Don River just outside Toronto. Evidently, one or both of the owners traveled the Erie Canal to New York to obtain paper-making equipment and supplies. They found Beekman St. buzzing with activity, and according to a receipt dated April of 1832, they obtained sheet felting at J. & S. Campbell's warehouse. This quarter also had available wire cloth for hand moulds, wire mesh for the cylinder-mould, and wires for the Fourdrinier machine, and so the next day the firm visited John M'Chensney, a wireworker at 108 Beekman St., and picked up nineteen feet of 16 gauge brass cloth and thirty-nine feet of 55 gauge brass cloth.[15]

Beekman Street intersected Front Street just two blocks down from Wall Street. In 1832, Charles Owen forwarded a letter to Thomas Hurlbut at an office located above a store at 22 Exchange St., this between Broadway and Broad St., and just one block from either Beekman St. or the Merchants Exchange Building. After the exchange burned down in the great fire of 1835 a new facility was established in a building on Steamship Row (1 Bowling Green), where the Alexander Hamilton Customs House (est. 1899) now stands. Following the fire Hurlbut's mail was directed care of Barstow's paper warehouse on Beekman Street, an area that luckily survived the catastrophe. Displaced stock and bond traders later acquired the site of the old Merchants Exchange Building, where they built the New York Stock Exchange.

The O&H brand sold extremely well in New York, as in 1836 the firm received an order from the firm of Myers & Minn exclaiming: "We are ... out of blue & white satin letter paper. We could have sold 20 reams this day if we had it. Please send by first opportunity. We should like to receive a general assortment of all your papers as spring sales have already commenced." An inventory sent by Owen to Hurlbut in 1837 describes three cases of paper, each containing forty reams of white and blue satin letter paper from the Phoenix Mill. The paper was destined for the New York warehouses of Herman & Mellen, A.D. Camp & Co., and George & Edgar Barstow. These shipments all went via by barge. In September of 1838 Owen wrote, "We sent today by Mr. Powers 3 cases of paper ... to be shipped by Hudson Towboat Barnes Line." This shipment contained a mixed assortment, including blue laid satin cap (No. 1, and medium), white satin letter, blue laid satin letter, blue vellum satin letter, and white vellum satin.[16]

As traditional papermakers, Owen & Hurlbut treated their writing paper with sizing, which was a gelatinous glue made from ground up hides, horns, and hooves of farm animals. This method is called "tub sizing" because the sheets were literally dipped in a tub of size. Traditional sizing featured largely in the manufacture of fine letter and ledger paper with the idea being to allow the pen to glide effortlessly across the sheet, while sizing also delayed ink saturation long enough to reduce blotting and even allow for minor error correction. About 1835 Owen & Hurlbut began the process of sizing and calendering to produce a fine mirrored finish they called "satin." Other manufacturers did the same thing under the brand of "superfine bath satin" or combinations thereof.[17]

Mr. Hurlbut's name had apparently been bandied all around Beekman St., and in 1838 a recent immigrant to the city sent a letter via Albany steamboat to the South Lee Mill with a business proposal, saying: "I take the liberty of penning a few lines to inform you that in connection with one of my brothers we have taken the store at No. 7 John St. corner of Gold for the purpose of doing a commission business.... [Would] you be inclined to send us an invoice of your fine letter & foolscap & some of your laid printing.... Call on me when you come to the city.... P.S. As you are postmaster [of South Lee], I send this for free."[18]

In 1840 Owen & Hurlbut installed a new paper machine that was used for writing, printing, and plate paper. In September of 1841 they receive an ecstatic request from A.D. Moore's New York paper warehouse: "I sold nearly all of your 22" × 32" and expect to close the balance tomorrow.... I enclose samples of your 22" × 32", it is a good paper & I can obtain 12.5 cents per sht. for the sizes I have named." This may not the astronomical $100 to $125 per ream commanded by Tileston & Hollingsworth's plate paper, nevertheless, at $62.50 a ream these sales must have been music to a papermaker's ear. Moore continues: "I have sold only 100 reams of 24" × 36" as the quality is inferior for the price. I expect Mr. Hunt [a printer?] will take the 100 reams of the 24" × 36" medium at $4.... I can sell 500 or 600 reams of this medium pretty soon at $4, but cannot get more. I should like 200 reams 21" × 31" of the quality of the 22" × 32", also 100 reams 26" × 37" ... a very good article as good as the 22" × 32" will answer. Also, 200 reams 24" × 28" like the 22" × 32". Also say 100 reams 24" × 38" of the same quality." What may be gathered from this exchange is that O&H rapidly adapted to making paper by machine, and had begun producing high quality print paper at a very competitive prices.[19]

In 1842 a publishing convention was held in New York and the organizing committee valued the nation's annual paper production at $5 million. The growth in the publishing industry over the next generation remained strong. In 1847, the year Hoe & Co. introduced its "lightning" press, the New York *Weekly Tribune*, which got most of its paper in Lee, Mass., was the largest newspaper in the city with a circulation of 200,000. By comparison, the *New York Herald*, dubbed the world's largest newspaper (physically), boasted a circulation of 77,000 in 1860. The railroads and telegraphs now allowed for just-in-time inventories, for example, the *Journal of Commerce* reported in 1848 that a major paper in New York could place an order (by telegraph) at 9 o'clock in the morning and have it by 9 o'clock the next day. The publishing trade had become the most capital intensive business in the city, and by 1860 the presses of New York alone consumed over $5 million of paper in the production of books and newspapers valued at $11 million.

Not surprisingly, a number of inventions and improvements came out of New York over the years. In 1830 Ephraim F. and Thomas Blank patented a method of turning leather shavings into pulp and making leather board on a wet machine. At its peak about twenty-eight mills in the nation made leather board, which had a variety of decorative and practical applications. Between 1855 and 1857, Louis Koch patented improvements to the manufacture of paste board that would give rise to cardboard. In 1859 John Meyerhofer patented a rosin used in making the first waterproof paper.[20]

During the latter half of the 1850s Owen & Hurlbut still did great business in New York. In July of 1858 W.H. Parsons' paper warehouse at No. 16 Beekman St. requested samples of O&H's wove blue ruled letter valued at $1.75, $2, $2.25, and $2.50 per ream, and wove blue ruled foolscap at $1.75 and $2.25 a ream, but the buyer was still chaffing over the price increases, writing: "We regret we cannot offer you more for the paper of which we wrote. But, we mean to do more in the plan of paper. You may make & may possibly send an order of respectable size [but] we wrote for foolscap & letter samples from a large western house." Foolscap continued to be a big seller for the South Lee Mill; for example, in 1859 O&H received an order for one case of 16 pound flat cap at $2.25 a ream from the warehouse of H.C.M. Hulbert, a customer since 1842.[21]

Cyrus Field, Esquire

During the 1840s one of the most successful commission warehouses in New York was owned by Cyrus Field, who went on to earn a place in history in conjunction with laying

5—The Growing Empire

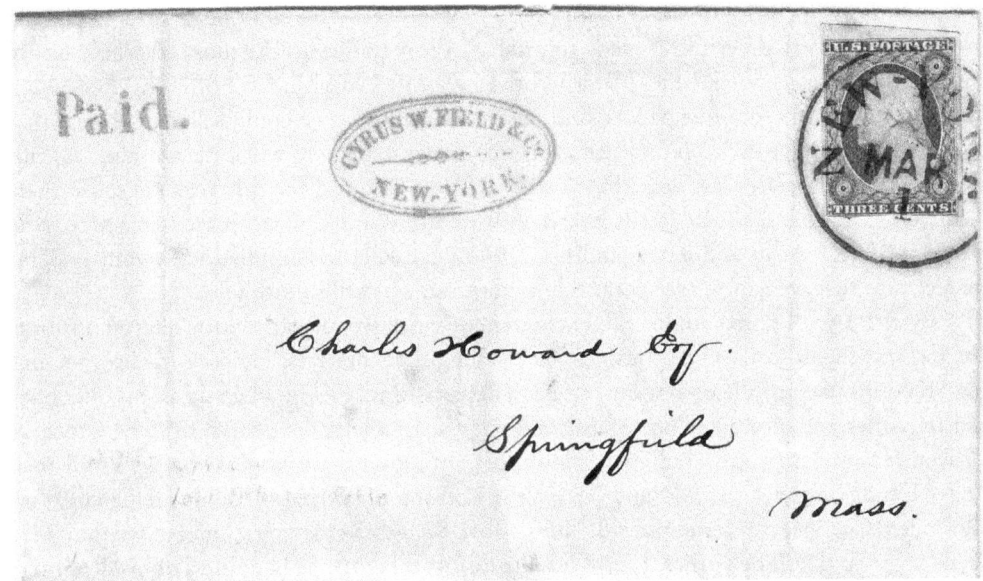

A letter of 1853 from Cyrus Field addressed to Charles Howard, Springfield, Mass.

the first transatlantic telegraph cable. Born in Stockbridge on November 30, 1819, Cyrus West Field began with little more than a grade school education when he went away to New York on eight dollars given him by his father. A fifteen-year-old errand boy for a hardware and dry goods store on Broadway between Murray and Warren, Field quickly worked his way up to storeroom clerk at $300 per year. Meanwhile, older brother Marshall went to work in a Berkshire paper mill, rising to the position of superintendent. Marshall Field formed a partnership with George H. Phelps in 1836 to buy the Columbia Mill in Lee. After Phelps & Field took over operations of the mill, young Cyrus was hired as a bookkeeper at the salary of $250 a year plus room and board. in October 1838 Cyrus Field wrote Owen & Hurlbut concerning a $381 shipment of lead that had come in from Wells & Woodruff of Hartford. Lead had a variety of uses in the paper mill, everything from sealing pipe joints to balancing weights on the waterwheel. Obviously Field's knowledge from this short stint at his brother's mill would serve him well in later years.

The Phelps & Field mill succumbed to fire in 1840 with losses of $20,000. Marshall Field stayed on to oversee reconstruction; however, by 1849 he had left the business and moved to Chicago, where he established Marshall Field & Co., a leading department store. Out of work, Cyrus took a minority interest at a paper mill in Russell, Mass., but soon after accepted a position at E. Root & Co., noted paper dealers on Maiden Lane in New York. E. Root & Co. was badly weakened in the Panic of 1837, and despite the best efforts of Cyrus Field the firm failed in 1841.

With financial help from his brother-in-law, Joseph F. Stone, Cyrus Field now started a commission warehouse of his own in New York. Reprising his role as traveling salesman and purchasing agent, Field established a paper warehouse at 11 Cliff St. near John Street, a rag warehouse at 58 Cliff St., near Beekman, and offices at 9 Burling Slip where he could expedite rag shipments. The firm published a catalog of goods and their letterhead of 1845 describes the availability of rags, felting, bleaching power, wire cloth, alum, oil of vitriol, and "paper of every description made to order at short notice on the most favorable terms for cash or approval notes." Field later wrote the South Lee Mill: "We often have inquires from our customers for your papers, and I have no doubt ... we could sell a large quantity in the course of a year. If you will consign our paper to sell we will cash all sales less the interest. Our cash sales last year (1845) were $221,484.21 and for this year are more than they were last. Our sales the last twelve days were $13,051.99." Field placed an order with O&H in 1848 for cap and letter, adding, "If you can keep us supplied with your papers, we feel sure that we can render you such account sales as will please you."[22]

By the age of thirty-three the entrepreneur's net worth came to a quarter million dollars, and Field now left the business to pursue other interests. He gave his brother-in-law $100,000 to continue operations, and as a departing act Field looked up all his old debts and liabilities and cleared them regardless of status. In a letter of March 1853 he wrote to Charles Howard in South Hadley to announce: "In looking over my old books I find that I have a judgment against you and Mr. Wells Lathrop occurring in the supreme court of Hampden Co., Mass., September 3, 1850, [for] $2,975.54. I have this day written Mr. Wells Lathrop that if he wishes it, or if it will add anything to the happiness of his declining years, I will cancel this judgment. I will say the same to you hoping that you may live many years in health, be prosperous, and happy, is the sincere wish." Howard & Lathrop's mill was closed down in 1848 to make way for a new dam at Hadley Falls. On the other side of the ledger, Field also looked up creditors of the defunct E. Root & Co. and paid all these old debts in full with seven percent interest.[23]

It may be possible that Field had personal reasons for leaving New York at this time. In a note to the Lee Bank in 1854 he writes, "It is impossible for me to say when I shall return to the city [New York], for I shall not leave here until my dear little boy Arthur is better." A few months later Field wrote the same on mourning border stationery indicating a death in the family.[24]

Throughout his career Cyrus Field prided himself for answering all written communications within twenty-four hours, so conducting business with the London warehouses must have been frustrating with all the long distance letter-writing. In the mid–1850s Field became involved with a venture to lay telegraph cables between New York City, St. Johns in Newfoundland, and the coast of Ireland. Commanding the propeller ships HMS *Agamemnon* and HMS *Niagara*, Field completed the sixteen hundred mile span across the North Atlantic in October of 1858. The unprecedented achievement prompted a telegraph from Queen Victoria to then President James Buchanan. Unfortunately, the underwater cable snapped after just three weeks of use, and it would take engineers nearly five years to develop a multi-wire armored cable that achieved lasting success. Following the cable-laying voyage of the propeller ship SS *Great Eastern* in 1866, Cyrus Field went on to receive the coveted Congressional Gold Medal for his contributions to transatlantic communications.

The Madras Mission

During the 1830s and 1840s, U.S. paper exports exceeded imports by fifty percent. Although still a fraction of production, American-made paper was slowly beginning to make its way around the world. It seems everywhere there was an embassy or mission, opportunity presented itself. Seeking to boost government revenues, the U.S. State Department had become very pro-business, a major impetus in the opening of the Japanese market by Admiral Matthew C. Perry in 1854.

Hoe & Co., with offices on Gold Street in New York, manufactured the steam-powered Napier and lightning presses. The publishing industry was largely responsible for success of the paper machine, and Hoe presses helped fuel the growth and expansion of the paper industry. The company also sent printing presses overseas whenever the opportunity presented. Evidently Henry Hull of the American Board of Commerce in Boston had tipped them off about the needs of one Reverend M. Winslow at the American mission in Madras.

Madras was the presidency (capital) of the British Empire in India, and the center of publishing on the Indian subcontinent. After sending a printing press to the American mission, Hoe & Co. must have offered to assist in any way with ink or paper, and local papermaker C. Yaneat Chelty took up the invitation, writing to Hoe & Co. in 1854: "I have established here paper mills & manufacture paper & therefore ... request ... a diagram and state the cost & charges for a paper making machine ... giving full particulars as to the power required and drive and ... quantity of paper that could be manufactured."[25]

The Indian papermaker went into great detail about the kind of machine, saying: "In some mills paper is taken out of the moulds by hand. Please suggest a plan by which the mould with its deckle could be dropped, and paper made by a small machine which should give both horizontal and circular motion ... when the paper is moulded & comes up to the surface of the pit [vat] ... a boy standing near may take the mould off the machine & substitute another ... the ... couch moulded paper ... taken out to a felt. If you favor me with drawings & specifications of the same with a memo of the cost & charges if possible by return mail. I require a dozen machines of the description."

The request appears to be for a chain-mould machine of the type Nathan Sellers had begun work on some thirty years previous. This machine has its roots in France and England around 1811, and according to Clapperton it was none other than St. Ledger Didot who registered the initial patent. The basic element of this machine was a chain drive that submerged the mould into the vat and shook it on the way up. In the Didot version the machine emptied the mould onto a conveyor that automatically couched the paper as well. A key advantage was that existing hand moulds in either laid or wove format could be used. If the chain-mould was ever commercially used Dr. Hunter makes no mention of it, so it seems to be yet another invention that fell by the wayside.

6

The Demise of Handmade Paper

Frank Taylor, a Washington, D.C., paper merchant, wrote to Owen & Hurlbut in May of 1842 to inquire, "I suppose you were probably making the paper I wanted, but I have heard ... that you have given up making hand paper and mean to confine yourselves entirely to machine." If true, this would be a remarkable turn of events, as handmade paper had been O&H's stock in trade for nearly twenty years. This event gives rise to the question whether this had been a recent decision, or something in the planning for many years.

Going back to the turn of the century, the elders of Pittsfield had long desired a paper mill be built in Berkshire County. The region possessed many natural advantages in the way of waterpower, but not until the arrival of a young papermaker from Worcester in 1801 did the first mill become reality. After construction of a paper mill in Dalton, a local landowner named Samuel Church, who once worked as a vatman at a mill in East Hartford, Conn., established a small two-vat operation on the Housatonic River in the southern reaches of Lee. Encouraged by leading members of the community, and with a generous grant of assistance from the town of Lee, Church erected another mill in the northern reaches of Lee Town Center. This plant was called the Union Mill, and here Church would be joined by two sons, Joseph and Leonard. To further raise the family's fortunes the South Lee Mill was expanded from two to four vats, and in 1819 another of Samuel's sons, Luman Church, erected the Forest Mill in East Lee. The following year Luman converted an old carriage shop in Lee to papermaking, naming it the Enterprise Mill. Samuel Church then sold the South Lee Mill in 1822 as he began work on the Castle Mill on Laurel Lake Stream in East Lee.[1]

The paper machine came to Lee largely as the result of an accident. During work on the Erie Canal the firm of Walter & Winthrop Laflin and Riley Loomis of Southwick, Mass., erected a blasting powder mill in the northern reaches of Lee. Demand by the gigantic project so exceeded capacity that the same erected a second plant on the Beartown Mountain Stream (later known as Powder Mill Brook) in South Lee. In 1823 an accident at the South Lee plant claimed two lives. The following year the other plant in Lee exploded, causing severe damage to nearby houses. The townsfolk were now so opposed to the plant that Laflin & Loomis were forced to make alternate plans. A new company called W.W. & C. Laflin was formed in 1826 and the Crow Hollow Mill was converted to papermaking with four rag engines and a cylinder-wire machine supplied by John Ames. Sitting next to the river, the head-race ran 140 rods above the mill, while the tail-race covered 30 rods back to the Housatonic. The renamed Columbia Mill employed 60 people and produced 40 reams a day, making it one of the larger mills in the country. The venture's success was assured when it contracted with Horace Greeley to supply newsprint for the *New York Tribune*. To keep up with the newspaper's demand, W.W. & C. Laflin then erected the

6—*The Demise of Handmade Paper*   77

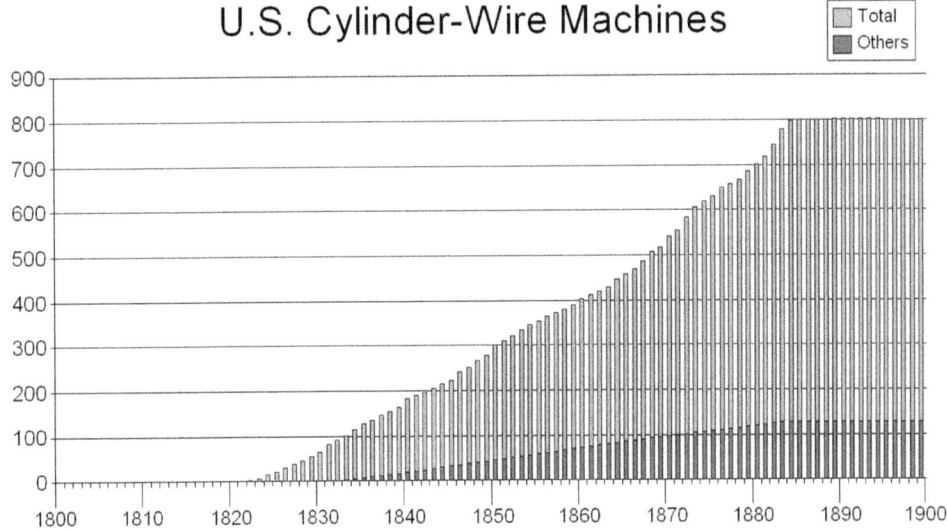

Housatonic Mill about ³/₄ mile upstream with a further capacity of 60 reams of printing paper a day.[2]

Riley Loomis (formerly of Laflin & Loomis) served in the state legislature where he made the acquaintance of Stephen Thatcher, the representative from Lee. Thatcher had grown up on the family farm in Lee, where his father, Jethro Thatcher, supplemented the meager farm income by working as a cooper. In his father's metal shop young Thatcher learned to make copper pails, butter churns, cider barrels, and wire. Thatcher's rise to success came after witnessing the success of Berkshire's early paper mills. In 1825 Thatcher worked with Zenas Crane of the Red Mill in Dalton to build a paper mill on the farmer's property in East Lee. Crane assisted in outfitting the mill, even loaning one of his vatmen to the operation, and within two years Thatcher paid off all his partners and took sole ownership. Now in a burst of ingenuity Thatcher heated a copper plate and stamped out a damp sheet of yellow-tinted paper creating an imitation straw hat. The farmer-turned-papermaker now commenced the production of paper hats and bonnets in white, black, and straw colors, sending his son-in-law, Jared Ingersoll, on sales trips up the Erie Canal in New York and elsewhere. High quality Navarino bonnets retailed for as much as five dollars in New York, while the cost of manufacture was a mere nickel. Demand was so strong that Thatcher soon engaged his neighbors at the South Lee Mill to assist in production.

Riley Loomis likely introduced Thatcher to his former partner, Walter Laflin, who agreed to manufacture the paper bonnets on a much larger scale. Bonnet sales at the New York house of Arthur Tappan, for example, grew to 50 dozen a day. Collaboration on the project eventually brought Owen & Hurlbut in contact with W.W. & C. Laflin, and as the mills in Lee consumed great quantities of rags, it was only a matter of time before the two collaborated on the New York exchange. W.W. & C. Laflin later wrote the South Lee Mill, "Please send us your note for mill rags & [quote] by your letter N.Y. Apr 30/1834 as we have made no entry on our books here, and only a day book in N.Y." In 1833 Owen & Hurlbut joined W.W. & C. Laflin in placing an order for the first sheet-calenders manufactured by Coleman Sellers of Philadelphia. Sellers sent detailed assembly instructions complete with six diagrams, but the Laflins couldn't seem to get the machine working properly. To resolve the problem the machine manufacturer sent his son, Charles Lee, to Lee to set up calenders for both papermakers and to train personnel on their maintenance and care. Apologizing for the inconvenience, Sellers wrote, "We would in all cases prefer sending a man to put them into operation, but that would yield so small a profit that we could not afford it."[3]

Owen & Hurlbut were first contacted by D&J Ames through their business dealings with W.W. & C. Laflin. Since D&J Ames also had a large trade in rags, it was not long before the two were engaged in the trade. In March of 1831 D&J Ames wrote Owen & Hurlbut: "Yours we duly rec'd by the last mail and note the contents. We will send you a wad of the rags on the terms of your letter. They will be of the former quality as those you have had of us lately. We know of no difference in all our No.1 rags, they are sorted mostly for our own use. We should like to sell [to] ... you on the former terms and we will receive all your shavings." There must also have been some discussion about the paper machine, as the letter goes on to say: "We shall charge you one hundred and fifty dollars for our patent right for a one engine mill and two hundred for a two engine. You will have a great deal in the use of them, and all those that work fine stock as you do ought to have these to save stock."[4]

Owen & Hurlbut had begun to use a strong alkaline agent to wash rags, and in 1831 the firm received a letter from D&J Ames saying, "If you want to save expense of transportation we think your teamsters could sell lime here at from 50 to 57 cents a bushel." The active ingredient in lime is calcium carbonate, and this was very effective in washing rags. It so happens that Lee, Mass., sat on a great bed of marble and limestone, and in rebuilding the capitol building in Washington during the 1850s architects traveled to Lee, where they found the fine white grain dolomite they would later use. However, Berkshire farmers found the stones more of a nuisance than anything, and so constructed kilns in small shacks for the burning of marble and limestone to make lime. The results of really clean rags were quite noticeable, as in 1832 William Parker of Troy wrote the South Lee Mill: "The flat thick post arrived from New York a few days since is a very old fashioned No. 2 ... all of it old and smoke dried, and looking quite different from the kind you make now."[5]

Following up on their previous conversations, David Ames Jr. wrote Owen & Hurlbut in his familiar left-handed style to say: "It will cost you one hundred fifty dollars for a one engine mill and two hundred dollars for a two engine mill. This will make the finest paper ever, this is the finest machine ever." South Lee had more than enough waterpower for a machine mill, but Owen & Hurlbut would need to sacrifice one of their plants to do it. Eventually the firm decided to rebuild the original Church mill located east of the bridge over the Housatonic River. In 1833 Owen & Hurlbut allowed their insurance with Aetna to expire and hired a new insurance agent from Boston by the name of C.W. Cartwright. Drawing from experience in the industry, the agent wrote the papermakers regarding the perils of steam boilers: "Yours of the 30th was duly received stating your intention to make some alterations in your paper mill.... This company will not object to your planning a steam boiler in your mill, sitting in mason block(s) as you describe, provided ... you will have no wood work to come nearest the brickwork around your boiler more than one foot....

If you intend to build a new chimney for the boiler flu it should be at least 8" thick, and 9½" would be still better.... These precautions are determined absolutely necessary from our own as well as the experience of others. The danger to be guarded against is the great concentration of heat in a small chimney."[6]

A steam boiler at the South Lee Mill could only mean that D&J Ames provided a steam-heated dryer section to go with the paper machine. John Dickinson had invented the steam dryer (patented in 1817), which alleviated the need for loft drying. Essentially it dried the paper by passing the wet web over a large rotating drum. Steam was admitted to the cylinder through the journal bearing, while the exhaust, or condensation, exited via the bearing on the opposite end.

Detail of earliest multi-cylinder dryer showing steam pipes connected through journal bearings. A hood kept soot from landing on the web as it dried.

Owen & Hurlbut now possessed the best of both worlds. They could make news, wrapping, and wall paper on the machine, continuing to specialize in high quality writing, ledger, and book paper at the Phoenix Mill. As for their good friends in Lee, fire struck the Columbia Mill with a loss of $20,000. After rebuilding, W.W. & C. Laflin sold the plant to the newly formed firm of Phelps and Field.

The Worcester Engineers

Owen & Hurlbut had operated their paper machine in South Lee for two years before a faulty regulator caused the auxiliary boiler to overheat. The paper-makers were lucky the boiler didn't explode, but super-heated steam did end up warping the main drying cylinder, rendering it unusable. The paper machine now had to be shut down, leaving the papermakers to weigh their options. Searching for a competent millwright, the firm settled on a new company from Worcester by the name of Howe & Goddard. As South Lee was still a remote location, much of this business was conducted by mail. Assessing the situation in early 1837, Henry Howe touted his improved dryer, writing: "The D&J Ames dryer [works] well [so] the fault must be in their boiler.... I know that D&J Ames has ... a plan to dry something like mine, but they did not have as large a space for steam [and] it was not conveyed in the same way."[7]

Henry P. Howe was a millwright from Shirley, Mass., who worked in paper mills in the Boston area and is best known for the fire dryer. In 1836 Howe took on an enterprising young partner named Isaac Goddard, and the firm set up offices in Worcester, a rapidly growing railroad hub just outside Boston. Although Howe & Goddard's earliest clients were in Milton and Newton, Worcester seemed the ideal location to allow them to expand their territory westward. In explaining his current availability, Howe informed the South Lee papermakers: "[I] have ... received a letter from Newton where I am to put in this steam dryer. They are putting in a new chain wheel and I think it doubtful about their [being] ready ... to operate this dryer [in] less than 4 or 5 weeks. Therefore, I do not wish to have you wait as it will take 2 weeks to finish yours."[8]

Owen & Hurlbut would not have to wait long, as in February of 1837 Howe wrote again: "Your boiler with all apparatus except the force pump & pipe was delivered at the railroad Monday for Worcester.... One of my calender rolls was ... bad ... & I had to get another cast.... The dryer felting I shall send [from] Boston ... this week. I ordered it made 2 months ago & I think it is at Boston by this time." The following week, Howe reported:

> I this day have recv'd a letter from Mr. Seth Adams [of Boston] concerning the boiler ... being ready ... and [put] board the railroad for Worcester this week. I requested Mr. Adams make a pattern for the grates & send with the boiler. I shall finish your dryer next week.... I can get you a man ... here in Worcester that is acquainted with putting up steam pipes, boilers, & machines ... and knows how to tend steam dryers ... for one dollar per day.... I wish you would write me by return of mail & draw a plan of the room in which the machine & boiler will set [and] on which side ... your paper machine [sits].[9]

Dryer felting is the conveyor belt that guides the web around the dryer section, holding the paper tight against the drum to keep it from curling, cockling or blistering. Evidently not just any belt would do, as Howe later wrote: "The dryer felting is here but I am going to Boston this day & I shall carry it back & get something better." The original belts were made of wool, thus the term "dryer felting." It was later discovered an all-cotton belt better withstood the heat, so in the future dryer belting was made of heavy cotton duck.[10]

The railroad at this time extended only as far west as Worcester, so the new equipment, including the eight foot diameter drying cylinder, had to be shipped to western Massachusetts by ox-drawn sled. Howe updated the customer on its progress: "One of the calender rolls ... did not arrive here till Saturday being blocked up on the Railroad.... I shall send a man this week to assist in setting the Boiler & Dryer." Evidently the job went smoothly and a few months later Howe followed up with the papermakers to ask if he could use their name in his upcoming circular.[11]

Goodwin Rag Dresser

Imported rags accumulated a good deal of dust and dirt during their long sea voyage, and papermakers employed a labor saving device called a "duster" to remove these impurities.

The duster was essentially a cylindrical wire cage driven by an auxiliary gear off the water wheel, and as cage rotated the rags tumbled around, shaking loose miscellaneous debris where it collected below in a tin pan. The duster had been around for a while, for instance, Zenas Crane had supplied Stephen Thatcher's mill with a duster back in 1824.[12]

In January of 1837 Owen & Hurlbut receive a broadside announcing a valuable new labor saving device. Someone evidently combined Beach's rag cutter with the common duster to create a "rag dresser." According to the circular, the rag dresser allowed papermakers to dust the rags and prepare them for the beater in one continuous action:

> Hartford: Having purchased of the original inventor, the patent right to a valuable improvement in paper making ... I take the liberty to lay before you a brief statement of facts.... The improvement ... is designed to supersede the necessity of hand labor in dressing and cleaning rags for the engine.... It consists mainly of the common duster, heretofore used in paper mills, with the solution of a shaft running through the center of the same, in which shaft are inserted a number of knives or beaters, and both shaft and duster are so adjusted by means of hollow axles and bands, or gearing, as to revolve by distinct and several motions.
>
> After a thorough trial, this invention has been found a complete substitute for the old process of dressing and cleaning rags by hand. With this improvement ... a single person may thoroughly dress and clean the rags for the engine, at the rate of from 75 to 150 pounds per hour.... My rag cleaner has been in use several years, with approbation, in the ... establishments of Messrs. H. Hudson, of Hartford, and A.C. & W. Curtis of Newton, Lower Falls, Mass.; also in the mills of Mr. James Donaghee, and Messrs. Candee, Page, & Lester, New Haven; Mr. John H. Walsh, Newburg, N.Y.; more recently in Mr. Henry Burgess' Mill, Hartford; and several others.
>
> At an early period, one of my rag cleaners was purchased by Mr. John Butler, and used in his two engine mill in Manchester. Subsequent to this, the foreman of Mr. Butler, Mr. George Carriel, constructed a rag cleaner, which was put into operation in a four engine mill, then recently erected by Mr. Butler. This machine I consider a palpable infringement of my patent.... Under these circumstances, it is presumed no honorable manufacturer, apprised of the facts in the case, will countenance an attempt to evade my patent....
>
> I will dispose of the right to use my rag cleaner, at a reasonable price; and should be pleased to receive and answer any communication on the subject. Yours respectfully, Geo. Goodwin Sr.[13]

As previously reported, most papermakers preferred to dress their rags into strips or squares by hand and only makers of inexpensive newsprint or wrapping paper used the mechanized rag cutter. Traditional papermakers assert that hand dressing preserves the integrity of the fibers, making for a stronger paper. Still, this is not to say that rag cutting machines were entirely restricted to certain mills; for example, the aforementioned A.C. & W. Curtis of Newton, Henry Hudson of Hartford, and John Walsh of Newburgh were all makers of high quality paper. Even so, simply possessing the machine is not necessarily an endorsement, as for example McGaw reports that the Cranes of Dalton had a rag cutter, but upon seeing the result for the first time, the owner ordered the machine disconnected and never run again.

As mentioned in the Goodwin broadside, John Butler made fine writing paper at his mill in Northampton, Mass., a town about twenty miles north of Springfield. Butler was a papermaker from Manchester who came to Northampton in 1836 to build a single-machine mill served by four rag engines. According to an 1838 letter to a New York banker, Butler now sought a $10,000 loan to expand this mill. The owner reported that the mill and stock were insured for $20,000, while surrounding property of forty acres was valued at an additional $20,000.[14]

Many inventions from this early period would see only limited use and quickly fell by the wayside. Goodwin makes the case for the rag dresser in part by making it appear worthy of theft. His broadside seems more aimed at slandering Butler than disseminating any useful information, and is a fine example of the kind of chicanery that ran through the industry at this time.

The Tandem Dryer

Owen & Hurlbut had already made preparations to replace their steam dryer when news reached them of a more advanced dryer recently installed at a mill in Needham owned by John L. Rice and Luthor Crane. Rice & Crane were rebuilding after a recent fire and imported a new paper machine from England that came with an advanced dryer section. Luthor Crane's brother, Zenas, owned a mill in Dalton, and this is likely where Owen & Hurlbut heard the news. Sometime during work on the South Lee Mill the papermakers quizzed Henry Howe on the subject and received the following answer: "Mr. Crane's steam dryer goes beyond any dryers in use. I received two orders yesterday for this kind of steam. One of them had ordered [a dryer from] Phelps & Spafford but he has seen Mr. Crane's operate ... and has now stopped his order for one of this kind. This man lives in Newton near Mr. Crane where he has a good chance to know."[15]

To get out of this awkward position Howe & Goddard now offered to swap out paper machines and dryers at the South Lee Mill for twelve hundred dollars, adding: "We have one cylinder[-wire machine] of this dimension done w/dryer & rolls.... We will warrant this dryer to dry the handsomest paper of any dryer in use in this country except D&J Ames patent pending dryer." O&H would have been satisfied to simply trade dryer assemblies, and they couldn't understand the need to buy a new paper machine too. It would have been a waste of time to retrofit O&H's aging machine, and so Howe replied:

> We should rather not take yours. But, if you should like to make a change with us at a fair rate & let us make you a whole machine complete w/driers, & take your drier & rolls that I built, we would try & trade.... We should like to hear your views.... Mr. Goddard, my partner, is now in Rochester N.Y. setting up a drier & will be at home in two weeks from this day. I expect he will come by the way of N.York City.... If you write me before that ... I will write him to come by your place & see what can be done by the way of a trade. If we build anything we wish to build the vat, pulp dispensers, brass cylinders, machine frame, felt washers, one set of prep rolls, two five foot cylinder driers, two sets of calender rolls, reels & cutter, with all the gearing shafting & pulleys (except your first shaft & gears). Mr. Goddard would go & help put the machine in operation by paying his expenses to and from home. We charge nothing for putting up machines unless hindered by other work.

Henry Howe then goes on to describe the operation of the new dryer assembly:

> Our machine will consist of: two cast iron cylinders 60 inches wide 5 foot diameter, two pair iron calender rolls, one pair iron press rolls ... the frames to be cast iron with cutter, vat, pulp dressers & all the shafting & gearing & pulley except the first driving power.
>
> The way they are built thus: The cast iron cylinders five feet diameter each with heads put in perfectly tight. The cylinders sit within 3 inches of the flow & the paper is put into the top & a pair of calender rolls sits between the two cylinders so that the paper after it has passed round the first drying cylinder it then passes through the calender rolls to the next

cylinder which finishes the drying & then it passes through another set of calenders which by this operation all the impression of the felt is taken out & the paper is smooth both sides. Calendering when it is hot & partly dry eliminates wrinkles by having two so large cylinders to dry on. Both sets of calenders must sit so to receive the paper very soon after it leaves the cylinder [-wire machine] before it, in order to have calenders operate well....

We are now building steam driers on an improved plan and found out it makes the best paper ... liked ... by printers. We finished one a week ago & am now building one ... to go to New Hartford.... Besides making much handsomer paper ... this kind of machine will take the lead soon [over] any other steam kind.[16]

Owen & Hurlbut now received some good news, as Henry Howe found a home for their old machine. Evidently while in New York, Goddard met with John Butler of the Northampton Mill, and Butler is said to have been in a great hurry and had already dispatched a team to South Lee to collect the machine. Howe now traveled to Northampton in August of 1839 to supervise setting up the old machine before heading to South Lee. The new cylinder-wire with tandem dryer went into the South Lee Mill in early 1840. Coincidently, the Northampton Mill went on to manufacture inexpensive newsprint for the *Pittsfield Eagle* in 1841, but the South Lee Mill captured this business a year later.[17]

Shuttering the Phoenix Mill

In February of 1839 Charles Owen went on a trip to Springfield to visit with D&J Ames. While there Owen may have stopped to look at the new railroad station. Owen & Hurlbut owned stock in the Western Railroad and anxiously awaited its arrival in Berkshire County. Simultaneously, a second line running north through Connecticut was also being built, this known as the Housatonic Railroad since it traveled as far north as the Housatonic River before turning west to its terminus on the Hudson River. The Western Railroad would continue through Pittsfield before turning south to connect to the Housatonic Railroad at West Stockbridge. Since both lines bypassed the town of Lee, an interconnecting link called the Stockbridge & Pittsfield line was built. Both Charles Owen and Thomas Hurlbut had served in the state legislature, and they would make sure the line ran through South Lee and serviced the South Lee Mill.

Following his return home the papermaker received a letter from John Ames, who wrote: "On the other side please find copy of my specification of trimming reams. I believe it to be a copy of my specification or patent on said machine. I found a copy of the patent among my papers soon after you left. That I made known of when your Mr. Owen was here." The patent reads:

I, John Ames ... have invented a new and useful improvement for cutting or trimming paper in the ream, and the following is a full and exact description of the construction and operation of the said machine as invented and improved by me.

I construct the frame ... with suitable beams.... The knife carriage is of cast metal about half an inch thick.... The said knife is fixed with its edge parallel with the lower edge of the carriage. ... motion by water, steam or other power, imparts a perpendicular motion to the knife carriage. The parts in operation resemble the inverted action and motion of the sweep and saw-gate of a saw-mill....

When the ream to be trimmed is firmly adjusted in the press it is brought under the knife in motion which trims the edge of the paper on one side at a stroke...

John Ames' letter of 1839 to Owen & Hurlbut discussing the latest trimming machine.

>   I do not claim the knife or the separate parts of said machine or said press as my invention: but what I claim as my invention and improvement, is the combination of machinery in the manners above described for the purposes aforesaid. [signed] John Ames ... third day of December, 1833 A.D., before the subscriber, a justice of the peace in and for the said county ... William Bliss[18]

It seems most unlikely Owen traveled all the way to Springfield in the dead of winter just to look at a ream trimmer. At this time the machine at the Water Shops Mill would have been configured with a number of accessories licensed to Ames, including: a hog box with a Sellers pulp dresser, a set of prep rollers and felt washers to eliminate pulp build-up on the press felt, the tandem dryer, and a finishing section with two sets of stacked calender rollers, a reel-up, and automated paper cutter. The reel-up was a recent invention that allowed the web to be collected at one speed and discharged at another, making the job of slicing and cutting much less hectic.

The tandem dryer more than lived up to its billing, and the New York market was soon flooded with high quality machine-made writing paper. Owen & Hurlbut were fortunate to be among the first to have one, and writing paper samples from the South Lee Mill dating between 1842 and 1844 reveal a range of calendered, engine-sized articles in pale blue and off white. Having a hand mill of their own, Owen & Hurlbut were instantly aware of the threat to the handmade article, so the decision came to close the Phoenix Mill sooner, in 1840, rather than later. The mill's closing created new opportunities for the firm, and in 1845 they built an addition to the South Lee Mill along with a new mill dam to claim the unused waterpower from the former Phoenix Mill. About 1847 the real estate and former mill buildings to the south of the bridge were converted into an iron foundry called the South Lee Manufacturing Co., where workshops turned out small parts such as angle irons, strapping, bell hammers, and shafting, and worked in strategic association with the paper mill.

Part Two

The Moving-Wire Machine

7

The Machine Manufacturers

The invention of the first paper machine came within ten years of Christophe-Philippe Oberkampf's invention of the wallpaper-printing machine in 1785. Wallpaper was big business in France, its history going way back to 1481 when Jean Bourdichon painted 50 rolls of paper with angels on a blue background for Louis XI. The king had the paper mounted on panels to use as a portable background when moving from one castle to the next. To service the aristocracy's growing appetite, a paperhangers' guild was established in 1599. In 1675 Jean-Michel Papillon developed engraved block printing using continuous patterns, and it was this process that Oberkampf's machine later automated.

In the early stages of the French Revolution seven members of the Didot family in Paris were engaged in various forms of publishing, and while some dabbled in political scandal sheets, St. Ledger Didot (1767–1829) was in the business of printing classical works for the literary elite. About 1790 he hired a twenty-eight year old proofreader who had recently emerged from the army. St. Ledger's father, Pierre-Francis, owned a paper mill in Essones, about twenty miles outside Paris, and as Pierre-Francis was getting on in age, St. Leger sent his new clerk to the mill to assist with the bookkeeping. The following year his father passed away and the bookkeeper was temporarily put in charge of the paper mill.[1]

It must have come as quite a relief to Nicholas-Louis Robert to receive a bookkeeping assignment outside of Paris just prior to commencement of the Terror. Quite unexpectedly, he was elevated to position of manager of the Essones Mill following the death of its owner in 1793. The Essones Mill made paper of all kinds, including wall paper. The manufacturing of wall paper was particularly labor intensive, a standard 34 foot roll could only be made by pressing individual sheets together and gluing them along the edges. At one point the former sergeant-major clashed with the papermakers' guild over wallpaper construction, and soon realized he had little chance of imposing his will. Stung by the experience, Nicholas-Louis began thinking about the feasibility of a machine to make long sheets of paper to avoid the expensive and time-consuming process. St. Leger Didot published quite a number of technical and scientific works at his shop in Paris, and borrowing from these, Robert had all the information he needed to create just such a device.

Nicholas-Louis' concept was a machine that imitated the shaking motion of the vatman. St. Leger Didot soon warmed to the idea and granted Robert full access to the mill's workshop. Here Robert completed the first working model in 1797, and the results were so encouraging he went on to build a full-scale version the following year. The first web of paper from the machine didn't pass its trial printing on Oberkamph's press, but after experimenting with different grades of pulp, Robert finally developed an acceptable sheet of wall paper made of 100 percent cotton. Following this success Didot encouraged Robert to apply for a patent, which was granted on 9 September 1798. The French Bureau of Arts and Trades was so impressed it issued a special declaration announcing that "Citizen Robert" was "the first to imagine a machine capable of making paper from the vat.... This machine forms paper of great width and of indefinite length ... of perfect quality and thickness.... [This] advantage cannot be derived from ordinary methods of forming paper by hand, where each sheet is limited in size."

The Fourdrinier

The paper machine was perhaps the most significant development in the history of paper since its invention in China eighteen hundred years previous. For his effort St. Leger Didot offered Nicholas-Louis Robert 27,400 francs for rights to the patent, and Robert agreed to payment in installments. Didot went on to file for a separate patent in January of 1799 covering the manufacture of wall paper up to twelve feet wide and fifty feet long. Still, the timing couldn't have been worse, for the French Revolution was wreaking havoc among the business community, and despite the best of intentions, the Didot family couldn't continue to come up with the money to pay Robert. So, lacking the necessary cash or credit, St. Leger Didot had little choice but to sell the patent, in effect, to pay for it. The nation was currently at war with England, but Didot's brother-in-law, John Gamble, happened to be in Paris working on the prisoners exchange commission. Gamble agreed to help, and Didot lent him Robert's model along with several drawings to take back to England. High on the list of potential investors was the London stationer Bloxham & Fourdrinier. The Fourdrinier brothers, Henry and Sealy, were most enthusiastic, and after Gamble registered the plans at the patent office (#2487, April 1801) the brothers paid £1300 for rights to build the machine.

John Smeaton was an English wheelwright best known for his work on Newcome's steam engine and the Eddystone lighthouse. About 1800, Smeaton interviewed a bright young engineering student and recommended him as an apprentice to a millwright in Dartford, then a small town about 15 miles outside London. The millwright, John Hall, had previously done work at a paper mill about ten miles away in Bermondsey, and owners of the same now brought to his workshop a model of a French paper machine. Hall deduced the machine needed a lot of work, and probably more time than it was worth to him; however, his new apprentice seemed very much up to the task. The papermakers engaged Hall to build them a new foundry and workshop at their mill in Bermondsey (Fort Place), and following its completion the apprentice agreed to stay on and work on the machine.

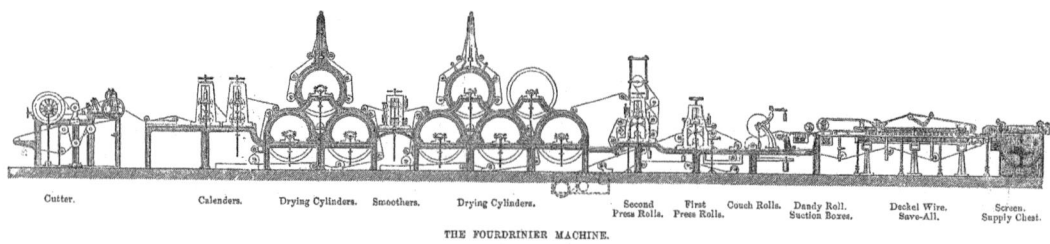

The Fourdrinier machine of 1825–1835. Note the old-style two-stage multi-cylinder dryer.

The Fourdrinier Brothers' new workshop was completed in 1802 and John Gamble was put in charge of building the machine. Experiments with the first machine proved so satisfactory that the Fourdriniers cast their eyes on a mill near Hertfordshire (at Two Waters), which was then available for £2,000. Gamble was directed to build two new 48-inch wide machines for the Two Waters Mill, with which the Fourdriniers intended to make a fortune in the thriving London market.

It was Bryan Donkin's knowledge of physics and problem solving ability that landed him a job with John Hall at Dartford. After witnessing the paper machine in Hall's workshop, the young engineer saw a number of possible improvements and may have been instrumental in convincing the owners to take the next logical step. Then, under the guidance of John Gamble, the aspiring wheelwright made remarkable progress. The frame of the machine was lengthened to twenty-seven feet, and numerous rollers with up-and-down motion were placed along the path allowing the wire to run smoother and drain faster. The overextended frame now required a new backwater collection system, which Donkin pulled off flawlessly. Flexible leather straps in contact with the edge of the wire were used to keep the stock from overflowing the edges, while a new and improved vibrating frame caused the wire to dance as it moved along. When the work was all done Donkin dubbed it the "moving-wire" machine. At the Frogmore Mill, Donkin widened the waterwheel to provide added power for the machine, and following initial trials the new machine performed so well that John Gamble insisted on taking out a further patent (#2708, 1803). Work now began on two machines for the Two Waters Mill, and as these were capable of making 1200 pounds a day, Donkin designed several new and larger capacity hollanders. Following success at the Two Waters Mill, Bryan Donkin was elevated to plant manager and given lease on the tool and dye workshop at Fort Place, where he went on to build eighteen more machines over a ten year period from 1806 to 1816.[2]

John Dickinson invented the cylinder-wire machine in 1809 and set up his first machine mill the following year. Dickinson's machine held a number of advantages over the Fourdrinier; it had fewer moving parts, was cheaper to build and maintain, and was easier to operate. Dickinson's machine also had twin stuff chests that allowed the machinery to continue running from batch to batch. The biggest knock against the Fourdrinier was it only made paper from cotton rags, while Dickinson's, it was said, "could make paper from a hedge." If the Fourdrinier Brothers wished to preserve their investment then certain accommodations had to be made with Dickinson. To begin with, Dickinson was given a long-term lease on a moving-wire machine for his own use. The next year, 1814, Dickinson was given a royalty-free lease on the Two Waters Mill for £100 a year. No mention was ever made of a quid pro quo, but Dickinson, for his part, kept the cylinder-wire machine off the market until after the Fourdrinier patent expired in 1822.

The give and take with John Dickinson gave Bryan Donkin the time he needed to make further refinements to the Fourdrinier machine. In 1811 Donkin invented a staggered couch-roll that allowed the machine to run considerably faster. Speed would become the main advantage of the moving-wire machine, and following this innovation Donkin went on to sell another 48 machines before the Fourdrinier patent ran out. As a lasting testament to his work, one machine Donkin made for a mill in Berlin, Germany, ran continuously until 1877. The old machine was then acquired by the firm of Kraft & Kunst, who installed it at another mill where it continued in operation for another thirty-five years.

Once St. Leger Didot sold the rights to Robert's machine, he became interested in a new invention called the chain-mould. Between the wars he traveled to Bermondsey to visit with Gamble and Donkin to obtain help with this latest mechanical device, but his timing was such that Donkin was then working on staggered couch rollers (1811). As Donkin's grandson relates the story: "One evening at his lodgings the following idea occurred to him; that by placing the couch rolls at an angle instead of vertically one over the other ... the effect would be that the sheet would be ... less liable to break in passing.... The trial was made at once the same night, as it was not wished that a certain person should be present. With great exertion this was effected, and the mill started, the noise however, awoke the said individual who at once appeared on the scene."[3]

After 1822 the cylinder-wire machine made great inroads in the market. Attachments such as presses, dryers, and cutters were soon available, while similar devices for the moving-wire took years to develop. One story is told of Bryan Donkin, who in 1822 struggled with the design of the steam-heated dryer section for the moving-wire machine. Having heard about Dickinson's work in the area, the wheelwright simply showed up one morning at the Apsley Mill and was given a tour by an unsuspecting worker. Ann Dickinson later recorded the fallout from the visit in her diary, saying: "Mr. Taylor told Mr. D[ickinson] that Elliot had [brought] a tramping paper-maker all over the mills ... which so enraged Mr. D that he gave Elliot notice ... and order'd poor Francis the machine man and Gates the steam roll man to be dismiss'd." Five days later Dickinson forgave the workers and put them back on the job—such was the nature of the business in the early days.[4]

Connecticut Papermaking

The paper industry in Connecticut was very active from an early date, given readily available waterpower and a busy publishing industry in Hartford and New York. Hartford was a regional center of publishing, the fourth largest in the country. Most of the paper mills in the state were located along the Connecticut River, but many also took root in eastern and western reaches of the state due to ready access of costal shipping or overland routes to New York. As early as 1800 the state had sixteen paper mills, the first erected in 1766 by a merchant from Norwich named Christopher Leffingwell. Leffingwell sent a trusted aid named John Bliss to a paper mill near Philadelphia to learn the trade, and upon his return started up a mill on the Yantic River at Norwich Falls just north of New London. The *New London Gazette* carried a notice on December 10, 1766, that said: "The paper on which this Gazette is printed was manufactured at Norwich. ... proof that this Colony can furnish itself with one very considerable article which has heretofore carried thousand of pounds out of it." The Leffingwell Mill was financially troubled from the start, but the enterprise was deemed necessary by state officials, who began subsidizing the plant at the rate of forty pence a ream on writing paper and twenty pence per ream on printing paper. During the Revolutionary War the Leffingwell Mill proved indispensable, and for his contribution to the cause George Washington later appointed Christopher Leffingwell captain of the Port of Norwich.

Hartford sits on the Connecticut River about sixty miles upstream from Long Island Sound. Inland schooners and flatboats navigated the river as far north as the falls at Hadley, which were so imposing it was actually cheaper to ship by way of Boston than to port goods around them. The state of Connecticut finally authorized construction of a canal around the falls in 1792, setting off a succession of canal building activities on the upper Connecticut River in the communities of Turners Falls, Mass., and Bellows Falls, Vt. Eventually, riverboats could travel up the length of the Connecticut River all the way to New Hampshire.

Hartford County became home to several paper mills, most situated on the Hockanum River around the town of Manchester. In 1789 the brothers Simeon and Asa Butler erected a mill in a suburb of Suffield called Union Village. The name of the plant was the Eagle Mill, in honor of the recently ratified national bird, and like other enterprises in the region it made paper for the book binders of Hartford. A former teamster at the mill recalls carting paper to Hartford and picking up a keg of rum before returning by way of Windsor to collect a load of rags. The mill's claim to fame came in 1820 when it secured the stationery contract for the U.S. Senate, though the papermakers may have been assisted to some degree by the postmaster general, Gideon Granger, who also was from Suffield.

The firm of Bissell & Pease started another paper mill in Suffield around 1803, this plant perched on Stony Brook, a tributary of the Connecticut River. Here the firm developed a heavy grayish wrapping paper used to ship tobacco. Following approval by customs inspectors the wrapping paper replaced the old tobacco cask, saving weight and reducing the cost of shipping. The mill was named in honor of Benjamin Franklin, and by the late 1820s it was running fine writing paper for the New York and Hartford markets. The hand mill eventually closed in 1848 and was sold to Joseph Daniel Stowe, a machinist from Newton Lower Falls. Stowe went on to refurbish the plant and install cylinder machines for the manufacture of manila paper.[5]

The cylinder-wire machine started a wave of new activity in Connecticut between 1829 and 1851. Several new machine mills went up in New London and New Haven, while Hartford experienced the most growth with eight new machine mills constructed between 1830 and 1854. The first straw mill in the state is believed to have been built by the firm of Smith & Bassett around 1834 in the town of Seymour, New Haven County. In 1847, C.H. Dexter converted his mill in Windsor Locks to manila. Dexter began by using gunny sacks and jute butts, but eventually found all the mill's needs on the New York Exchange. Another mill in Windsor Locks was called the Hibernian Mill because most of its workers were recent immigrants of Ireland. This plant caught fire in 1856, but it was completely rebuilt in 90 days' time at a cost of just $2,000.

Beach, Pickering, and the French Connection

Joseph Pickering owned a hand mill in North Windham, Conn., and in December of 1827 imported a Fourdrinier machine. He had initially gone to England, but instead found a used machine in France. The machine was shipped to the workshop of George Spafford, a talented mechanic in South Windham, Conn. Spafford's partner was a millwright named James Phelps, who made all the necessary waterpower improvements to the Pickering Mill. Local historians report the machine went into operation in March 1828.

The first Fourdrinier machine in America reportedly went into a mill in Saugerties in October of 1827. The firm of Beach, Hommerken, & Kearney purchased the Henry Barclay mill in Saugerties and installed the machine, although the identity of the millwright is not known. It seems most unusual the future newspaper publisher is so silent on the matter. For that matter, Beach was also fairly restrained in his circular that came out in June of 1828. In that case, however, it's not what he says, but how he says it. The circular is datelined Springfield (Mass.) and on paper watermarked "AMES." Since Beach's association with D&J Ames goes without saying, it gives the appearance the latter wanted to keep their name out of the news.[6]

Solomon Curtis came to Newton Lower Falls in 1789 and established a mill powered by the upper dam of the Charles River. Curtis was succeeded in business by sons Allen, Crocker and William. The new firm of A.C. & W. Curtis went in for waterpower improvements in 1825, taking water from the pond at a higher elevation. Three years later they imported two 60" moving-wire machines from Donkin & Co. of London. These were the first moving-wire machines in Massachusetts, and as a security measure iron bars were installed all along the outside windows of the mill. The original structure soon outlived its usefulness, and in 1834 a new plant of stone construction was built over the existing site.

The extent of Bryan Donkin's involvement regarding the first moving-wire machine in America is another mystery. Donkin is said to have made two 60" machines for A.C. & W. Curtis and one for Beach; however, Donkin's records reveal only two machines were ever shipped to the U.S. In 1827 there was still considerable opposition in Parliament to exporting the paper machine, so Donkin may have used an intermediary. A woodcut engraving of the Didot Mill in Essonnes reveals an impressive facility where Nicholas-Louis Robert built the first paper machine. Donkin evidently made a machine for St. Ledger Didot Mill in 1815, and this machine is believed to have been sold to Pickering in 1827. According to Donkin's records the London manufacturer built a 60" machine for Fermin, Didot & Sons just two years earlier in 1825. Since there were no other paper machines in France at the time, it's certain that Fermin, Didot & Sons were involved at least to this degree. Donkin's records also show Didot made a repeat order in 1827 for another 60" Fourdrinier, and one can only speculate whether this was the machine that ended up in Saugerties. If Beach's machine did take such a circuitous route, this may explain his silence on the matter.

The South Windham Engineers

It took Phelps & Spafford about three months to get the Pickering machine up and running. The old machine was in need of new belts, wires, and other spare parts from Bryan Donkin & Co. of London. In the course of construction George Spafford came to believe the paper machine could best be built in this country and at much lower cost, so he created detailed drawings of every part during reassembly. The drawings were hastily made on smooth pine boards that were carefully planed into regular strips. To begin building the machine, Phelps & Spafford needed a tool and die shop, so they purchased an old school house and moved it to a site below a small dam in South Windham. Here the firm constructed a primitive lathe and boring mill, all powered by a small waterwheel. Machine parts were made on site, while cast parts were ported in by ox-cart from a foundry twenty miles away. Spafford also took on a 19-year-old apprentice named Charles Smith, who proved an excellent machinist.[7]

George Spafford liked to believe his moving-wire machine would outperform the original. The first Phelps & Spafford machine went into the old Leffingwell Mill in Norwich Falls. The firm put together a second machine for the mill of Henry Hudson in North Manchester. Back in '02 Henry Hudson converted an old grist mill in the village of Oakland and began making writing paper for the New York market. To advertise this latest achievement Hudson ordered custom machine wires from London. One wire had a seated Britannia, while a second featured an English coat of arms with scales and chevrons. It might have taken the papermaker some time to dispel the impression these were imported papers, and in later years Hudson used watermarks with distinctively American themes.

Phelps & Spafford went on to make machines for mills in East Hartford and Bloomfield, New Jersey, and later took out a patent for the belt-driven steam dryer in 1830. The belt dryer held the paper closer to the cylinder for more even drying, and had originally been patented in England in 1822. While the firm built a going concern, they were caught off-guard by the Panic of 1837 and forced into receivership. Phelps evidently maintained his connections within the industry, and a few years patented the rag washer, which was duly recorded by Munsell:

> 1843. James Phelps, of West Sutton, Mass., made improvements in the washing machine, which consisted of an adjustable, rotating water elevator and strainer, which could be raised or lowered in the vat of the washing or beating engine. Also a rotating prismatic screen, or strainer, for straining the water from the paper-stock, in the vat of a washing or beating engine, in combination with devices for discharging the strained water, being not only more efficient than a Cylindrical screen, but also admitting of more ready repair.[8]

The rag washer was a specially built hollander that strained the pulp with fresh water to remove any residual bleach. The rag washer processed and bleached the pulp into half-stuff, and from there it would be sent to the standard hollander to finish the job. Hunter reports the invention originated at a mill in Breton, France, between 1835 and 1840. In 1847 Phelps issued a circular (including a list of customers that included nearly all the leading papermakers in the country) that read:

> To The Manufacturers of Paper,
> The subscriber takes this method to inform his friends who are engaged in the paper business, that he is the sole inventor and proprietor of the celebrated rag washer, so extensively used and so generally admired. The great number which he has supplied, and the continued demand for them, induce him to believe that no washer has hitherto been used which will compare with this in point of utility and economy. He has now made such arrangements that he will be able to supply with greater dispatch than heretofore all who may be disposed to patronize him. Having for many years been engaged in the paper business, and also in the manufacture of various kinds of paper machinery, he feels full confidence in offering his washer to the public as the best which has been tried.
> His letters patent bear date November 24, 1843, and though many changes have been made in other Washers, and many attempts to evade this patent, he has good authority for believing that none of the artifices resorted to for the purpose of such evasion, can possibly succeed, and he notifies all manufacturers and other persons that the shall be obliged, in the defense of his rights, to prevent the building, vending, or use of such Washers as shall appear to infringe his patent, and wishes that they will consider this such notice as shall put them on their guard. He feels grateful for the very many favors which he has thus far received from this class of manufacturers, and trusts that his work will meet their continued approbations.
> James Phelps, West Sutton, Mass, July 1, 1847[9]

TO THE MANUFACTURERS OF PAPER.

The subscriber takes this method to inform his friends who are engaged in the paper business, that he is the sole Inventor and Proprietor of the celebrated **RAG WASHER**, so extensively used and so generally admired. The great number which he has supplied, and the continued demand for them, induce him to believe that no Washer has hitherto been used which will compare with this in point of utility and economy. He has now made such arrangements that he will be able to supply with greater dispatch than heretofore all who may be disposed to patronize him. Having for many years been engaged in the Paper Business, and also in the manufacture of various kinds of Paper Machinery, he feels full confidence in offering his Washer to the public as the best which has been tried.

His letters patent bear date *November 24, 1843*, and though many changes have been made in other Washers and many attempts to evade this patent, he has good authority for believing that none of the artifices resorted to for the purpose of such evasion, can possibly succeed, and he notifies all Manufacturers and other persons that he shall be obliged, in the defence of his rights, to prevent the building, vending or use of such Washers as shall appear to infringe his patent, and wishes that they will consider this such notice as shall put them on their guard. He feels grateful for the very many favors which he has thus far received from this class of manufacturers, and trusts that his work will meet their continued approbation.

The following are some of the gentlemen who have patronized him, and to whom he is permitted to refer:

TILESTON & HOLLINGWORTH, Boston, Mass.
A. C. & W. CURTIS, Newton L. Falls, "
CREHORE & NEAL, " " "
THOMAS RICE, Jr. " " "
WALES & MILLS, " " "
BENJAMIN FARLYS, " " "
PLATNER & SMITH, Lee, "
BENTON & GARFIELD, " "
ALEX. WHYTE, " "
OWEN & HURLBUT, South Lee, "
CRANE & CO. Dalton, "
W. CARSON & SONS, " "
JESSUP & BROTHERS, Westfield, "
BURBANK & FALES, Russell, "
SILAS GODDARD, Millbury, "
RICHARDS & HOSKINS, Gardiner, Me.
DAY, LYON & CO. Portland, "
R. & A. H. HUBBARD, Norwich, Ct. Ct.

DAVID SMITH, Granville, Ct.
CULVER & MICKLE, Granville, "
GOODWIN & CO. Hartford, "
PLATNER & PORTER, Union Ville, "
PERRSE & BROOKS, New-York City, N. Y.
GAUNT & DERRICKSON, " "
HENRY BARCLAY, Sangerties, "
J. H. WALSH & SON, Newburgh, "
THOMAS HOWLAND, Troy, "
SAVAGE & MOORE, Saquoit, "
KNOWLTON & RICE, Watertown, "
STODDARD & FREEMAN, Rochester, "
H. V. BUTLER, Paterson, N. J.
WM. KAY, Morristown, "
J. M. WILLCOX, Philadelphia, Penn.
JASPER HARDING, Philadelphia, "
MAGARGA & COPE, " "
GEORGE KING, York, "

West Sutton, Mass. July 1, 1847.

JAMES PHELPS.

A James Phelps rag washer circular of 1847 lists thirty-six of the most prominent paper mills and paper warehouses in the country. Phelps, formally of Phelps & Spafford, had developed the first American-made moving wire machine.

About 1840 businessman Harvey Winchester bought the failed South Windham business along with all its patents. The aforementioned Charles Smith, who had made a name for himself in the industry, was duly elevated to full partner. This was the beginning of a long-standing enterprise known as Smith, Winchester & Co., which went on to become a leading manufacturer during the heady days of the 1850s and 1860s at a time when the moving-wire machine finally came of age. In the latter half of the century Guilford Smith and Arthur Winchester carried on the name of the firm; Guilford proved a natural leader, while Arthur was most adept at finances. Smith-Winchester sent machines across the country

Engine room showing the octagonal shaped rag washer attachment to the hollander in action (left center) (from *Harper's Illustrated*, June 1887).

to California, south of the border to Mexico and South America, and even to Great Britain, which would have made George Spafford most proud.

In 1851 Smith-Winchester received an order for dryer felting from Hoffmanville, Md., to which William Hoffman penned the following comment, "I have had my cylinder machine improved by putting in more suction ... to make better paper." This reference is to the new centrifugal pump that Smith-Winchester used in all its new cylinder-wire machines. The centrifugal fan pump provided a more regulated source of vacuum than the old piston pump, and evidently Smith-Winchester had taken to retrofitting a number of mills in the course of their business.[10]

In 1850 Smith-Winchester returned to the old Leffingwell Mill to perform some upgrades. This had been the first mill in the state, as well as the site of the second moving-wire machine. The now twenty-year-old machine was run by the firm of Culver & Mitchell, who went in for upgrades that included a new stuff chest, doctor plates, and a No. 2 pulp regulator. The regulator, or pulp meter, measured the amount of pulp needed for webs of different thickness, thus eliminating the old trial-and-error process. The first regulators for moving-wire machine came out in 1849, so here again Smith-Winchester is seen retrofitting machines with the latest technology. Smith-Winchester also put in a new set of press rollers and knives for the paper cutter. Culver & Mitchell purchased a Phelps rag washer and put in a wider water wheel too. It seems Smith-Winchester sold a good number of rag washers that, over time, earned James Phelps some $77,000 in royalties.[11]

Howe, Goddard, Rice, Barton, and Fales

Machinist Henry Howe and millwright Isaac Goddard started the firm of Howe & Goddard in 1836. From the beginning the firm specialized in fire dryers and steam dryers, and eventually grew to become a manufacturer of cylinder-wire machines as well. When the first pulp regulators for the cylinder-wire machine came out in 1842 Thomas Howland of Troy wrote Owen & Hurlbut, "I have to inform you that the regulator that you speak of that Howe & Goddard put in my mill I am very much pleased with it, and would advise you to get one for it will regulate the making of paper."[12]

With the retirement of Henry Howe in 1845, two new partners, George M. Rice and George S. Barton, were admitted to the firm. Now incorporated as Goddard, Rice & Co., the firm hired a number of French machinists who were experienced in the manufacture of the moving-wire machine. By 1862 the plant manager, Fales, rose to partner, and he began a program of modernization and retooling. The firm now added an iron foundry to make cast parts, and the firm name changed to Rice, Barton and Fales Machine and Iron Company.[13]

Key to the firm's success was the acquisition of important patents such as the Hulton wire guide, Gavitt cutter, Kneeland layboy, and Van de Water undershot waterwheel. The firm made a major acquisition in 1873 when it bought rights to the Harper Fourdrinier. James Harper was a papermaker from East Haven, Conn., and in 1862 he patented an improvement to the moving-wire machine that borrowed from a technique used on cylinder-wire machines. A special set of rollers under the primary wire and couching belt allowed improved contact with the suction roller, gaining valuable efficiency and allowing the web to traverse at much higher speeds. Harper produced just eight machines before selling the rights to Rice, Barton & Fales.

Factory floor showing the standard moving-wire machine at left and the scarce "Harper Fourdrinier" at right. Stacked paper at center rear is being wheeled away from the finishing section (from *Harper's Illustrated*, June 1887).

The key to identification of machine-made paper is in the wire marks. The cylinder-wire machine runs fairly slowly, its speed strictly governed by a secondary gear off the main drive. The slow speed combined with high suction leaves a distinctive wove pattern throughout the sheet. These are true watermarks, meaning the press felt cannot erase them. The moving-wire, on the other hand, does not rely on suction, and has a much higher rate of travel. As a result wire marks are somewhat randomly distributed throughout the sheet, and in later years the machines run so fast it is difficult to discern any wire marks at all.

Pusey, Jones & Co.

In 1848 two mechanics with paper-making experience erected the Nonatum Mill on the Brandywine River in Delaware. The Gilpins were the last in this area to build their own machines, and Joshua Pusey and John Jones now established a machine shop in conjunction with the mill for the same purpose. In 1851 two iron foundry workers from Wilmington, Edward Betts and Joshua Seal, joined the firm and erected a forge where they could make casting parts for the paper machine. Given the remarkable growth in the paper industry at the time, Pusey, Jones & Co. continued building paper machines for others. The firm sent a line dated July 3, 1871, to a papermaker in Suffield Conn., saying: "Dear Sirs, Yours of 28th is at hand, the included specification of a 62" Fourdrinier paper machine which we will make for you ... for the sum of $14,000, line 3$1/2$ months. A machine built upon a specification such as we enclose as a perfect one & strictly first class in all its appointments & entirely the cheapest machine on the market. Should you be pleased to have your order we have in the neighborhood one of the same class running and would take pleasure in showing you."[14]

Pusey-Jones also specialized in dryer assemblies that consisted of six 36" tandem pairs instead of the usual three. With the ever-faster moving-wire machine the strategy was to regulate the temperature of each tandem pair to allow the paper to dry gradually. The temperature of the first pair was relatively low, while each pair in succession was set slightly higher. Manufacturing machines of custom widths was another strategy of the firm; of course this didn't happen until certain advances in wire-working were available. In 1867 Pusey-Jones built two new 200 feet-per-minute machines of 86" width for the Jessup & Moore Mill in Wilmington. The fast-moving machine was a far cry from models that appeared early in the century and was some twenty times faster than the cylinder-wire. Over the next twenty years some of the more popular machine widths were 66", 72", and 76", each seemingly designed for a specific purpose. Pusey-Jones machines went into mills in Roaring Springs, Pa., Louisville, Ky., Ballston Spa, N.Y., Palmer Falls, N.Y., Lawrence, Mass., and Holyoke, Mass., to name a few. The machinery business was so good to the firm they sold their paper mill.

Papertown, USA

In 1852 Walter Laflin wrote to a machine manufacturer in South Windham, Conn., to inquire about the sale of his old cylinder-wire that came out of the Housatonic Mill in

Lee Town Center. Laflin had sold the mill and the new owners installed a moving-wire machine. It was the beginning of a new era in Lee, a time when papermakers increasingly turned to the high-speed paper machine. The moving-wire machine was used primarily for newsprint since its introduction in 1827, but now with the addition of the tandem dryer it could also make book and writing paper. At this point the moving-wire machine was preferred for all new rag mills, while the cylinder-wire continued in the manufacture of straw and manila paper.[15]

East Lee Village would eventually outgrow Lee Town Center in numbers of mills. The village got its first mill in 1819 when Luman Church built the Old Forest Mill. This plant had a mammoth 30-foot-tall overshot water wheel and specialized in fine writing and printing paper. The plant was sold to Jared Ingersoll and Caleb Benton in 1831, and Ingersoll later sold his shares to Harrison Garfield, a farmer and owner of the local country store. Garfield possessed one of the finest greenhouses and graperies in the county, and also presided over the National and Savings Bank. Now working with Caleb Benton, the two expanded the Forest Mill and started up a paper machine in 1835. It was also about this time the secret of "engine sizing" first appeared in the United States. The technology was brought over by master papermaker Joseph Krah, who had recently immigrated to Baltimore. Krah's skills were highly sought after, but eventually he came to work for Garfield & Benton in East Lee. The firm now developed the property next door, and in 1846 set up a moving-wire machine in what would be called the Upper Forest Mill.[16]

Next to arrive in East Lee was Joseph Allen and a group of investors who erected the Washington Mill in 1836. The mill, however, only made wrapping paper, and it subsequently failed in 1840. The firm was reorganized as Allen & Co. and the mill re-started, but failed

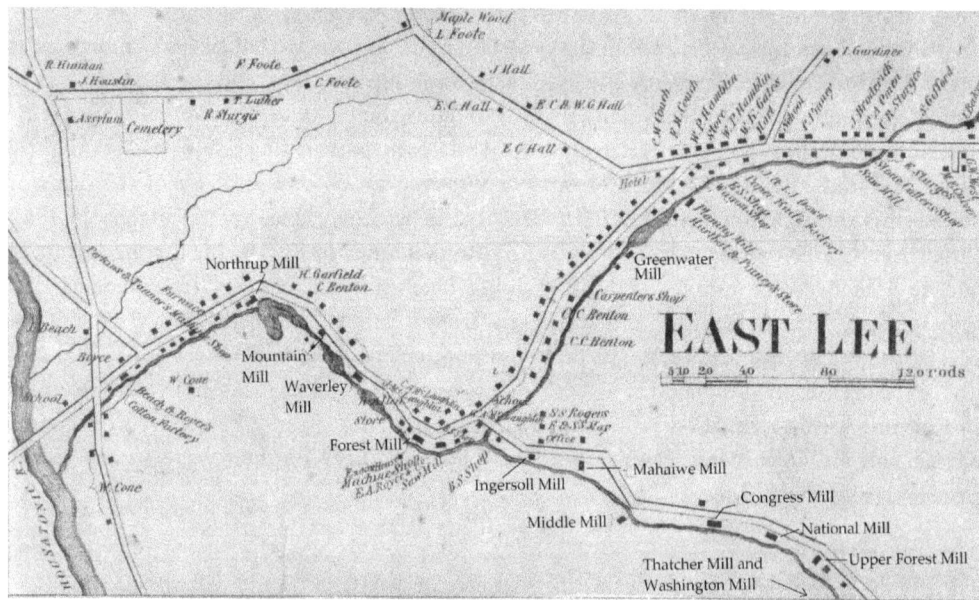

Map of East Lee, Mass., depicting paper mills. This region had the highest concentration of paper mills in the nation prior to 1860 (adapted from Henry F. Walling County Map of Berkshire County, Massachusetts, Smith, Gallup & Co., publishers, 1858).

7—The Machine Manufacturers

A staff of twenty poses in front of the Old Forest Mill in East Lee, Mass. At left is the boiler chimney and wood pile. Typical machine mills built during the 1850s no longer featured a drying loft.

again in 1847. The facility now came into the capable hands of Platner & Porter of Unionville, Conn., who refurbished the plant and sold it to Lunman Phiney & Co., with the "and company" being none other than Harrison Garfield and Caleb Benton.

Stephen Thatcher's mill in East Lee had been built in 1824 with help from Zenas Crane. In 1840 Thatcher and his son-in-law, Jared Ingersoll, refurbished the facility and erected a paper machine. Thatcher & Ingersoll then bought the last remaining Church family mill in East Lee known as the Waverly Mill, and converted it to machine production the manufacture of fine book and writing paper. The old mill was well situated at the confluence of Lake May and Greenwater Pond, which provided some sixteen foot of head, more than enough waterpower for three rags engines making 1000 pounds of pulp per day.

Jared Ingersoll seemed to be everywhere in East Lee, and in 1833 he partnered with George Platner to purchase the Turkey Mill on Hop Brook in Tyringham. This five-year-old mill had been the second machine mill in Berkshire County, and through his many acquisitions Ingersoll gained enough experience to build a mill of his own. He subsequently sold his interest in the Turkey Mill and built a new plant in East Lee on Lake May Stream. Ingersoll's Mill unfortunately burned in 1839, and with all the knowledge gained over the years working with the likes of Joseph Krah, Ingersoll now decided upon a career selling engine sizing and paper colorants.

The ruins of Ingersoll's Mill was sold to the firm of Edward and Sylvester May. Sylvester

May had come to Lee in 1834 to work as foreman of the Columbia Mill. His brother, Edward, owned a woolen factory in Walpole, N.H. Upon buying the Ingersoll Mill the firm converted the plant to the manufacture of straw wrapper, opening the first straw mill in Berkshire County. Sylvester now took interest in a sawmill at the outlet of Lake May reservoir. With cooperation of other local manufacturers he expanded the reservoir to one and one-half miles long and three-quarters of a mile wide. This now supplied a full thirteen foot head of water at the site of the old sawmill, which the Mays now converted to straw manufacture (Middle Mill). In 1853 the May Brothers built the Mahaiwe Mill on Lake May Stream where they made high quality writing, printing, and manuscript paper.

The intensity of papermaking in East Lee changed even more following the building of a depot for the Pittsfield & Stockbridge Railroad. The availability of unused waterpower combined with ready access to the New York market set off yet another wave of mill construction. Five new mills were built between 1852 and 1855, including the Congress Mill, Mountain Mill, National Mill, and Northrup Mill. The ill-fated Northrup Mill was the last one built in the area, and so ended up in a poor location. Over the next thirty years it burned down on four occasions and was rebuilt thrice, while ownership filed for bankruptcy on no less than five separate occasions. The mills of East Lee, when combined with those of South Lee and Lee Town Center, now represented the largest concentration of paper mills in the nation, and for this reason local historians would refer to it as "Papertown, U.S.A."

8

Ohio and the West

At the turn of the century Ohio had no paper mills and wasn't even a state, yet by the end of the Civil War Ohio's paper industry would rank fourth in the nation. The first mill west of the Allegheny Mountains was the Redstone Mill, built in 1796 about four miles outside Brownsville in Fayette County, Pa. This mill supplied paper for the *Pittsfield Gazette*, and became the inspiration for the first paper mill on the Ohio side of the border called the Beaver Creek Mill, which was built in 1806 in Columbiana County. The latter mill made first class writing and book paper on wove moulds measuring 13 by 16 inches.

Dard Hunter, a native of Steubenville, researched much of the early history of Ohio papermaking and discovered that the early steam engine played a large part in the development of hand mills in the region. The town of Steubenville sits on the banks of the Ohio River about ten miles west of Pittsburgh, and according to the *Pittsburgh Almanac* a mill was established here in 1813 that employed a small steam engine to power the beaters. It seemingly all began in 1811 when the first steamboat plied the Ohio River near Pittsburgh. Its rudimentary boiler hadn't enough steam pressure to propel the craft against the strong river current, but other uses were soon found for the sturdy engines. The Bishop Mill in Pittsburgh was probably the first of its kind, employing a sixteen horsepower engine to run the hollander, thus alleviating the need for raceways, waterwheels, or even riverfront property for that matter. The steam engine seems a costly investment for a pioneering mill, as they required some 10,000 bushels of coal per year to operate; still the Bishop Steam Mill seemed to be profitable, employing 40 people and grossing $30,000 a year.[1]

The Clinton Steam Mill in Steubenville began producing wall paper in 1819 for Thomas Cole, an artist who migrated from Chorley, England. The mill added a cylinder machine in 1830 to make paper for the popular *United States Spelling-Book*. Such was the state of affairs when Charles Owen visited Pittsburgh in July of that year. Owen & Hurlbut had engaged in the manufacture of paper bonnets in conjunction with two other mills, and Owen had come to the region to promote their sales and distribution. In making the rounds he evidently made the acquaintance of Christopher Lamden, an apprentice at the Clinton Steam Mill. Christopher's father, who owned a mill in Wheeling, Virginia (now West Virginia) had sent the young man to Steubenville to see what could be learned about the paper machine. It seems immediately after Owen's visit, Christopher went to work at the South Lee Mill, and it was also about this time the Lamden Mill began the manufacture of paper bonnets. Fire struck the family's mill in 1835, and Christopher immediately returned to Wheeling to assist his father in rebuilding.

Completion of the Erie Canal in 1826 provided Ohio with ready markets in the East. Trading centers on the Great Lakes such as Cleveland and Toledo now grew into large commercial ports, and a number of paper mills sprang up around the region at this time. Many

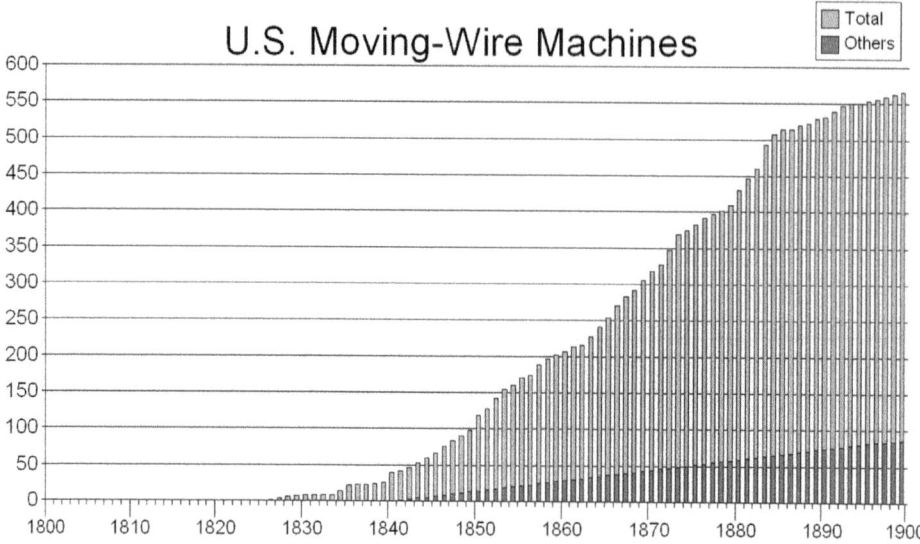

of these plants were steam-powered, such as a mill in Cleveland owned by J. Kellogg that was destroyed by fire in 1831 with a loss of $7,000 and no insurance. The first cylinder-wire machine in this part of the state was established at a mill in Newton Falls, about 10 miles south of Cleveland, and it produced paper at the rate of 425 pounds a day. The mill was leased in 1841 to Frederick Trendley for $1,000 per year. To supplement the mill's income Trendley did small jobs for neighboring mills, such as calendering paper for a mill in Cuyahoga Falls about 30 miles to the west. By 1850 the Cuyahoga Falls Mill was being run on manila wrapper and was selling off its better rags to writing paper mills. By chance Owen & Hurlbut bought some of the rags in the spring of 1851, and papermaker T.L. Miller writes from Cuyahoga Falls to report that all the No. 1 rags had been sent to South Lee by express via East Albany (and the Erie Canal), adding, "My No 2. [rags] are worth more to me to work here than are to ship."[2]

Despite the expense of having to ship all its print paper over the Appalachian Mountains, the state of Ohio had a dozen newspapers by 1810. Publishers in Cincinnati worked to entice experienced papermakers from the East who were willing to migrate. One such recruit was Christian Waldschmidt, son of a master papermaker of Gengenbach, Germany, who had immigrated to American in 1786 to work at a hand mill in Montgomery County, Pa. In January of 1810, Waldschmidt erected a hand mill on the Little Miami River in Sycamore Township, advertising in a local newspaper for rags at three cents a pound cash, or four cents a pound to be paid in blank books or stationery. Waldschmidt's vatman, Johan Schmidt, came from a pioneering mill in Maryland, while another employee, Ebenezer Hiram Stedman, came by way of a pioneering mill in Kentucky.

Zanesville was an important crossroads town on the Muskingum River, a tributary of the Ohio. In 1819 Ezekiel Taylor Cox closed his print shop in New Jersey and moved to Zanesville, where he bought a local newspaper called the *Muskingum Messenger*. As circulation grew the publisher sought to establish a paper mill, and in 1828 Cox & Co. converted a sawmill north of town for this purpose. The hastily constructed mill burned down in 1836, but the enterprise had proved so successful that a new brick plant was built in its place. Cox set up a cylinder-wire machine that supplied all of his interests, with enough leftover capacity to service hungry markets in the central and western reaches of the state.[3]

Ezekiel Taylor Cox's sons, Horatio J. and Jones L., joined the firm in 1846, and they established a paper and rag warehouse in Zanesville's courthouse square. The *Zanesville City Times* reported in 1857 that the Coxes sold printing, wrapping, and roofing paper, but no mention was made of the emerging specialty of telegraph paper. This product was a long narrow strip of soft white paper used to register the dots and dashes of the telegraph machine. Zanesville's telegraphic paper came to be the most well-known brand in the country, and during the Civil War the War Department contracted with the mill, boosting its gross sales to $150,000 a year. A special paper cutter made for the purpose was the inspiration of C.R. Hubbell, a long-time employee of the Zanesville Mill. Rolls of paper were taken from the paper machine and rewound into seven-inch diameter spools that were mounted on the cutter and severed into one-inch segments.

Construction of the Ohio and Erie Canal was authorized by the Canal Commission in 1820. The waterway ran from Lake Erie south through the Cuyahoga and Scioto Valleys, and enabled a network of canals that crisscrossed the state. The canal finally arrived in the central Ohio town of Chillicothe in 1832. Chillicothe had hosted the Ohio Constitutional Convention in 1802, but struggled with Zanesville to hold on to the seat until 1816, when the capital was permanently fixed in Columbus some forty-five miles to the north. In July of 1831 the brothers Hezekiah and Isaiah Ingham abandoned their hand mill in Bucks County, Pennsylvania, and moved to Chillicothe, where they erected a machine mill. Called the Hydraulic Mill, this plant used waterpower supplied by the canal and ran on cap, post, handbill, and wrapping paper for the central Ohio market.

The Miami & Ohio Canal ran directly from Toledo down to Cincinnati, effectively short-circuiting the meandering Ohio & Erie Canal in the race to the Mississippi. Along its banks the town of Middletown would grow into a center of textile manufacturing. The first paper mill was established here in 1827, and Middleton would go on to support eight more paper mills, making it the largest such complex in the state. The town of Dayton was then a crossroads community connected by no fewer than nine turnpikes. With the coming of the canal in 1834 a paper mill was built here by F.J. Diem & Co. Dayton would also host the first straw mill in the state, this erected in 1859 by William Clark. Perhaps the most famous Ohio papermaker was Daniel Mead, who built the first of his three rag mills in Dayton in 1846. The Mead Corporation eventually closed down operations in 1890 and moved to Chillicothe, where it had built an enormous wood-pulp facility.

Following developments in Ohio, paper manufacturing rapidly spread to Indiana. In 1826 Isaac Mooney, who had learned the trade at a mill on the Little Miami River, established a two-vat mill in Big Creek, just north of Madison. The following year John Sheets, of Warren County, Ohio, built a two-vat mill on the Indian-Kentuck Creek east of Madison, and in 1831 a small one-vat mill was set-up in the town Richmond by the firm of Leeds, Jones, & Bissell. The first paper machine in the state was brought from Cincinnati in 1834 and erected in Brookville, Franklin County, by the firm of Phillips and Speer. The Wabash & Erie Canal linked Fort Wayne with Lake Erie in 1835, and soon thereafter paper mills were established in the towns of Delphi, Logansport, and Lafayette. The earliest manila mill in Indiana was established in Rockton in 1854.[4]

Cincinnati and Hamilton

Brothers James and Joseph Graham had been working at the Redstone Mill outside Pittsburgh at the time the owners installed an Ames cylinder-wire machine about 1825. The brothers then moved west to Cincinnati with plans of building a machine mill of their own. They established a mill on Decker Creek powered by water from a subterranean channel running from the east side of the narrows. To avoid potential lawsuits they built a cylinder-wire machine in secret, sending patterns and castings disguised as unrelated equipment to various different foundries and workshops. The Grahams hired plant manager Samuel Jackson, a relative of Andrew Jackson, who was given strict orders to keep the mill closed to all visitors, and that included stockholders. Having the only key, Jackson duly locked up the mill every day at lunch when he went across the street to eat at a local tavern, where he was promptly served regardless of other patrons.[5]

In 1825 the Graham brothers built a second plant on the Ohio River at foot of Western Row, now Central Avenue. Again they assembled a cylinder-wire of their own making, but the night before going into operation the plant caught fire and burned to the ground. It was later determined that local competition, fearing the threat of the new machine-mill, had organized the carnage. A new and even larger mill of 130 by 36 feet went up, this called the Phoenix Mill, as it had been raised from the ashes. This plant also employed an auxiliary steam engine to power the beaters, and when it finally went into operation in June of 1826 it employed some seventy-two workers. One of the specialties of this mill was a very smooth book binder board made by rubbing with polishing stones brought from France especially for the purpose.

To service the town's growing needs the Graham brothers also erected a paper and rag warehouse. In 1835 the brothers purchased 300 acres of land just north of town around Hamilton. To establish water rights they constructed a brush dam across the Little Miami River, this built by laying entire trees side by side with limbs or brush ends up and pinned across the butt end to form a crib that was filled with stone and gravel. Here the brothers made writing, bonnet boards, and wrapping paper. Hard times forced the closure of the Grahams' warehouse in 1851, and the Graham Mill was sold at sheriff's sale in 1857 to M.P. Alston, the mayor of Hamilton. Alston wisely converted the plant to straw wrapping paper, saving money by using broom corn, clean cuttings, and waste paper. In 1866, and again in 1868, floods changed the course of Little Miami River, leaving the mill high and dry. The site was abandoned, and about 1877 a machine tender from Gedds, Mich., came to Hamilton to view the abandoned property. Upon inspection the machinist found much of the equipment still buried, and purchased the property for $150. The machinery was removed to an abandoned sawmill in Hamilton, and after refurbishment the hollanders and related equipment were put to work at a mill at Delphos, Ohio. Here was established two 62" cylinder-wire machines with 36" triple-tandem copper dryers, a reel-up with stop cutter, and two Knight dryers. The former machine tender later sold the mill and realized seven thousand dollars of profit in the deal.

James Brown Graham left his father's warehouse in Cincinnati and moved to St. Louis in 1855 to become a dealer of paper, rags, and paper-making supplies. In 1859 James was joined by his brother Benjamin. The Graham Paper Co. went on to become the largest distributor of finished paper in Missouri.

As Cincinnati grew a number of other paper mills were established in and around Hamilton. John Erwin erected the Hamilton Mill in 1848 with machinery removed from the now defunct Phoenix Mill in Cincinnati. To this was added tandem dryers, an automated cutter, and two Knight dryers. Knight dryers were hollow cast iron rollers 12 inches in diameter, set in groups of three or five. Used for tub-sizing, the polished cylinders were heated by steam around which the paper passed, creating a super fine finish. This mill was later sold to J.C. Skinner & Co., which converted the plant to manila in 1878, renaming it the Hamilton Tissue Mill.

In 1848 Calvin Reilly and Adam Laurie, the latter a papermaker from Glasgow, Scotland, began an ambitious project to build a two-machine mill in Hamilton before running into serious financial difficulties. Laurie turned to a wealthy lawyer in town named William Beckett, who loaned the enterprise $12,000. Business was slow at first, the mill making one ton of wrapping paper a day valued at just $155. However, during the Civil War the Miami Mill contracted with *Cincinnati Commercial Gazette* and became quite profitable. The price of rags kept going up during the war, so Reilly & Laurie looked to an added source of raw material in the form of old and unsold newspapers. Engineers knew no method of de-inking so the pulp ran grayish; also from time to time there were reports of readers finding small bits of undigested text within the lines of the newspaper.

Chicago Style

Chicago was first settled around 1837. At the time, Oliver Morris Butler was working at a small wrapping paper mill in Hubble Falls, Vt., before transferring to the South Lee Mill. Two years later Butler took part of his pay in finished paper and moved to Chicago where he and his brother, Julius, established a paper warehouse. The warehouse sold all kinds of print paper, such as medium, double-medium, super royal, and imperial. J.W. Butler Paper Co. also stocked first class letter, extra fine map, lithograph, and plate paper, while wrapping paper, banknote, and tissue could be made to order. During the 1850s the firm relocated to 42 and 44 State St., and Mathew Laflin joined the firm at one point, making it Laflin, Butler & Co. In 1869 the warehouse moved to 114-116 Wabash Ave in the prominent Drake block. A fire in 1870 led to heavy losses and forced a move to Monroe Street Laflin sold out just before the Great Chicago Fire of 1871, and by 1876 the firm reincorporated again under the old title of the J.W. Butler Paper Co.[6]

In 1839 a hand mill was set up in an old barn using a windmill to power the hollander just like in the old country. Still, the windy city didn't provide enough power to keep the beaters turning, so the papermaker, M. Devitt of Pennsylvania, moved operations thirty-five miles to the west to St. Charles. Devitt set up the machinery in an old blacksmith shop powered by an ancient paddle wheel in the Fox River. Oliver Butler and Joseph Hunt bought the Devitt Mill in 1842, and after securing water rights on both sides of the river built a new dam and erected a 36" cylinder-wire machine ordered from Thomas & Woodcock of Troy. The machine made 400 pounds a day and made wrapping paper except for every fourth week, when it made newsprint. Over 1847-48 Butler & Hunt built a large brick mill on opposite bank called the West Side Mill. Here was installed a new 36" cylinder-wire machine with three 36" tandem dryers, a calender, and stop cutter. The firm made newsprint in two gigantic sizes; 29 × 43 in. called Long John, and 31×46 in. better known as Elephant. Both mills, employing a total of eighty persons, eventually came under the ownership of J.W. Butler Paper Co. When the West Side Mill was destroyed by fire in 1866, the Butler brothers gave up on manufacturing and focused entirely on their warehouse. Soon after the Great Chicago Fire, Oliver Morris Butler withdrew from the firm to become president of the Lockport Paper Co. The West Side Mill was eventually sold to Brownell & Miller, which refurbished the plant with three new water wheels and a 56" cylinder-wire machine that made two tons of wrapping paper a day.

In 1848 the Illinois and Michigan Canal connected Lake Michigan to the Mississippi River via the Illinois River, opening up ship traffic through Chicago for the first time. Chicago now became the most important hub in the Midwest and began experiencing unprecedented growth. The city's paper supplies came from mills in Indiana, like the one at Lafayette established in 1842 by Daniel Tondes. The rapid change in demographics caught some papermakers by surprise. In 1850, E.R. & C. Beardsley established a mill at Elkhart, Indiana, near the border of Michigan. This mill was originally outfitted with a 56" cylinder-wire for the manufacture of wrapping paper, but it was soon replaced with a 62" moving-wire machine running on newsprint for the Chicago market.

9

North and South

Kentucky Territory was in the vanguard of the western movement, and in 1793, just a year after statehood, the Rev. Elijah Craig started the first paper mill by converting a fulling mill just north of Louisville. Between 1800 and 1805 two mills were built on the Elkhorn River near Lexington by Isaac Yarnall. Another mill went up at this time in Logan County near Bowling Green, and two steam mills were built in Lexington itself. In 1817 a mill was erected in Louisville to supply the *Louisville Courier*, and in 1821 Amos Kendall built the Franklin Mill on the Elkhorn River. As the story goes, Kendall was speculating on land he believed the government wanted for a new federal arsenal and the paper mill was simply an afterthought. Still, the plant did very well for itself and Kendall sold it five years later at nice profit. Kendall took a job with the government in Washington, and went on to become postmaster general and treasury secretary under President Andrew Jackson. The Franklin Mill was bought by E.H. Stedman, who installed perhaps the first cylinder-wire machine in the state. In 1836 a machine mill went up in Louisville to supply the *Louisville Advertiser*. The newspaper ended up buying the mill in 1840 for $9,000, but sold it a few years later for $14,000. Success in the paper industry seemed assured when a flour mill in Mayville, 17 miles outside Louisville, was converted to papermaking in 1833. The owners installed a cylinder-wire machine supplied by two rag engines, but operations proved unsuccessful and within two years the facility reverted back to making flour.[1]

The first paper mill in Missouri was also called the Redstone Mill, this built in 1808 by Craig, Parkers & Co. of Georgetown, Kentucky, which made newsprint for the *Missouri Gazette* until 1814. The first machine mill in the state was established in 1834 by Lamme, Keiser & Co. in Columbia, Boone County. The plant manufactured newsprint for the *Missouri Intelligencer*, offering three cents a pound for linen or cotton rags and a penny a pound for jeans.

William S. Whiteman of Philadelphia came to Tennessee in 1806 and erected the first mill in that state on Middle Brook Creek about four miles outside Knoxville. Following his death in 1840 his son, W.S. Whiteman II, collaborated with leading citizens of Nashville to build a machine mill on the Cumberland River to supply newsprint for the *Nashville Banner*. In 1849 Whiteman II completely dismantled the old Cumberland Mill and moved everything to White's Creek Pike about eight miles outside Nashville. The papermaker evidently had some outside help, as in December of that year a millwright by the name of Thomas C. Levop wrote to Owen & Hurlbut from Nashville on an unrelated matter to say, "As my business called me away ... on short notice I left [your] statement ... with David Davis of Fishkill Landing." Whiteman II also set up a pulp mill at Paradise Ridge that sat atop an artesian spring called Loggin Springs. This facility allowed the papermaker to branch into the manufacture of bond, banknote, and other plate papers of the highest quality.[2]

In 1850 Nashville was a rapidly growing community with railroad service connecting to Charleston, S.C., via Atlanta, Ga. Despite the presence of the Whiteman's mill, supplies of paper often ran low and merchants naturally turned to sources in the north. In 1857 a merchant of that city named H.J. Yeatman wrote the South Lee Mill regarding a case of writing paper to be sent on the usual terms, adding, "You can sent it ... care of A.C. Schaffer, Baltimore ... or S. Wyatt, Charleston S.C."[3]

The earliest mill in North Carolina was built near Salem in 1791; however, by the 1850s the only going concern in the region was the Buffalo Paper Mills at Shelby. This plant was a large two-engine mill making book and news. In 1856 a former employee named Augustus Curtis Wiswall accepted the position of supervisor and moved his family down from Newton, Massachusetts. Wiswall managed the mill for about four years, but during this time he also sought to establish a mill of his own. In 1861, with real estate assistance from two local farmers, A.C. Wiswall & Co. completed construction of a single-machine mill in the town of Lincolnton, some twenty miles Northeast of Shelby.

Around 1846 two paper mills were established near Greenville in South Carolina. The *Savanna Republic*, some two hundred miles away in Atlanta, Ga., received all its newsprint from these mills, but finding the cost of freight excessive, the publishers went on to help finance the building of two new mills in their state. In 1847 William Whiteman II's uncle, James Byrd, opened a plant in Marietta, Ga., while the firm of Chase & Linton put up another machine mill in Athens, Ga. Between the two production was about 2,000 pounds a day, more than enough to supply the newspaper. The Athens mill burned down in 1856 and again in 1861, but was quickly rebuilt each time.

In Virginia, Conway Robinson and Thomas Ritchie of Richmond founded the Franklin Manufacturing Company to make iron and textile goods for the local market. Robinson and Ritchie were also publishers of the *Richmond Enquirer*, and in 1845, with capital of $50,000, the firm established a modern two-machine mill at the south end of 8th St. The Franklin Mill was an imposing three-story brick building with two moving-wire machines and eight rag engines, and employed 40 to 50 people making news, letter, printing, envelope, and wrapping paper. Formerly known as Prior's Garden, the location had long been a farmers' market, and after the mill went up a local historian (Mordecai) lamented that the formerly peaceful terraced gardens had become the source of "roaring and whistling of steam and the rumblings of water wheels and machines."[4]

In Louisiana the *New Orleans Picayune* placed orders three to six months in advance from Tileston & Hollingsworth of Boston. In August of 1845 the newspaper's publisher, Lasserden, Kendell & Co., wrote to ask that several bundles of "hardware" paper from the winter order be forwarded immediately by steamship. Newsprint may have been referred to as hardware paper, the hardware in this case being moveable type. The high cost of shipping to Louisiana drove the publishers of the rival *New Orleans Bulletin* to set up a cylinder-wire machine of its own in 1847. As the city of New Orleans hadn't enough rags to sustain the venture, additional raw materials were barged in from communities along the Mississippi River.[5]

The Jordan

The tightening of the rag market during the 1850s caused newsprint mills to move towards increased efficiency. The engine room became the focus of attention since the moving-wire machine consumed ever-increasing amounts of pulp, and mills needed to add more and more hollanders to keep up. In 1855 the firm of J. & R. Kingsland of New Jersey began pre-beating the pulp into half-stuff using a set of rotating blades mounted in a small enclosure. Kingsland's refiner cut beating time in half, thus eliminating the need for one or more hollanders. The refiner was patented in 1856 and came to be manufactured and distributed by Cyrus Currier & Sons of Newark.

News of the Kingsland refiner may have given the superintendent of a neighboring mill some new ideas. Joseph Jordan had worked for Tileston & Hollingsworth before taking a job with the Ivanhoe Paper Co. of Passaic Falls, N.J. Scavenging discarded equipment from the mill yard, Jordan mounted a set of rotating blades inside a cone shaped casing to grind the pulp to a finer degree than the Kingsland. His goal was to replace the hollander entirely, and to this end the idea was patented in 1858. Jordan contracted a machine shop in Hartford to begin building the new refiner, but may have jumped the gun a bit, as it was soon discovered the device couldn't process pulp in a single pass like the Kingsland. The hollander had grown very efficient in the two centuries since its invention, and Jordan's refiner ultimately proved too small and inefficient to perform the job any better.

It seems another good idea would go by the wayside until papermakers at the Hartford Mill in North Manchester, Conn., began experimenting with the refiner for a slightly different purpose. Engineers were working on methods to inject sizing into the rag engine while still running, and to this end they affixed a hopper to the base of the Jordan and filled it with alum and sizing soap. Now as pulp passed through the refiner it gradually drew the sizing into the stream, while the rotating blades smoothed out any lumps before returning it the hollander. The papermakers now experimented with inexpensive filler materials such as 2 percent starch and 5 percent clay for the manufacture of book paper. Boswell Keene Co., the mill owners, began selling the new book paper through the warehouse of Grant, Warren & Co. in Boston, and word quickly spread through the industry about the Jordan. It was subsequently found that newsprint could be fabricated with up to 50 percent starch, effectively cutting the cost of materials in half. This activity now relaxed the pressure on the rag market, and after nearly a decade paper prices began to fall once again. Another advantage of refined-rag papers was that starch in the matrix eliminated post-printing shrinkage, and this now launched a new age of multi-colored printing (and advertising). Smith, Winchester & Co. bought the rights to the Jordan in 1860 after most mills already had one, but the refiner's popularity continued to be very strong through the rest of the century.

The invention of the Jordan not only reduced the price of newsprint, but of book and letter paper as well. As the marketplace became cluttered with inexpensive papers, industry leaders met in Pittsfield, Mass., intent on implementing price supports. A similar cartel achieved success back in 1853 when the newspaper manufacturers of the Miami Valley regulated prices in the markets of Chicago, Cincinnati, St. Louis, and Indianapolis. The Pittsfield cartel called themselves the "Writing Paper Manufacturers of America," and their first action would be to trim production by one-third for a three-month period beginning in March of 1861.[6]

Trials and Tribulations

The actions of the Pittsfield cartel were quickly reversed by mid–1861 when civil war broke out and demand for paper went through the roof. That year alone the price of letter and book paper went from 17 cents a pound to 45 cents a pound. The Writing Paper Manufacturers of America held meetings periodically during the war to help stabilize prices, and afterwards they continued to regulate prices until the practice was eventually outlawed by the Sherman Antitrust Act of 1890.

Demand for war news sent newspaper publishers scrambling as the price of ordinary newsprint climbed from 8 cents to 22 cents per pound. Congress was obliged to eliminate the duty on rag imports to support the sagging market, but this had seemingly little effect, as by 1863 the price of newsprint peaked at 28 cents per pound. Dailies in New York and Chicago switched to straw paper for the duration, as the paper industry came under increasing fire in the press. The crisis led to the formation of the "Paper Manufacturers Association of New York," which began regular meetings at 7 Beekman St. with the goal of implementing price controls. The one positive development to come from the crisis was the invention of the cylinder press, which allowed newsprint to be fed, printed, and cut in one continuous operation. In 1862 the Niagara Falls Paper Company contracted with a New York daily to deliver paper in reels equaling 2000 sheets.[7]

Recycled paper became another important commodity during the war, launching a rather sad chapter in American history, as many historical documents were lost in the process. The drive apparently began in June of 1857 when the United States Bank in Philadelphia sold off ten tons of surplus ledger books, letters, drafts, and checks, many dating back to the Revolutionary War era. Paper warehouses bought all the offerings and sold them to paper mills. Unlike clean shavings, scrap paper couldn't go directly into the beater because the inks change the color of the pulp. Mills were thus left with the problem of how to de-ink mass quantities of written and printed matter. Beater engineers initially filled an iron tub with water, cut paper, and soda ash, and soaked the mixture for a few weeks. This mode of operation proved inadequate over the long run, so the process was transferred to the rag boiler. The rag boiler came into general use during the 1850s when the tightening rag market caused newsprint manufacturers to seek more inexpensive rags. To de-ink used paper the iron sphere would be loaded with water, paper, and caustic soda (with a bit of lime to increase causticity), and rotated under steam pressure for several hours. Newsprint mills, in particular, were able to process more used paper than ever before, and some even put in a second rotary boiler just for the purpose.

At the height of the Civil War scrap paper went for as high as nine and a half cents a pound (about what new paper sold for a ten years earlier), and subsequent raids on the back rooms and attics of the nation netted tons of old newspapers, account books, and correspondence for the mills. The stellar price eventually drove collections efforts in Europe, and U.S. shipping firms made extra money buying bales of scrap paper at one cent per pound to use as ballast on return voyages. Following the war the scrap paper industry continued strong up until about 1880, when wood pulp became the primary source of newsprint.

At the outset of the war the paper mills of the Southern states had a combined daily production of about 75,000 pounds. Unfortunately, total paper consumption was 150,000 pounds a day, so after suspension of North-South trade many a rural newspaper simply collapsed. The *Macon Telegraph* had been printed on paper imported from Belgium, but following the blockade the publisher switched to newsprint made at mills in Georgia and South Carolina. As the war lingered on some newspapers resorted to unconventional means, for example in 1863 the *Louisiana Opelousas Courier* ran several editions on the reverse side of unused wall paper.

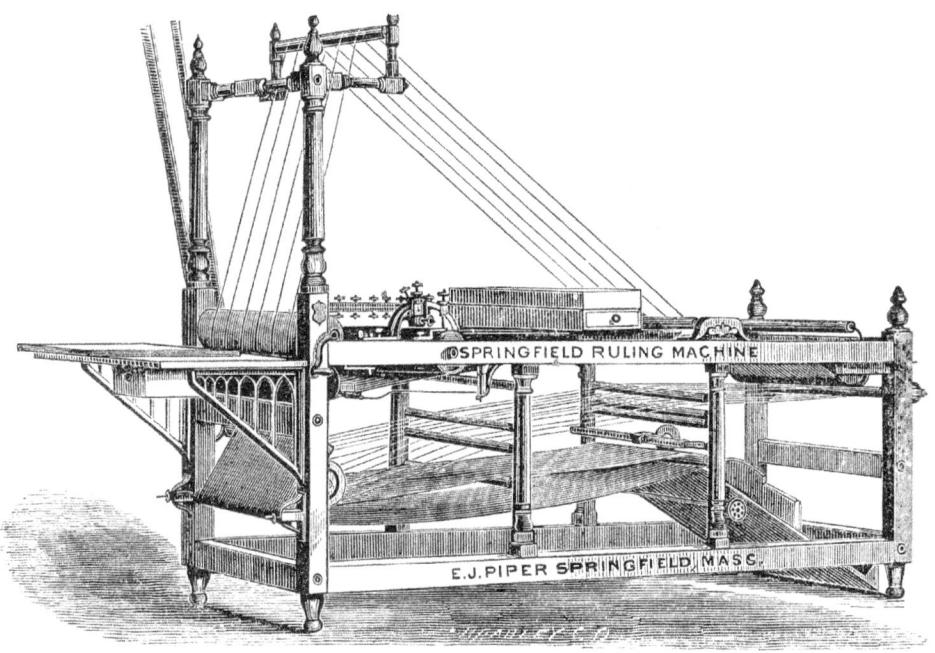

Ruling machine advertisement of 1884 (from advertisement in *Paper Mills of the World*, 1885).

The merchants of Louisville were unsure which side to support, so Kentucky affirmed its neutrality. Confederate President Jefferson Davis broke the stalemate by sending in 20,000 troops that drove as far as Bowling Green, leading Union President Abraham Lincoln to exclaim, "I hope to have God on my side, but I must have Kentucky!" At the former Kendall mill in Elkhorn, Ky., the Confederate government placed a substantial order for book paper, but after failing to make payment, the workers simply walked off the job. A Northern victory at the battle of Perryville in 1862 put Kentucky firmly in Union hands, but economic conditions outside Louisville had deteriorated to the point where most, if not all, the state's paper mills went out of business.

The state of Missouri, situated as it was on the border of Iowa, Illinois, Kentucky, Tennessee, and Arkansas, saw some of the most prolonged and widespread fighting of the war.

Governor Claiborne Jackson wanted to secede, but the Missouri State Guard was quickly routed by Union forces, forcing the state government to abandon its capital in June of 1861. Missouri was formally admitted to the Confederacy in November of that year, but never allowed to leave the Union camp. Once again, economic decline and the abandonment of workers forced the closure of all the state's paper mills.

The most prominent papermaker of the South was William Whiteman II, who made security paper for the Confederate government and other Southern institutions. The White's Creek Mill operated non-stop seven days a week, pausing only to replace belts or to clean boiler scales. The next generation of Whitemans also come of age at this time, when William Whiteman III established the Old Stone Fort Mill near Manchester in 1860. During the war this plant made news, book, and wrapping paper. The property sat adjacent to a powder mill owned by Whiteman II that supplied the Confederate War Department. Following the fall of Fort Donnelson in 1862, Union commanders reported the powder mill was blown to "kingdom come." The incomparable White's Creek Mill would later be rebuilt by Judge W. Pickerson.

The city of New Orleans was captured by Union forces in April of 1862, and cut off from material resources on the Mississippi, the local paper mill was closed. The firm of Chase & Linton had become the largest producer in the South following the loss of the Whiteman mills, but much to the papermaker's dismay, the Athens Mill burned down in the middle of 1863, an event deemed a national calamity, as it silenced most remaining periodicals in the South. The other Georgia mill at Marietta was destroyed in the face of Sherman's march after Confederate forces withdrew from positions around the town in July of 1864. Later that year the mills of South Carolina suffered a similar fate at the hands of marauding Union troops.

A New England Papermaker in Confederate Service

The war had been the source of great hardship to papermakers in the Southern states. One of the rare surviving wartime diaries of a Southern papermaker comes from Augustus Curtis Wiswall, who in early 1861 commenced operations at a new mill in Lincolnton, N.C., about twenty-five miles northwest of Charlotte. From the outset of the war the Lincolnton Mill suffered from a lack of supplies and materials such as bleach, lime, soda ash, and vitriol. The shortage of material would be matched by a shortage of labor, and in order to exempt two or three skilled workers from military service Wiswell had to agree to contract with the Confederate government for book and handbill paper. To meet his obligations the papermaker also found it necessary to lease slaves from a nearby plantation.[8]

Paper prices in Richmond were ten times higher than in New York, with book paper bringing 50 Confederate dollars a ream and sized writing 77 dollars a ream. The Lincolnton Mill had a cylinder-wire machine with a ruling attachment at the winder and was capable of making 1,500 pounds of writing paper a day. The mill was also equipped with an envelope cutter, and the job of gumming and folding was contracted out to a local women's group. A four-mule team went on regular routes visiting merchants in Charlotte and Morganton in N.C. and Columbia, S.C., to deliver paper and collect rags, which at one point in 1864 were so scarce they reached 30 cents a pound. In September of that year papermaker supplies became so scarce that Wiswall reportedly paid 2,970 Confederate dollars for a cask of soda ash weighing 1188 lbs., and 1,890 dollars for a cask of lime weighing 540 pounds.

Throughout the ordeal the sentiments of the New England papermaker were never in doubt. When several Union prisoners, who escaped from the military prison in Charlotte, N.C., ended up at his door, Wiswall gave them food, clothing, refuge, and most importantly, maps to Union lines. In appreciation a group of army officers signed a letter declaring Augustus Curtis Wiswall a "Union Man." Fortunately, Wiswall never had to use the letter as the war ended with General Joseph E. Johnson's surrender near Durham, N.C., in April of 1865. The Shelby and Lincolnton mills were thus the only two paper mills in the South to survive the conflict.

Wiswall sold the mill a few years later and returned to Newton, Massachusetts, in July of 1869. During his absence, his old employer, Thomas Rice Jr. had purchased the Foster Mill in 1857, assigning the property to his brother, Charles, under heavy mortgage. However, once Wiswall's story became known, Thomas took the property back and sold it to Augustus for nine thousand dollars. Wiswall went on to replace the mill's wet machine with a 48" cylinder-wire and began the manufacture of manila paper.

The Franklin Mill in Pictures

The Franklin Mill in Richmond, Va., happened to be the only plant in the South to employ the moving-wire machine, and obtaining replacement wires and felts was a distinct problem. The Fourdrinier Brothers received royalties on replacement wires and felts for

Ruins of Franklin Mill showing an overshot waterwheel with a twenty-foot drop, 1865. Shafting and pulleys at right rear powered the engine room on the second floor.

many years after their patent on the paper machine had run out, but the monopoly in the U.S. was finally broken in 1847. William Staniar, an apprentice at the largest wire-weaving establishment in Manchester, England, was hired by the firm of Stephens & Thomas of Belleville, N.J., and he brought over a model from which was made the first loom for a 62 inch wide by 24 foot long primary wire. The first of the American-made wires was successfully tested at the plant of J. & R. Kingsland of North Belleview, N.J. Wire-weavers in New Jersey, New York, and Connecticut would later take up the trade, and manufactures were eventually able to produce wires of any given width, not just the standard 62 inch gauge.[9]

In the early stages of the Civil War the lines between North and South were highly

Ruins of moving-wire machines at Franklin Mill, 1865. Machine room of half-timber construction was a separate wing of the plant. Press rollers on the machine at left are clearly visible.

porous, and blockade runners ran wagons laden with replacement wires and felts through the mountainous terrain of Virginia, Kentucky, and Tennessee. Land routes were eventually shut down as the war progressed, and the only remaining source of spare parts was Great Britain. The textile manufacturing city of Manchester dispatched special blockade running ships to bring back raw cotton in exchange for strategic supplies, including banknote paper, printing plates, and even finished postage stamps, but there is no record of any belts or wires brought back for the Franklin Mill.

While the both North and South had capable weavers, the moving-wire machine required a special two-ply press felt that could only be made by advanced textile mills in Great Britain. Finally, in 1864 textile manufacturer Johnson, Fuller & Co. leased a moving-wire machine at a mill in Camden, Maine, to experiment with a homespun version of the press felt. Their first belts made paper well enough, but wore out rather quickly. At that time press belts had to be replaced about four times a year depending on usage, but the down time was very costly to the paper mill, and so belts that didn't last very long would find no market. The following year the Ohio firm of Schuler & Benninghofen did the wise thing and hired a former weaver from a prominent woolen mill in Great Britain who brought with him the secret of making a belt with great strength and longevity. Essentially, the base weave consisted of linen warp and cotton filing fibers, while the web side was made of fine wool so as not to leave any marks in the paper. The Schuler & Benninghofen belt rapidly

An artist's reconstruction of Franklin Mill, Richmond, Va. (courtesy Frances Alcorn).

caught hold and effectively priced the imports out of the market. Other textile manufacturers entering the business after 1870 were Waterbury & Sons, Acme Felt Company, Akron Woolen and Felt Company, Lockport Felt Company, Megunticook Woolen Company, Appleton Woolen Company, Green Brothers, Rumford Falls Woolen Company, L. Heathcote, and Weiss & Son.[10]

The proximity of the Franklin Mill next to the Richmond armory fairly sealed its fate during the evacuation riots of April 1865. In the war's aftermath the photographic chronicler Mathew Brady traveled to Richmond to snap a few fleeting images of the Confederate capital, and one of his subjects happened to be the Franklin Mill. Remarkably, these were the first photographs taken of an American paper mill. One picture shows the approximately 20 foot tall and 6 foot wide overshot water wheel with an iron gudgeon and wooden slats that shone brilliantly in the sun after years of use. The photo also reveals what's left of the engine room where eight rag engines once hummed with activity. Along the basement floor lay the twisted remnants of shafts and pulleys that had once driven equipment mounted on the floor above. In another photograph twin pairs of press rollers from the wet end of a moving-wire machine are clearly visible. The two moving-wire machines sat rather close together supported by a single retaining wall running between. This was not an unusual configuration, as paper machines were often located in a separate wing of the plant due to high humidity and the need to evacuate waste water.

10

Crane vs. Willcox

Thomas Willcox established the Ivy Mill on Chester Creek near Philadelphia, which was named for a slip of English ivy brought over from the Old Ivy Bridge in Devonshire, England. The mill was of typical European half-timber construction, and when completed in 1729 it manufactured press board for weavers back in England. The pulp was all ground by hand, and paper-making utensils mentioned in the lease of 1730 are somewhat revealing; these include mortise and hammers, vat and pott, two stuff tubs, rag knife and block, paper press mould, a pair of shop paper moulds, 26 fulling paper felts, 77 shop paper felts, two press paper planks, a press and rag wheel, a screw and box, glazing engine, two paring knives, two pails with iron hoops, two pairing frames, one paring bench, two troughs, one winch, two trestles, six posts and eighteen rails for hanging paper, a hanging stool, one pad, a box for paper, three rag washers, 160 tap pots, and 20 cogs.[1]

About 1760 Thomas Willcox redeemed an indentured servant named John Readen to modernize his plant, and along the creek bank Readen erected an engine house with a waterwheel and a hollander. It was about this time the mill obtained rags from Benjamin Franklin and began making paper for the *Philadelphia Gazette*. The papermaker's son, Mark, took over mill operations in 1767, and he became the first in the country to employ the wove mould (as demonstrated by the state of Maryland hand bills of 1775). The mould came from England, but whether Franklin had anything to do with it is not known. Willcox was making high quality banknote paper for the state of Pennsylvania in 1776 when war broke out, and following the British withdrawal from Philadelphia he went back to making high quality writing and print paper. Willcox became noted for the purest white linen banknote paper such as that made for the Bank of North America in 1781, and it seems the papermaker could hardly wait for peace to break out, as records show the mill ordered double foolscap and double post moulds from London in 1783. By 1791 Willcox was back to ordering banknote and letter moulds from Nathan Sellers of Philadelphia, some of these continuing in use as late as 1848.

Younger brothers John and James Willcox joined the firm in 1808, and the family now opened a warehouse in Philadelphia where they offered (in order of increasing size): folio cap, lattice, demy, post, medium, royal, extra royal, super royal, imperial, manslaughter, and atlas. The paper was sold in bundles of 20 quires, or 480 sheets, and in the days before the ream label each quire had an imperfect (or broken) sheet covering the top and bottom. Willcox remained in good standing with the U.S. government, and was contracted to make all the treasury note paper during the War of 1812.

James M. Willcox took over management of the mill in 1826 and incorporated the ivy leaf into the watermark for the first time. Three years later he made improvements to the mill, erecting a much larger facility next door to the original. In 1835 he purchased the former Sharples Iron Works with extensive water rights from heirs of the long-time owner.

Letter addressed to J.M. Willcox Co. from Grant, Dennis, & Co. warehouse of Boston, 1848.

The property was located on the main branch of Chester Creek about two and a half miles from the Ivy Mill, and here the papermaker established a new plant with a large overshot water wheel and a 62 inch moving-wire machine for the manufacture of book paper and newsprint. Business at the Willcox Glen Mill was so good they added second moving-wire machine in 1845. Following James M. Willcox's death in March of 1852 his eldest son, Joseph, took over the Ivy Mill, while younger sons Mark and James Jr. oversaw the Glen Mill.

The Old Red Mill

Stephen Crane was born in Milton, Mass., in 1766 and began work at the Neponset Mills just as the Revolutionary War was beginning. Paul Revere was a copper-plate engraver and printer in the city who was engaged by the state of Massachusetts to make their currency notes. According to an entry in the Neoponset Mill's day book, Revere had placed an order for banknote paper, and master Crane is said to have delivered the stock under armed guard. Crane eventually worked his way up in the paper-making business, and when the Neponset Mill was sold in 1793 he took a job as vatman at a nearby mill in Watertown. The next year Crane entered a partnership with Edward Jackson and William Hoggs of Newton Falls to build a new mill in the town of Needham. Crane went on to manage the facility for two years, but the mill was sold out from under him by certain Boston investors who had transferred their shares to William Hoggs & Son.[2]

Stephen Crane had five younger brothers, two of whom, Zenas and Luther, had also grown up in the paper trade. Zenas was born in Milton in 1777, and also worked at the

Neponset Mill. By 1796 Zenas had taken a job at the Burbank Mill near Worcester, then one of the leading mills in the state. He later traveled west to Springfield to look into establishing a mill near the county courthouse, always a good location for a paper mill, but was beaten to the punch by Eleazer Wright, who had already begun building the mill that would later be sold to Colonel Ames. The back-up plan was to continue traveling west to look into potential mill sites along the Housatonic River in Berkshire County.

The Housatonic River is comparable to other major waterways in New England such as the Naugatuck and Mohawk rivers. It begins as a fast rushing mountain stream descending six hundred feet over its first eighteen miles, and upon reaching the town of Dalton two tributaries join the stream to create the strong and swift-running river. The Housatonic next tumbles over a waterfall just north of the town of Lee before flowing at a more leisurely pace towards East Lee Village. Here the river is joined by the powerful Lake May Stream that had descended some 500 feet in just over two miles to reach this juncture. The Housatonic then turns west, passing in turn through the settlements of South Lee Village, Stockbridge, and Housatonic Village.

After seeing the river for himself Zenas heard some predictions from local business owners that Berkshire County would one day attract such a host of textile manufactures as to rival Manchester in Old England. But despite all the natural advantages, Boston was the only regional paper market, and nothing short of a miracle could attract investors out to the Berkshires. But miracles do happen, and the young papermaker now met a prosperous farmer in Dalton who happened to own an artesian spring on prime real estate next to the river. Pure spring water was highly desirable for the manufacture of bond and banknote paper, and Crane instantly recognized the potential for a mill site. The farmer had other uses for the spring, but was willing to part with a neighboring parcel on the upstream side of the farm. Crane's determination kicked in at this point, and while the decision to build was something of a compromise, he would still be in a good position to acquire the spring when the time came. Zenas managed to attract a single investor from Boston named Henry Wiswell, and for the rest of the investment he turned to Daniel Gilbert, an old friend of the family who had served with Zenas' father during the Revolutionary War. Gilbert was a prominent firearms manufacturer for the federal government, and the industrialist generously loaned $4,000 for construction and a further $6,000 for working capital.

Zenas Crane & Co. now established what would be called the Berkshire Mill, a simple one-vat affair with a daily capacity of six reams. Mill employees consisted of an engineer at $3 per week, vatman and coucher at $3.50 a week, and a layboy at 60 cents a week with room and board. Zenas Crane himself served as both superintendent and general manager at a salary of $9.00 per week. Daniel Gilbert was the first to be paid off in 1803, and Henry Wiswall sold his shares to Crane four years later. But, Wiswall wasn't completely done investing, and he now partnered with David Carson, another vatman from the Burbank Mill, and the pair managed to convince the farmer with the fourteen acres and artisan spring to join them in a fresh venture. The key property was acquired for a mere $194, and the enterprise soon had a one vat facility that used wooden pipes to convey spring water the short distance to the mill. This was called the Red Mill, presumably after the color of the building.

After all his careful planning Zenas Crane found himself mired in the management of the Berkshire Mill and took no part in the establishment of the Red Mill. It seems his interest in the property never wavered though, so he traded David Carson one-third interest in the Berkshire Mill in exchange for the latter's shares in the Red Mill. Carson seemed more interested in having a mill of his own, so Crane sold him his remaining shares in the Berkshire Mill, then used the cash to buy out the farmer's share in the Red Mill. Upon gaining control of the Red Mill Crane added a second vat, but even with that it would take until 1826 to pay off all his accumulated debt.

In 1823 David Carson teamed with another local farmer to build a third plant in Dalton called the Defiance Mill. Carson continued to expand his interests in the area, and just two years later partnered with Zenas Crane to buy the Castle Mill in Lee. Crane was just emerging from the burden of debt on the Red Mill, and he now partnered with a farmer in East Lee named Stephen Thatcher to build a new plant at the base of Lake May Stream. Crane sent one of his trusty vatmen to work the mill and ensure success of the venture, and this farmer seemed to get the hang of things rather quickly, paying off all the investors including Crane within three years' time.

Crane's original goal for the Red Mill was to sell his product in Boston where the crisp white paper brought a premium. Following completion of the Erie Canal, however, the trade in Manhattan picked up significantly, giving the papermaker another viable option. Sales of Crane paper at Johanthan Seymour's New York warehouse in 1828 amounted to $2791, and nearly doubled over the next year. Meanwhile, the paper machine made its first appearance in Lee in 1827, and Crane followed this new development quite closely. Production at the Red Mill was running at capacity, and a decision had to be made soon whether to expand or modernize or do both. Crane soon divested his interest in the Castle Mill and began improving the Red Mill, beginning with the installation of an overshot waterwheel with an iron gudgeon (bearing). Crane found the cylinder-wire machine capable of producing the same kind of linen-cotton composite papers he'd been accustomed to making, so in 1831 he abandoned hand papermaking and installed an Ames machine. Neighbor David Carson soon followed suit, erecting a machine at the Berkshire Mill the following year. Crane went on to add a multi-cylinder steam dryer in 1834, but continued using the Red Mill's extensive drying lofts in the production of tub-sized writing paper. Crane also produced plate paper of the highest quality, retaining the old screw press for the smoothing process. A veteran of the Old Red Mill later described this work as follows:

> The paper that was between these form boards was put in the press, that would be screwed down just as hard as possible. They would secure these presses, and then along toward night after they had got all the presses full, one of the men from the finishing room would go around the mill and call out "Press!" The men would know that they were to go in the finishing room....
>
> We had a long pole with an eye on the end of it, and we would stick one end of that pole into the hole in the screw press. Then at the upper end of the room there was a windlass with a big rope an inch in diameter, and a hook on the end of it. They would take that and pull it out, and hang it on to the eye of the pole. A man would then start and put it around just as far as he could and back. It was a slow process. In those days it was hard work — a whole lot of it, and slow.[3]

Crane Family Mills

About the time Zenas Crane began working with the paper machine his brother Luthor went into business with John L. Rice to buy the Longfellow Mill in Needham. Following a fire in 1834 Rice & Crane rebuilt the plant and installed a new 62 inch cylinder-wire machine they imported from England. This machine turned out to be the first to come with the new tandem dryer, and its success drove machine manufacturers such as Howe & Goddard, D&J Ames, and Phelps & Safford to patent an American version. Rice & Crane went on to manufacture book paper and newsprint, but Luthor Crane retired a few years later and the Rice family continued in the business thereafter.

Zenas Crane retired in 1842 and the Old Red Mill was entrusted to sons Zenas Marshall and James Brewer, who formed Crane & Company. That same year the Western Railroad came to town and Crane & Co. divided their property and began construction on a second plant called the Stone Mill. Here they installed a moving-wire machine, the first in Berkshire County, and the mill proved so successful that by 1846 Crane & Co. had an inventory alone valued at $20,000 with accounts receivable of a further $9,000. The next generation of Berkshire papermakers was also fairly active at this time, with Seymour Crane and Lindley Murray Crane taking interest in a mill in Ballston Spa, N.Y. It is perhaps no coincidence that other sons of Berkshire manufacturers went to neighboring New York at this time; David Carson's sons bought a mill in Newburgh, while Walter Laflin's sons took over a plant in Herkimer.[4]

After construction of the Stone Mill, James Brewer Crane found a natural spring in the town of Westfield, Mass., and here the papermaker built a new plant and piped in water from Wolf Pit Springs some three-quarters of a mile away. J.B. Crane also brought over his favorite mill superintendent, F.A. Thompson, from the Old Red Mill, and began making up to two tons of fine writing and ledger paper daily on a 60 inch moving-wire machine. Crane was also something of a tinker, and he invented an early version of the layboy, so named after the worker who tended stacks of paper at the end of the line. Sixteen automated layboys were then sold, mostly to friends in Berkshire Co. and mills in Herkimer, N.Y., and Ware, Mass. Crane later added a 90 foot double cylinder for the manufacture of manila, and at one point in making paper collars he took the process one step further and began making manila belts to replace the leather belts used to drive auxiliary equipment. Manila fibers were found to be very strong and durable, and the secret to the process lay in pressing together thick strips of paper on the double-cylinder.

Following his retirement in 1847, J.B. Crane was succeeded by sons James A. and Robert B. Crane, who formed Crane Brothers, Inc. Expanding on their father's manila belt patent, the Crane Brothers began manufacturing all sorts of household items such as paper pails, shipping trunks, guitar cases, and a line of disposable bowls for photographic baths and pharmaceutical works. The firm also went into business with E. Waters & Sons of Troy, N.Y., which patented the paper boat. These watercraft were composed of one-quarter-inch manila paper over a fifteen-foot keel of soft pine. The skins were delivered to the boat yard in rolls made from five thicknesses of manila card about one-sixteenth of an inch thick, and from these Waters produced racing shells from single-sculls up to eight-oar craft. He also made larger boats up to 42 feet, but these cost considerably more than conventional timber construction and so their sales never materialized. Still, the racing shells proved very popular and in annual races at Saratoga in 1886 the teams of Harvard, Yale, and Columbia all raced Waters' models to victory in separate events.

Something even more surprising than paper boats were paper railroad wheels. A locomotive engineer named Richard N. Allen of the Cleveland and Toledo Railroad had been

Making paper car wheels (from *Harper's Illustrated*, June 1887).

looking into an alternative to iron-slag wheels, which had the tendency to develop micro fractures when exposed to high vibration and extreme cold. Allen approached the Crane brothers, who supplied him with manila paper disks of the desired thickness. Allen took the disks and bolted on a cast iron hub, then pressed the assembly into a steel ring that served as the shoe. The paper wheels were tested for six months on a wood car for the Central Vermont Railroad, and following this success the Pullman Palace Car Co. placed the first order for one hundred wheels in 1871. Paper wheels were more expensive than the cast parts they replaced, but the manila lasted longer, ran smoother and quieter, and could be re-shod over and over. Within ten years the Allen Company's sales ran to thirteen thousand car wheels a year.[5]

With the success of their manila operations the Crane Brothers built a new three-story (40 by 90 foot) plant called the Japanese Mill. This plant came to employ 100 people, and the Cranes found even more uses for manila paper in the manufacture of wash tubs, buggy and wagon boxes, piano cases, and even structural domes. The first paper dome made for the observatory of the Rensselaer Polytechnic Institute in Troy weighed just 1,000 pounds. Following this a dome was built in 1881 for a building on the campus of the U.S. Military Academy at Annapolis, this consisting of thirty-six sections totaling just 2,500 pounds. Remarkably, that structure survived until 1959, when the building was finally torn down.

Treasury, Bond, Currency, and Stamp Paper

During the heady days of the 1840s the Red Mill came to specialize in plate paper and took the process even one step further in what was called hot-pressing. Polished metal plates were preheated in a steam chamber, and these produced a sheet of incomparable smoothness. For banknote paper Crane also employed light sizing, which was a thin application of external sizing used to fill in tiny irregularities in the paper. The market for the Red Mill's pure white bond and banknote paper was exceedingly lucrative, and so the Cranes wisely transferred the manufacture of all other kinds of their writing and printing paper to the Stone Mill. To further their interests in the marketplace, Crane & Co. also issued a circular printed on a half-sheet of high quality linen plate paper, this reading:

> Dalton, Mass. Nov. 1845.
> We take the liberty of sending you a sample, half sheet, of our manufacture of bank-note paper. It is made entirely of new linen. Among the several banks sold to, in this state, we can refer you to the Northampton, Lee, and Adams Banks, also to the Citizens, and Central Banks, Worcester. A copy of an extract from a letter of the cashier of the last of which, is annexed, and to which we respectfully refer you. Most of the late emissions of these institutions are, we believe, executed on our paper.
> Our paper has been worked by the principal bank-note engravers in New York and Boston. One reason of our success in selling and on which we rely for effecting future sales to the banks, is the extreme low price at which we determine to put it. Those who have heretofore bought have paid as high as from twenty to thirty dollar per one thousand sheets, while ours is placed at the very low price of twelve dollars; being a savings to purchases of one half, and a superior article in point of strength and durability, to any ever made. We are prepared to send any quantity to any part of the United States as may be ordered. Banks at a distance can have it left at any of the printers in New York, Philadelphia, or Boston.[6]

Not to be outdone, Willcox issued a circular of their own in 1850, this reading:

> Enclosed are three samples of banknote paper, each a quarter of a sheet, of different thickness, but made in the manner (handmade), and of the same material. The subscriber has been a manufacturer of bank-note paper for nearly a half century, and for the last twenty years has made it almost an exclusive business. After various experiments, he thinks he has succeeded in making a paper particularly adapted to that purpose. It is a material different from any heretofore used, being tough and flexible, having a surface not liable to crack or wear, rough and susceptible of receiving the finest steel or copper-plate impressions of which the samples are specimens. He also has in store a large quantity of bank-note paper that is well seasoned, being from eight to ten years old, of which samples will be forwarded without delay when required. Every sheet is warranted perfect, and equal to the samples. The price of this paper is $22.50 per 1000 sheets, and is for sale at James M. Willcox & Co., No. 7 Minor Street, Philadelphia, and at manufacturers prices by Messrs. Rawdon, Wright, Hatch & Edson, Engravers and copper-plate printers, New York, New Orleans, and Cincinnati.[7]

Both papermakers identified linen as the essential ingredient in bank-note paper that created its exceptional folding strength and longevity. Cotton fibers were used to lend softness and absorbency, improving both the depth and quality of the printing. Both papermakers also came up with strategies for paper security. In 1844, Zenas Marshall Crane embedded silk threads in currency paper made for the Northampton Bank. A single thread spanned the width of the $1 denomination, two threads on the $2, and so on in an effort designed to prevent forgers from promoting a lower valued bill. For their part, the Willcox

Paper Co. developed increasingly complex watermarks. The Willcox paper used for the 1843 U.S. treasury notes had a basic watermark identifying each of the denominations, while five years later a more sophisticated multishadowed watermark was used on the U.S. bond issue of 1848.

The main difference between Willcox and Crane was in the method of manufacture, and while Willcox espoused traditional papermaking, Crane touted the modern machine-made variety. To decide the issue a group calling themselves the New England Association of Banks for the Suppression of Counterfeiting sent samples of each of the most popular banknote papers to an independent testing laboratory to judge their relative merits. The association's 1855 report reads: "Two of the most extensive bank-note paper manufacturers offered specimens.... These papers were tested by Charles T. Carney of Lowell. Sheets were drawn at random from 500 sheets of each specimen, and their strength tested both lengthwise or by perpendicular strain, and crosswise or by transverse strain, also with and without sizing. The first experiment was with paper made by Crane & Co., weighing 14 lbs. to the ream. For the second experiment a paper made by Willcox & Co., 14 lbs to the ream was used.... The premium was awarded to J.M. Willcox & Co., Ivy Mills, Penn." So it was the traditional hand-made paper won this round, while the cylinder-wire paper, otherwise identical in composition, was found to tear more readily along the long axis. This is the condition the pulp dresser was supposed to address, but it evidently wasn't the perfect solution so many had been led to believe.[8]

Competition between Crane and Willcox extended to other plate papers as well. Crane's bright white bond paper made with pure spring water was hard to beat, so in 1846 Willcox introduced chlorine bleaching to the Ivy Mill. The use of plate paper also extended to U.S. postage stamps, which used fine engravings as a deterrent to counterfeiting. The first U.S. stamps were printed on a bluish Willcox plate paper made on a ribbed mould having approximately thirty-one wires per centimeter. Ribbed moulds are essentially the same as laid but without chain lines, although in this case the laid lines are so fine they leave little or no marks in the paper. For the contract of 1851 the government desired a more opaque paper, meaning the contractor had to find something a little thicker. The stamp paper contract alone was valued at about $40,000 per year, and evidence of Crane's involvement at this time may be found within the archives of the Crane Museum in Dalton, Massachusetts.

The contractor selected by the postal service was Toppan Carpenter & Co., banknote printers of New York, Philadelphia, Boston, and Cincinnati. The printer obtained their stock from the McWilliams Warehouse in Philadelphia, and in this case they selected a medium-thick linen-cotton composite ribbed paper made by Willcox. It seems once the reduced postal rates took effect, demand for the new stamps quickly exceeded supply, and as production climbed the printer found it could no longer obtain paper from a commission warehouse in the quantities desired. The printer must have looked into contracting directly with Willcox, but it's believed the death of James M. Willcox in March of 1852 may have delayed any formal agreement. In the meantime, Zenas Marshall Crane went to Philadelphia discuss matters in person, leaving several samples of plate paper to test with. However, Crane ultimately lost this contract because he couldn't under-bid Willcox on price, nor could he offer free shipping to Philadelphia as Willcox did.[9]

The tables began to turn in 1857 when Congress authorized another issue of treasury notes and the printer, once again Toppan & Carpenter in Philadelphia, selected Crane to supply the paper. However, there were serious problems with the quality of the stock, and in July of 1858 Zenas M. Crane was handed a note from the printer saying: "We have a great complaint of the shrinkage of your bank-note paper ... the Treasury department has advised us that there is a difference of between ¼ and ½ in. in length for our $100 Treasury Notes.... This is very serious evil, and we beg that you will ... not only account for it, but inform us how we can prevent this. The same complaint has been made at our New York office ... where a great variation in the size of bank-notes from the same plate is found to exist. It naturally creates doubts about the genuineness, and therefore is a serious evil which we must correct."

Shrinkage has always been a factor in rag paper, though it didn't become an issue until the government began scrutinizing its high value issues. It should be noted that plate printing was a wet process whereby the sheets were moistened beforehand to keep the oil-based ink from blotting, and shrinkage becomes noticeable once the printed sheets dry out. Cotton fibers were the source of the problem, and Crane & Co. was certainly aware of this dynamic, which is why their plate paper was made from either 100 percent linen, or a linen-cotton composite. Barring other possible explanations, it appeared the source of the problem may have been in the shortage of linen imports.[10]

While linen rags were in short supply, a number of paper mills turned to raw flax in a process much like that used for manila paper. In 1852 a flax supplier sent the following to Owen & Hurlbut:

> Enclosed is a sample of flax sent to you for examination with a request that you would give me your opinion as to its adaptation for manufacturing into paper. The article is produced by a patent machine taking the flax from the field without knotting. The process is so rapid that the article could be produced at a moderate price. The inventor proposed to furnish it in bales similar to bales of cotton. It is easily bleached as white as the finest cotton.... If you think it would work well into paper, say about how low ... you think it should come to make it an object of paper manufacture.[11]

Crane did not reject every new trend that came down the line; for example, during the Civil War the company quick to adopt vulcanized coverings for couch and press rollers, which helped eliminate rust spots in the paper. Still, Crane remained very conservative in their methods; for example, they eschewed bleaching, and continued to rely on traditional wood-lined hollanders long after most mills converted to the more durable iron beaters. Crane also avoided the Jordan, and that decision would come back to haunt.

The stamp paper contract went to Crane very briefly over 1857-1858, and the loss of the business evidently had the desired effect, as Willcox was jolted to action. The Pennsylvania papermaker installed a Jordan at the Glen Mill and began working with the printer to get around the dependence on linen. It was subsequently found that starch in the matrix virtually eliminated shrinkage without sacrificing impression quality. The inexpensive stamp paper was readily adopted by the printer, which pocketed the savings. In 1861 Congress authorized a new issue of stamps to prevent the Confederacy from cashing in on large stocks of U.S. postage stamps in Southern post offices, and this time the printer would be the National Banknote Co. of New York, one of Crane's best customers. Still, despite the built in advantage, Crane found it could not compete on price or performance with Willcox's new refined-rag paper.[12]

Stamp paper would be the least of Crane's problems. In 1861 the Confederate government froze all payments to creditors in the North, leaving the papermaker with many unpaid accounts. The news got even worse, as to pay for the war effort President Abraham Lincoln nationalized the nation's currency, eliminating the remaining market for banknote paper. Crane & Co. continued making writing and book paper at the Stone Mill, but the Red Mill was effectively insolvent. Perhaps taking a cue from the Crane Brothers Mill in Westfield, the Old Red Mill was now converted to the manufacture of collar paper.[13]

Younger brother Henry B. Willcox had taken over management of the Ivy Mill in 1859 and had been instrumental in winning the paper contract for the new U.S. greenbacks in 1861. Production of currency paper was quickly ramped up at the Ivy Mill, but the Treasury Department became concerned with the pace of production and so issued a grant to the Willcox Brothers to experiment with the manufacture of banknote paper on the Fourdrinier. Hitherto, a linen-cotton composite could not be made on the moving-wire because dissimilar fibers tended to separate into homogenous lumps. The Willcox Brothers eventually developed the "localized fiber" process (patented 1864), and this paved the way for production to move to the Glen Mill, where more banknote paper was made in a day than the hand mill could in a month. The nation's currency paper was closely guarded through every stage of manufacture and at the height of the war the Treasury Department maintained a force of forty officers at the Willcox mills employed in both police and detective work. Moreover, they did the job well, as over the next fifteen years hundreds of millions of sheets of banknote, bond, and treasury paper were made at the Glen Mill without incident.[14]

Operations at the Ivy Mill limped along until 1866 when the old mill closed its doors for the final time after 137 years of glorious service. Years later, Dard Hunter visited the ruins of the Ivy Mill and reported, "As late as 1884 some of the buildings of this mill were still standing.... When I visited the site of this old mill ... nothing remained but the stone foundation, which stood in soggy ground almost covered with rank vegetation. A remnant of the old ivy vine ... still clung to the ruined walls. At the base of one of the crumbling masonry walls, deeply embedded in the marshy ground, lay a rusty and broken screw press fallen upon its side. It has been one of the indispensable implements of the old mill, and between its platens ream after ream of paper had been pressed."[15]

The Treasury Contract of 1878

The Ulysses S. Grant Administration had been riddled with charges of corruption, leading to the election of Rutherford B. Hayes in 1877 on a reform platform. Soon thereafter the new secretary of the treasury, John Sherman, asked the Willcox Paper Co. for production data on the manufacture of banknote paper. It was subsequently revealed the wartime price of 75 cents a pound had never changed, so Willcox agreed to an immediate reduction to 70 cents per pound. By this time America's banknote paper included small strands of jute added as a security precaution, and Willcox felt the high cost was justified. Secretary Sherman, however, asked for production details to make that determination for himself, and when Willcox stonewalled, the Treasury Department simply wrote a new specification and put it out for competitive bid.

Another firm vying for government business at this time was the Hudson-Cheney Paper Co. of North Manchester, Connecticut. Henry Hudson had been among the first in the nation to install the moving-wire machine, and following this conversion the Oakland Mill made writing, book, and handbill paper, and held stationery contracts with the United States government. During the Civil War, Hudson ran into financial difficulties over the price of rags, and in 1864 partnered with silk manufacturer F.W. Cheney. Cheney made substantial investments in the Oakland Mill, which was completely rebuilt and re-equipped with the latest machinery, including a Jordan. In 1869, following the government's ill-advised papermaking venture in the basement of the Treasury Department building, officials put out for bid their own brand of silk-threaded security paper known as "membrane." Cheney fairly jumped at this opportunity and the Oakland Mill was soon contracted to the government for revenue and tax stamp paper, presumably made using the localized fiber process. Henry Hudson retired in 1873, and Cheney continued to pursue government contracts such as the new postal cards of 1873. However, in this case Cheney may have overstepped his bounds, as the Oakland Mill lacked specialized equipment to make high quality card stock, and so the job had to be subcontracted to the Russell Paper Co. of Boston. Despite the minor setback with the postal service, in 1878 when Secretary Sherman opened bidding for the Treasury Department's banknote paper contract F.W. Cheney fairly believed he was the heir apparent.[16]

In the postwar period Crane & Co. underwent serious reorganization. In 1865 Zenas Crane Jr., son of Zenas M. Crane, was assigned to run the Bay State Mill, a former woolen mill converted to papermaking by Seymour Crane in 1851. The mill had failed in 1857 and idled thereafter. In revitalizing the plant Crane traveled to Europe to study the latest coloring and embossing techniques. On his return the Bay State Mill was retooled for the upscale stationery market, and the business swiftly took off making Crane paper for popular upscale department stores such as Tiffany's of New York, Baily, Banks & Biddle of Philadelphia, Shreve, Crump & Low of Boston, and Marshall Field of Chicago. The enterprise paid off a $54,000 mortgage within three years, and re-established Crane & Co. as a marquee papermaker. The firm suffered setbacks after back-to-back fires at the Stone Mill in 1869 and the Red Mill in 1870, and in reconstruction Crane & Co. built a new plant called the Pioneer Mill. About this time W. Murray Crane and Frederick G. Crane, sons of Z.M. Crane and J.B. Crane respectively, were admitted to the firm. In 1873 Murray Crane was fresh from the Wilbraham Academy and Williston Seminary, and after a period of paper mill training he was sent out on the road as sales representative. He visited the Winchester Arms Co. in New Haven to look into a problem with sealing paper used in rifle shells, and returning to Dalton developed a specification for a "bullet patch" that gained rapid approval from the arms manufacturer. The Winchester repeating rifle went on to great popularity, and demand for the specialized ammo secured the success of the Pioneer Mill.

Crane & Co. had been out of the currency paper business for a considerable time when Secretary Sherman put the government's contract out for bid. Despite the loss of the Red Mill, the 26-year-old W. Murray Crane was dispatched to Washington, D.C., to bid on the contract. By the time he arrived in the city Willcox had already opened bidding with a figure of 55.5 cents for standard stock and 61.4 cents for their premier paper. Murray may have felt a bit out of his element, as the other firms were represented by senior lawyers and lobbyists who wined and dined the Washington elite to all hours. It was a typical hot summer so Crane kept to his hotel room, venturing out only occasionally for a walk in the park. Crane stayed in close touch with the home office, and at the appropriate time submitted a bid of 40 cents a pound. Murray Crane now found himself besieged by overseers and influential government officials who probed his knowledge and commitment and watched his every move. Bidding was supposed be sealed, but influential parties soon went to work, resulting in a bid of 39.75 cents by the Cheney Paper Co. Evidently using the same unofficial channels, Murray Crane learned of the competition's clandestine activities, and so made plans to protect his company's interests. He organized an impromptu going home party at his hotel room at the close of bidding, knowing his overseers would all be in attendance. At the eleventh hour, Murray slipped out of the hotel and ran to the Treasury Department with a new proposal in hand. With minutes to spare Crane submitted an updated bid of 38.9 cents to win the bidding. News of the last minute heroics caused an uproar, and Cheney representatives demanded a competency hearing, as was their right. The government obliged and at an open hearing lawyers leveled a series of accusations, but Murray Crane answered each question in turn and demonstrated his impeccable knowledge of the business. In the end the committee stood by the decision and the contract became Crane's for the taking.

Crane & Co. had, in essence, offered to build a mill to suit, giving the Treasury Department complete control over the process. About this time the government transferred the localized fiber process to Murray Crane, and with assistance from superintendent William S. Warren of the Pioneer Mill, the company developed a currency paper meeting all government specifications. Influenced by techniques developed at the Old Red Mill, Crane submitted samples of loft-dried paper, writing (Note: "double-sized" refers to a combination of engine and tub sizing): "This paper is a double-sized loft dried paper, which, while it will not print quite as readily as machine dried paper, possess much better sizing qualities, is much stronger and will wear very much longer. We have facilities for making it either way [loft dried or machine dried], but our experience tells us that the former is much better, and that it is the only way to properly dry a bank-note paper by that process [which takes] from three to four days ... and doesn't burn the fiber or kill the sizing qualities. [Machine dried paper] is smoother, lies a little flatter, and is softer (and perhaps will print a trifle better), but on the other hand the loft dried paper ... will hold any impression ... much better ."[17] Upon viewing the samples government officials scoffed at the prospect of loft dried paper, believing it would only slow down the process as it had at the Ivy Mill years before. Following this decision Crane & Co. purchased the failed Colt Mill in Pittsfield and equipped the plant accordingly (including a drying loft for the manufacture of bond and treasury-note paper). Production of banknote paper at the Glen Mill ceased as of October 1878, with the government accumulating stocks sufficient to supply them for a year. The following year Secretary Sherman approved the contract with Crane & Co., and Crane's new facility known as the Government Mill went into operation. Photographs of the interior show a moving-wire machine with a long bank of tandem dryers and a finishing room with a rotary press for plating.

As tonnage of banknotes shipped to Washington over the next twenty years tripled, Crane established a laboratory to find ways of improving life expectancy of the bills. One method was to abandon the use of recycled rags in favor of clean cuttings direct from textile mills. The lab also became involved in the stationery business. Crane's stationery at that time continued its popularity, cumulating in the printed invitations for the dedication of the Statue of Liberty in 1886. Crane labs now developed a rag substitute for parchment that found a large market in schools, colleges, fraternities, and societies, and scientists also developed a thin, light substitute for India paper, used primarily in the publication of Bibles. On top of all this the firm dabbled a bit in the manufacture of cigarette paper made from bleached tobacco.

W. Murray Crane went on to serve as governor of Massachusetts from 1900 to 1903, and U.S. Senator from 1904 to 1918. Politically, he was a moderate Republican who supported the establishment of the League of Nations. At his funeral in 1920, William Howard Taft, whom Crane had opposed both in 1908 and 1912, declared: "Murray Crane and I were intimate friends. I cherish his memory as that of a loved and loving friend. I count it a great privilege to be permitted to come here and take part in this affectionate tribute to one of nature's noblemen." President Calvin Coolidge is quoted as saying, "It is as a friend that those who knew Senator Crane will always remember him."[18]

Failure to gain the treasury contract was another in a series of blows that led the Cheney Paper Co. to bankruptcy, and the Oakland Mill was sold in December of 1879 to the Hurlbut Paper Co. of South Lee. The Willcox Paper Co. went on to contract with the government of Germany to manufacture banknote paper at a mill near Berlin using the localized fiber process, this paper receiving the esteemed "diplome d'Honneur" at the Paris Exposition of 1878. The Glen Mill struggled to get by, and in 1885 was converted to manila. By 1894 the Willcoxes were bankrupt and the old mill sold to Joseph M. Dohan. Dohan formed the Glen Mills Paper Company that installed a new 76" Pusey-Jones machine for the manufacture of manila and vegetable parchment.

A Contemporary Account of the U.S. Industry

In December of 1850 James M. Willcox was requested to provide a report to Congress on the progress of the paper industry since the founding of the country. Willcox opens his report by saying, "For want of merits and dates, my report of the rise and progress of the paper manufacture in the United States, must be very meager, as I have to rely on my experience and observation, and on conversations with my father [Mark Willcox]." In the ad hoc report Willcox touts his knowledge of the subject while deftly steering his readers away from potential competitors such as Crane & Co. An abstract of the Willcox report is provided here for comparison with the modern perspective:

Little progress in the way of machinery occurred until 1820. From 1820 to 1830, some efforts were made to introduce the paper machine from England, some very imperfect. The expense of the Fourdrinier, as well as the cost of importation, was an inducement to Phelps & Spafford of Windham, Connecticut, to begin manufacture of the moving-wire machine in 1830. Phelps & Spafford and Howe & Goddard of Worcester, Massachusetts, are the only two firms manufacturing the Fourdrinier machine to date. The cylinder-wire machine, introduced at an earlier date, is simpler and less costly, and in more general use. Paper made on the cylinder machine is not of high quality, and was primarily limited to the manufacture of news and wrapping paper.

The increased demand led to a deficiency of raw material [rags], and the country was obliged to resort to Europe for supplies, in particular, Germany and Italy. As the price of linen rags continued to rise, manufacturers increasingly turned to articles of cotton, which when much worn and reduced to rags, becomes a tender substance scarcely able to support its weight when made into paper. Imported linen rags, when mixed with 20 percent of domestic cotton, imparts the paper a strength and firmness which it could not have without it. The best qualities of writing and printing paper contain from 30 to 50 percent linen, from which we are entirely dependent on imports. However, the use of cotton for clothing is yearly increasing all over Europe, decreasing linen imports from five to ten percent every year. Over 1837-38 the Willcox Paper Co. developed an excellent substitute consisting of raw cotton mixed with rags of the same material when the price of raw cotton was as low as six cents per pound.

A vast improvement in machinery occurred between the years 1830 to 1840. This period was also marked by the application of chlorine gas, chloride of lime, alkaloids, lime and soda-ash for bleaching, cleaning, and discharging colors from calicoes, worn out sails, tarred rope, hemp, bagging, and cotton waste. Hemp bagging, an excellent material that gives paper great strength, is currently in high demand for the manufacture of newspaper. Such articles have been made into coarse news and printing papers, and consequently have risen 300 percent in value. A few mills possess machinery to process such materials into fine paper, for instance, one mill uses hemp rope to make fine letter paper.

Some improvement has been made in the finish of writing and printing papers by the introduction of iron and paper calenders. An imitation of old fashioned hand made laid paper is now made on the machine, but still lacks the toughness, firmness, and surface qualities of the handmade product. Nonetheless, great quantities of imitation laid paper [are] used by the public under the supposition they were hand made. I believe there are only two mills still in the nation that manufacture hand laid letter paper, mine and one in Massachusetts [L.L. Brown & Co.]. There is a limited quantity of particular papers that can be better made by hand, that being bank-note, laid letter, and parchment. Such papers that are much handled require great strength and durability.

The increase in production from steam-powered printers has greatly increased the numbers of published materials over the past ten years. The average machine mill now in service has an average daily production of 300lbs. [not withstanding] the loss of time and power from a deficiency of water in the summer season.

At present, the industry in Pennsylvania is greatly reduced. Ten years ago, eighty percent of supplies [rags] in Philadelphia came from east of the North River, and at present that number is less than twenty percent. Formerly, a great quantity of paper was shipped east of the mountains, and a large quantity of rags brought in return. In consequence, a greater number of mills have been established in the west, particularly in Ohio.

After the American Revolution the number of mills grew steadily driven by the demand for book and news paper. Since then the quantity of imported paper has steadily decreased. Before the revision of the tariff in 1846 the numbers had dwindled to no more than two percent. The exception has been wall paper of which great quantities continue to be imported from France. Since 1846 there has also been an increase in French letter paper and envelopes, but total consumption is still perhaps no more than three percent.[19]

11

Holyoke, City of Industry

The Commonwealth of Massachusetts had long been the leader in paper production due mostly to its strategic location between the largest markets in the country. However, the paper industry still lagged far behind iron and textiles in total revenue generation, and in view of recent advances in transportation, combined with the state's limited resources, it seems the commonwealth's best days were behind it. With the advent of wood pulp, rag mills grew increasingly scarce, although just when one begins to perceive the end was near something remarkable took place. Hitherto, canal building lay primarily in the realm of transportation, and if a mill (paper, textile or otherwise) needed waterpower, it built a raceway. State officials recognized that without new sources of waterpower, industries would relocate elsewhere. So, there began a movement towards the construction of waterpower canals with the intent of attracting new businesses to the region. Massachusetts, with its ready access to major markets, became a prime target for development; the only problem lay in identifying locations with low land values, something of increasing scarcity in the Bay State.

The Connecticut River runs south from New Hampshire through central Massachusetts on its way to Long Island Sound, and was the principal waterway of New England. The river had but with three main obstacles to navigation: Bellows Falls, Turners Falls, and Hadley Falls. The building of a canal around Hadley Falls began in 1792, this cut from a sandstone bluff and extending southwards on the east side of the waterway for two and one-half miles before rejoining the river just below the falls. The original facility could accommodate flat boats and barges up to forty feet in length, and not until 1824 would a system of locks be introduced to allow traffic to flow in both directions. A small town grew up here to maintain the facility and serve the needs of travelers, and for lack of a better name this was simply called Canal Village.[1]

Canal Village was the site of a large paper mill built in 1824 by D&J Ames. Over on the west bank of the river was Ireland Parish, which consisted of a largely pastoral region known simply as "the fields." In 1831 the Hadley Falls Company was formed for the purpose of building a cotton mill on the west side of the river. A wing dam was constructed just ahead of the falls to direct flow through a raceway that powered the new textile mill. Little changed thereafter until the 1840s, when the Northampton to Springfield Railroad arrived. The tracks were originally to cross the river north of the Hadley Falls and enter Canal Village, but midway through construction a route change sent it south through Ireland Parish, bypassing Canal Village altogether. With the coming of the railroad a real estate speculator from New York named George Ewing saw the potential to build a waterpower canal on the west side of the river where textile mills could be serviced by the railroad. With this end in sight Ewing bought the Hadley Falls Company and began acquiring all the water rights on both sides of the falls, including all of Canal Village and the roughly 1100 acres of real estate in "the fields." Only about twenty-two houses were scattered over Ireland Parish at the time, and while some owners proved exceedingly stubborn, the entire acquisition was eventually made at a cost of $119,000.

The New York speculator soon sold his investment to the newly formed company of Thomas H. Perkins, George W. Lyman and Edmund Dwight, who unveiled plans to construct a dam across the Connecticut River and build an industrial community of 2,000 people in Ireland Parish. The dam was completed in November of 1848, but collapsed after only several hours. Undaunted, engineers reworked the plans and improved the levy, which re-entered service in October of 1849. The new dam consisted of twelve huge gates, each 15' long by 9' wide, and powered by a large water wheel mounted in an abutment. Although minor leaks appeared in the wall from time to time, the structure was easily maintained and continued in operation without further incident. Water diverted from above the embankment was now directed to a canal that wound along parallel to the bank of the river for about a mile, bringing waterpower to any number of potential building sites.

In March of 1850 the first act of the new civil government was to name the town after Elizur Holyoke, a prominent resident of those parts. Meanwhile, developers had sold fifty-nine lots of stores and dwellings that came to be filled with an iron foundry, a machine shop, blacksmith, spindle factory, and pattern shop. The first business block to be completed was the Gallaudet-Terry block at the corner of High and Lyman streets. The founders also built a steam plant, and a gas main was directed to all parts of the business district to be used for both heating and power. Most important was the arrival of the first textile plant called the Lyman Mills, which went into operation in 1854. Many former dam workers would staff this textile mill, and seven blocks of housing, comprising over two hundred tenements, were now built for them and their families to live in.

The former Canal Village came to be known as South Hadley, and during construction

Hampshire Mill, South Hadley, Mass., circa 1875. Louvers were built into the roof to help maintain plant temperature and humidity.

of the dam the Howard & Lathrup Mill was closed and its equipment sold to the former plant manager, Joseph Carew. A new waterpower canal was built here to allow the businesses to relocate, and in 1848 Carew Manufacturing Co. broke ground on a new four-story plant where the old cylinder-wire machines were set-up for the manufacture of manila. A second paper mill, called the Hampshire Mill, was built next door in 1866, this running a 62" moving-wire machine for the manufacture of bond and stationery.

The Parsons Mill

D&J Ames divested of their mill in South Hadley around 1840, and the new owners, the Judd Brothers, converted the plant to manila. The Ames brothers continued in business in Springfield ten miles to the north, as evidenced by an 1849 letter addressed to John Ames at the U.S. Hotel in Boston requesting a further order of hardware paper for the publishing firm of L.W. Moody in Springfield. Although sales of cylinder-wire machines had greatly fallen off, D&J Ames was still very involved in equipment manufacturing. In February of 1852 the company sent Owen & Hurlbut an offer for a bill and pamphlet ruling machine similar to the one built for Southworth & Co. in Mittineague.

David and John Ames had no interest in building or acquiring new mills, but Joseph C. Parsons, their former plant manager at the Eagle Mill in Suffield, found business opportunities in Holyoke much to his liking. Parsons approached city planners with plans for a new paper mill to be outfitted with cylinder-wire machines. Parsons wanted to build on prime real estate near the South Hadley Dam, but commissioners denied the request for water power because they believed it best to hold out for another textile mill. Parsons eventually won them over with a grand design calling for a large paper mill with two moving-wire machines.

Parsons Mill on the upper level canal, Holyoke, Mass., circa 1875. Whiting Mill No. 1 is seen in the distance at center right (from promotional literature circa 1875).

Completed in 1853, the Parsons Paper Mill was dedicated to the manufacture of fine writing paper. The main factory would be three stories tall and house two 62" moving-wire machines built by Goddard, Rice & Co. Fourteen 250 pound rag engines and two rag boilers occupied a separate building, and as business improved in later years the existing beaters were replaced by sixteen 450 pound hollanders, and a third moving-wire machine of 72" was added. The auxiliary boilers and extended chimney were relegated to a separate fire-proof structure known as a boiler house, and another feature of this mill that became common in Holyoke was the placement of waterwheels in the basement directly under the factory floor.

Joseph Parsons built a separate office building on the corner of the lot with a view of the dam and river. The main floor housed the finishing plant, consisting of four sheet calenders, a web calender, two hydraulic presses, and associated envelope folding machines. The counting room occupied the second floor, while the top two floors were dedicated to office space. Almost overnight the Parsons Mill was one of the largest writing paper concerns in the country. By 1863 the facility would make two tons of paper and some 120,000 envelopes a day, and became so profitable that at one time its common stock paid dividends of 50 percent a year.

The Newton Brothers' Mills

The Hadley Falls Company suffered a financial crisis in 1857, and the business was sold two years later to the Holyoke Water Power Company for the sum of $350,000. Company holdings at the time included the dam, the South Hadley canal, two cotton mills, the Hadley Thread Company, and 1100 acres of Holyoke real estate. The new owners realized that to attract the more labor-intensive textile business the town had to grow, so, the Holyoke Water Power Company, backed by investors from Hartford, rolled out a redevelopment plan that included scores of new tenements, homes, businesses, parks, churches, and schools. The core of the project was three majestic waterways known as the 1st, 2nd, and 3rd level canals located at varying heights above the river to supply waterpower to dozens of new manufacturing sites. Two canals ranging from 20 to 28 feet above the river were cut parallel to one another running west through the town for a distance of about a mile. The 3rd level canal was 12" in elevation, and intersected the first two canals at the west end of town, conveying the streams south and east along the river before rejoining the original canal at a point near the Sergeant Street Bridge. Water levels in the canals were carefully regulated by a corps of engineers to insure the needs of every mill were properly met.

Another part of Holyoke's town planning was a system of railways linking each manufacturing site to depots in the downtown core. The Connecticut River Railroad ran south along the riverbank from Springfield and wound southeast through Holyoke to the Main Street Station. This line exited Holyoke to the south via a railroad bridge over the Connecticut River. A second railroad called the Holyoke and Western entered town from the west and immediately split into two branches, allowing trains to loop back they way they came. One branch proceeded between the first and second level canals before terminating at the Front Street depot, while the southern branch of the same traveled along the 3rd level canal before turning north at Dwight Street.

The plan to attract the lucrative textile business proved successful, as the Hampden Mill was established in 1860 on the banks of the old upper level canal. The Holyoke Water Power Company had just completed several new tenements and business blocks, and the new town was just taking shape when the Civil War broke out in 1861. The shortage of Southern cotton brought on by the war soon idled the Lyman Mill. At this time the waterpower company divested of the Hadley Thread Mill, which was reorganized as the Hadley Spool and Cotton Company, and it was beginning to look as if Holyoke would be in for some tough times. A federal program allowed for the purchase of Southern cotton from loyal planters, and as supplies to the north picked up the Lyman Mill and related concerns in Holyoke eventually returned to full production.

Travel across the Connecticut River to South Hadley had always been by ferry, but this arrangement grew increasing inadequate, as at times over fifty horse teams were in the streets waiting to board. A committee assembled in 1864 recommended a span emanating from High Street, but this proposal met with stiff resistance from the waterpower company, which was uneasy about granting easements so close to the dam. After much wrangling city commissioners finally approved a route running through an empty lot farther south. The site was surveyed in 1868, but approval from the state proved illusive. The *Hampden Freeman*, later renamed the *Holyoke Freeman*, covered the confusion, and in 1870 citizens were duly agitated and sent a petition to the legislature containing over 1500 signatures. A committee of legislators arrived by train in 1871, and as chance would have it, spring floods damaged the only ferryboat, putting it temporarily out of action. Given the state of confusion at the time, lawmakers had clear view of the problem, and the committee promptly elevated the transportation bill and it passed in the legislature later that year. A new span between South Hadley and Holyoke was completed in 1872.

In 1862 the Newton brothers (James H., John C., Daniel H., and Moses) came to Holyoke and founded the Hampden Paper Company, which built a mill between the 1st and 2nd level canals on Dwight Street. This plant employed 110 people for the manufacture of bristol boards, and in 1871 the Newton brothers sold the mill to Jared Beebe, but not before dividing the lot in two and retaining the empty parcel on the corner of Dwight Street. The following year Beebe partnered with G.B. Holbrook to form Beebe & Holbrook, which proceeded to convert the Hampden Mill to the manufacture of fine linen folio and writing paper. The configuration of this mill was somewhat unusual owing to the confines of the real estate. The main factory was a three-story structure with company offices, a workshop, and a separate room with two plating machines on the 1st floor. The 2nd floor held a finishing room and engine room with six heavy (500 to 750 pound) beaters, two of these fitted with rag washers, while the top story was fitted out as a drying loft. The main wing housed a 76" shaking wire machine with laboratory in the basement for the preparation of sizing. The rag room occupied a second wing, that also housing a bleach room with three $1/2$-ton rotary bleach boilers. The boiler house was situated on the remaining land between the two wings.

James H. Newton built the Franklin Mill in 1866 for the manufacture of collar paper. The three-story structure was situated on the western end of the old upper level canal overlooking the Connecticut River. The main floor of the plant housed the engine room with eight 450-pound hollanders and associated Jordans. The finishing room, which held trimming presses, Hammond cutters, and two Holyoke Machine Co. web super-calenders was on the south side of the main floor. Upper stories held the rag room and associated equipment like dusters, thrashers, and a Sturdevant blower. The plant's 62" moving-wire machine was placed in a wing on the north side of the mill. The boiler house, with four-foot steam boiler and three-ton bleach boiler, was located at the rear of the plant. After 1876 the facilities were converted to the manufacture of book, writing, and envelope paper.

The aforementioned "super-calender" had been another recent innovation. One of the problems with the early calenders was that the rollers often went out of adjustment. The next step in the machine's development was to mount two sets of rollers, one atop another. The sheets would make an "S" pass around and through the rollers, which were configured such that the top set only loosely pressed the paper, while the bottom set pressed it much harder. The rollers were much easier to configure and to adjust on the fly, and at some point

A system of belts is used to turn a rotary bleach boiler, visible in the background at right. Cut rags are dumped into the boiler from a hole in the floor above (not shown). After processing, the rags are deposited into carts to be wheeled to the engine room.

Illustration of the rear of Union Mill circa 1875, showing loading dock and railroad tracks.

a set of reversing rollers were added to the mechanism so the calender could smooth the paper on both sides in a single pass.

The other Newton brothers, John C. and Daniel H., went on a building spree of their own, erecting a number of shops and buildings in town, including the Elm Street High School, a wire mill, A.T. Stewart's woolen mill, and a machinery plant for the Holyoke Machine Company. In 1865 William Whiting, a clerk at the Holyoke Paper Co., began working with the Newton brothers to convert their wire mill to paper manufacturing. Whiting would eventually take ownership of the plant. The Whiting Mill was located on a prime corner lot between the first and second level canal at the intersection of the upper level canal. The main building was three stories high, 280 feet long, and 45 feet wide. Four 1200-pound rag engines and two 1500-pound rag washers were situated on the main floor of the plant. The second floor was divided between business offices and the rag room. The daylight basement held a finishing room with plating machines. The main wing housed two 62" moving-wire machines with tub-sizing attachments. The automated tub-sizing section for the cylinder-wire machine was invented in 1830, while Ransom and Millbourn's 1839 sizing section was the first made for the moving-wire machine.

Daniel H. and John C. Newton built the Excelsior Mill in 1872 for the manufacture of book and plate paper. By now the Newton brothers were experts at wedging paper mills into small plots of land, and this plant, located on a narrow strip of real estate running perpendicular to the river, would derive its power from the third level canal near the Sergeant Street Bridge. The main floor housed the engine room, which was equipped with four 800-pound rag engines. The upper stories were dedicated to the rag room, and also contained two Daniels rag-cutters and two dusters. A short wing that extended toward the river held two steam boilers and a 3-ton rotary bleach boiler. One 64" moving-wire machine occupied its own building that ran parallel to the main factory.

In 1879 James H. Newton and Edward T. Newton formed the Wauregan Paper Co., which built a mill on Dwight Street on the short plat next door to Beebe & Holbrook. Here the firm planned to manufacture writing, ledger, and envelope paper, and like other cramped downtown facilities, the mill configuration was somewhat challenged. The first floor was occupied by company offices and a finishing room with three stock cutters, one trimmer, and three Holyoke Machine Co. web-calenders. The 84" Pusey, Jones & Co. moving-wire machine with tandem dryers was mounted on the second floor, while the engine room, with six 1,000-pound rag engines and related Jordans, was on the same floor. The rag room occupied the floor above, and held two rag-cutters, a dusting machine, and a Sturdevant blower mounted in the attic.

Paper City

The growth of the paper industry in Holyoke led to a number of opportunities in support industries. The Prentiss Wire Mills had been established in 1857 with a meager force of eight people. Among other things, the factory made wire belts for the moving-wire machine, and as more mills came to town, Prentiss built a new factory in 1871 employing over fifty workers. The oft-mentioned Holyoke Machine Co. was organized in 1863. This firm employed fifty workers specializing in paper mill equipment such as their highly touted web-calender. New developments in waterpower — such as double wheels, register gates, and swing gates — were coming on the market. These developments led the Holyoke Machine Co. into the field of water wheels and associated gears and shafting. By the 1870s the firm became intrinsically linked with the development of water turbines led by a key employee named J.B. McCormick. McCormick had been a cabinetmaker and a music teacher for twenty-two years before coming to Holyoke in 1871 to work as a mechanic. Here he learned the business of waterwheels and gearing, and took his hand at making a water turbine. The water turbine would be ideal for mills in Holyoke because it allowed the plant to run more machinery without a corresponding increase in water rights. On the down side, turbines were subject to fouling, and that is why many mills chose to use a combination of wheels and turbines. McCormick's first attempt at a water turbine resulted in the "Holyoke Turbine" patented in 1875. This device yielded a twenty-eight percent power to diameter ratio, as verified by independent hydraulic engineers in Manchester, N.H. The Holyoke Turbine came to be built under license by a number of independent manufacturers around the country.

In 1881 the brothers J&W Jolly entered the machinery business in Holyoke after buying the five-person shop of Roby & Saunders. Success would bring the Jolly brothers to employ over 200 workers making custom iron and steel forgings and servicing nearly every piece of machinery in town. Paper mills were one of their specialties, and Jolly not only replaced wires and felts, but also built and repaired hollanders, Jordans, calender rollers, boiler pumps, shafting, gearing, stuff pumps, and Manning's combination winder. In 1882 they hired J.B. McCormick away from the competition, and put him to developing the double-bladed "Hercules Turbine Wheel." The turbine-wheel had an eighty percent power to diameter ratio, more than twice as efficient as the previous water turbine. Jolly's first turbine-wheel of four hundred horsepower went into the Parsons Mill, which had begun to lose market share to some of the newer and larger mills in town, and was desperate to add another paper machine. The first out-of-town order for the turbine-wheel came from the Government Mill in Dalton. Jolly also worked the European market, where they supplied a thirty-inch turbine-wheel for a plant in Glasgow, Scotland, and two horizontal and one vertical turbine wheels with a combined capacity of nearly one thousand horsepower for the Linden Paper Company.

Naturally Holyoke also had its own millwrights. No two paper mills were built alike, and each had specific needs depending on the products they were intended for, and the principal job of the millwright was to evaluate the available water power and come up with an equipment list that met the owner's requirements. On the corner of Main and Race streets was the architectural firm of David H. & Ashley B. Tower. The firm's files contained nearly eight thousand drawings of site plans, mill machinery, and waterpower improvement projects. True to their name, the Tower firm was also largely responsible for the architectural towers that graced the front of many of Holyoke's mills.

Success of the Parsons Mill generated heightened interest in Holyoke as a manufacturing center, and in 1860 the newly formed Holyoke Paper Company erected a three-story structure on Main Street for the manufacture of fine writing and bond paper. The ample real estate available on the 2nd level canal in this location allowed the owners to build big. The main factory was located in a 208 by 54 foot wing off the main building, housing two 62" moving-wire machines. Below this was the engine room with twenty 500-pound rag engines, eight fit with rag washers, and a complement of Jordans. The finishing room on the third floor held thirteen sheet calenders, five roll calenders, two plating machines, four trimming machines, four hydraulic presses, and eight ruling-machines. The upper story was divided between the sizing room and drying loft. Five auxiliary boilers were housed in a separate boiler house. Power for the plant was supplied by a two-story wheel house measuring 48 by 110 feet, housing eight twenty-foot water wheels that later were replaced by a similar number of water turbines totaling 500 horsepower. By 1863 the mill was capable of four tons of paper a day. At the Paris exposition of 1878 the Holyoke Paper Company received a gold medal for its ruled parchment and loft-dried linen bond.

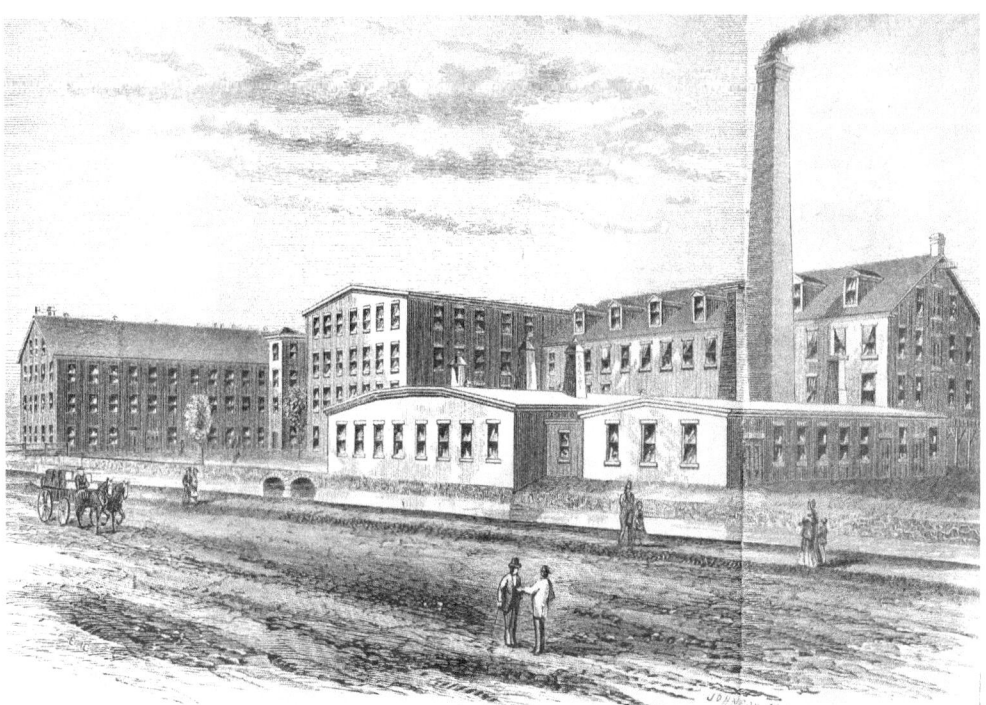

Holyoke Mill, circa 1875. A water outlet to the lower level canal is seen at center left.

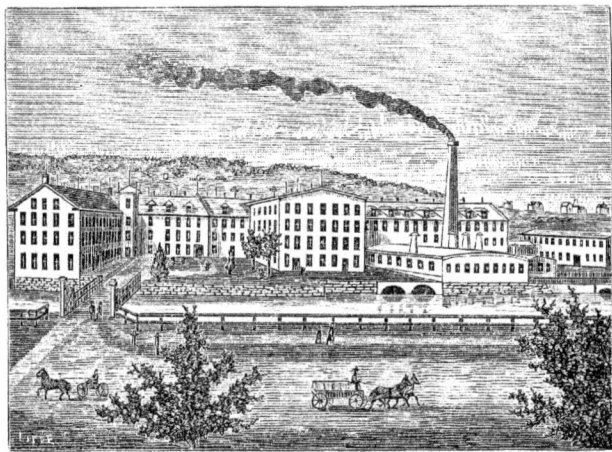

A Holyoke Paper Co. advertisement for high quality plate paper, ledger, and parchment (from *Paper Mills of the World*, 1885).

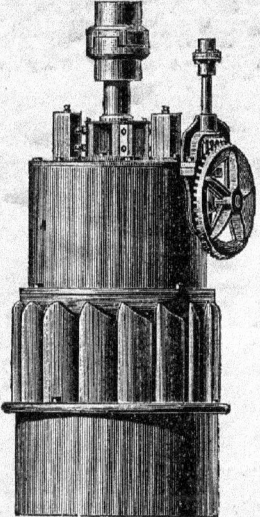

A Holyoke Machine Co. advertisement depicting J.B. McCormick's "Hercules" water turbine (from *Paper Mills of the World*, 1885).

It is something of a misnomer in that the bleach boiler was used for washing rags, while the rag washer was used for bleaching. The Holyoke mill had three rotary bleach boilers with a daily capacity of nine tons mounted on the main floor. Drainage bins for storing the pulp after bleaching were located in the basement directly underneath. Rag storage occupied the third floor, while the four rag dusters with daily capacity of ten tons were located on the floor below. A hole in the above flooring allowed the bleach boilers to be loaded from the rag room directly above. The bleach boiler, or rotary bleach boiler, was designed to remove dirt, grease, and grime from secondary grade rags, rendering them useable for fine papermaking.

C.S. Buchanan of Ballston Spa, N.Y., was the first to adapt the rag boiler for the manufacture of bogus manila. Over the years the device evolved into a long cylinder that rag mills would fill with several thousand pounds of rag clippings, to which were added water, lime, and caustic soda or soda ash. When the boiler-side doors were closed tight, steam was admitted to the maximum of fifty pounds per square inch, and the cylinder set to rotate. The tumbling action combined with the alkaline solution rapidly removed dirt and grease, and when the cycle was completed the rag clippings came out light tan in color, ready for the hollander.

Holyoke's building boom began in the waning days of the Civil War. In 1864 David M. Butterfield, a finisher at Parsons' mill, started the Valley Paper Co. that built a new mill in the southwest corner of town on the lot adjoining the proposed easement for the South Hadley bridge. The main mill was a three-story structure housing seven 450-pound rag engines on the main floor. The rag department and company offices occupied the second floor, while the third story was devoted to finishing and drying. A 62" moving-wire machine resided in a separate wing extending out toward the river while the boiler house was wedged into the angle formed by the main building and factory wing. The Valley Mill specialized in fine writing and envelope papers and business was so good that in 1877 a second plant with an architectural tower was built facing the upper level canal.

The Riverside Mill was built by Charles O. Chapin in 1867 for the manufacture of fine writing paper. This mill occupied the corner of Canal and Cabot streets, taking waterpower from the upper level canal. The plant consisted of a two and one-half story building with two long wings. A series of seven gables over the loft space was a distinctive feature of this mill. The daylight basement housed four Holyoke Machine Co. water wheels and four water turbines (a 60" American, a 25" American, a 40" Risdon, and a 15" Hercules). The main floor housed a bank of seven rag engines, while the second story served as the finishing department equipped with seven sheet-calenders and five ruling machines. A 72" moving-wire machine was housed in the main factory wing, and an 80" moving-wire machine was added in later years. The remaining wing housed the rag room, which included two dusters and a four-ton rotary bleach boiler. The auxiliary boilers and main chimney were built within the confines of the main building in a throw-back design.

The Hampden Cotton Mill was converted to papermaking in 1869 by the Albion Paper Company. A separate wing was built to house the paper machine where the owners planned to manufacture paper collars, but when that fashion faded in 1877 the mill was sold to Daniel H. and John C. Newton, who converted the facility to the manufacture of book and newsprint. The Newton brothers went in for improvements, expanding the original mill and building a twenty-foot tower similar to that of the neighboring Valley Mill. Within a few years an even larger factory building went up near the river; this connected to the main plant by the factory wing. To this was added a second tower topped by a French roof facing Canal Street, while a second wing would be added to house an 84" moving-wire machine. The Canal Street building would be filled with company offices, a repair shop, and finishing rooms containing two 36" nine-roll web super-calenders, two 72" Hammond cutters, and a Cranston trimmer. The upper floor and loft area held the rag room, equipped with two Daniels cutters, a duster, and a thresher. At the rear of the mill facing the river sat the engine house with five 1000-pound rag engines (three fitted with rag washers), and two rotary bleach boilers.

The Crocker Mill was built next door to the Albion Mill in 1871. This "T" shaped facility had been designed for the manufacture of paper collars, card stock, and bristol board. The mill sported ten 450-pound rag engines, a 76" shaking wire machine, and a cylinder-wire machine. A second wing was later added to house a Rice, Barton, & Fales 86" moving-wire machine. In 1878 Crocker acquired the Albion Mill, thereafter known as Crocker Mill No. 2.

The little-known Bemis Paper Co. operated a small plant on Canal Street between the Franklin and Riverside Mills. In 1870 the mill was sold to the Union Paper Company, a firm consisting of Henry and Edwin Dickinson and J.E. Taylor (formerly of Greenleaf & Taylor). Union planned to manufacture collars, but soon switched to writing paper. The original facility proved too small for their purposes, so a wing was added to house two moving-wire machines. The main plant of three stories was reorganized with the engine room containing ten 600-pound rag engines on the main floor. The finishing room, with a bank of sheet-fed super-calenders and two trimming presses, occupied the second story, while the upper story was devoted to a rag room equipped with two rag dusters. A boiler house was built on the river bank to house three 4.5-foot auxiliary steam boilers and two 3-ton rotary bleach boilers. A separate building for shipping and receiving with a small repair shop in the back was built adjoining the railroad tracks.

The firm of Greenleaf & Taylor was organized in New Jersey in 1853, and operated a mill in Huntington. Spurred on by the success of J.E. Taylor and the Union Paper Co., Greenleaf & Taylor moved to Holyoke in 1872 and formed the Massasoit Paper Company. Massasoit erected a four-story brick building fronting the first level canal near Dwight Street. The plant's first floor contained company offices, storeroom, sizing room, and plating shop with a large plater manufactured by the Holyoke Machine Co. The second floor held the finishing room with six sheet calenders, one Cranston trimming machine, two hydraulic presses, five ruling machines, and a stamper. The west end of the same floor had an engine room with eight rag engines, three fitted with rag washers. The third floor was divided between the rag room and rag storage, while the top floor and attic were used for drying. A main wing housed a 79" moving-wire machine capable of making three tons of finished paper a day. Massasoit was famous for its superior brands of linen with silky finishes known as Crown Leghorn, Imperial Parchment, and Lyons Parchment.

William Whiting had become one of the more successful papermakers in Holyoke. During the 1870s, Whiting made several civic improvements, including the Holyoke Opera House and the Windsor Hotel. Then in 1873 William Whiting opened a second plant on the corner of Dwight Street across from Beebe & Holbrook. Called the Whiting Mill No. 2, this plant made fine writing, bond, ledger, and envelope paper. The massive four-story building with a daylight basement had a wonderful mansard pattern roof and an eighty-five foot tower facing the second level canal. The main floor housed an engine room with ten 1000-pound rag engines. The second story was divided between the rag room and finishing room, while upper story and loft space were reserved for drying. Two wings flanked the rear of the building; one containing the auxiliary boilers and two massive (6,000 to 7,000 capacity) rotary bleach boilers, the other housing two moving-wire machines. Whiting, along with investment from the Jenks Family of Adams, had built another paper mill on the Chicopee River in Wilbraham, Mass. All told, the Whiting Paper Co. was the leading rag paper manufacturer in the country.

The Newton Paper Company was formed in 1876 by Moses Newton, James Rampage, and George A. Clark for the manufacture of building materials, heavy wrapping paper, and corrugated carpet lining. Named the Newton Mill, it had no association with the aforementioned Newton brothers other than having been erected on the third level canal next door to the Newtons' Excelsior Mill. The Newton Mill was a long, rambling two story affair running parallel to the river, with three wings that housing 42", 54", and 62" cylinder-wire machines. This was one of the largest concerns in the country and produced fifteen tons a day. The mill also featured a rather extensive shipping and receiving dock alongside the railroad spur.

In the far western reaches of town, at the intersection of the second and third level canals, was the Holyoke Manila Mill. Erected in 1875 by the firm of Robertson & Black, the plant was capable of making 2500 pounds of tissue paper a day. This facility was another rambling two-story affair with two wings. The main building housed three 500-pound rag engines with water wheels in the basement. The upper story had a rope cutter and an extensive area for storage. The west wing held a 62" cylinder-wire machine, while the east wing housed a three-ton rotary bleach boiler and two 4½ foot auxiliary boilers.

In what was increasingly becoming paper mill row, the Dickinson & Clark Mill was erected in 1878 on Canal Street between the Valley and Albion mills. This mill produced book and writing paper. The original mill was a modest appearing single-story facility with a stubby two-story tower looking more like a country manor than a paper mill, yet the facility housed four 2000-pound rag engines, two super-calenders, two trimmers, one 84" and one 88" moving-wire machines, and associated Jordans. In later years a 110" moving-wire machine and eight additional rag engines were added, extending the mill's capacity to a staggering twenty-three tons a day.

Holyoke by this time had so grown in stature it eclipsed Lee as the center of papermaking in Massachusetts. So many mills had come here that local newspapers began to refer to their community as a "paper city." Indeed, over time Holyoke became synonymous with fine rag paper, a fact made all to clear to the papermakers in South Lee when in 1885 an order of theirs from a stationer in Buffalo was accidentally sent to Holyoke by mistake.[2]

A Tour of the Whiting Mill

Sometime around the turn of the century the Whiting Paper Company commissioned a brochure entitled "How Paper Is Made — The Process as Seen at the Mills of the Whiting Paper Company." This piece is essentially a walking tour of Whiting's Mill No.1, the original mill built in 1865. We pick up the action immediately upon arrival:

11— Holyoke, City of Industry

An illustration of Whiting Mill No. 1 showing railroad tracks in the foreground. The machine-room wing is topped by a curved roof, while next door the boiler house features a very tall chimney. The bridge from Holyoke to South Hadley is in the background (from promotional literature circa 1875).

Let us enter a paper mill, and look about. I suppose we will find ourselves in the office in the first place and there see rows of desks and various clerks and bookkeepers, who scratch away with their pens very much after the manner of their class anywhere. There are the great brown bales of rags weighing from six to nine hundred pounds each, and here are heavy paper-wrapped bundles.... Reach into one and pull forth a piece Why! it's ... paper — a very heavy, coarse-textured white paper. But that is the shape in which it reaches the mills and it must be ground over again to make an article for the public....

The lower floors of this part of the mill are mostly given up to storage. We mount to the fourth story. Here the bundles of rags are being slashed open by a man with a big knife; the brown sacking and cords are removed, and the close-packed mass is pulled to pieces and thrown into a great hopper [thrasher]. There a swiftly revolving wheel catches the rags on its spikes and whirls them about so fiercely that you wonder to find any rags left after the process, to say nothing of getting the dust out of them.

From this hopper they are dropped through a hole in the floor, and if we follow them ... down the stairway ... we enter a large room where a little army of girls is at work amidst a great array of immense baskets heaped to overflowing with the white rags.... Most of them stand facing the windows before a wide continuous table divided into apartments and floored with a coarse wire screen. In front of each girl is a heavy upright knife like a broad-bladed scythe, which they use to cut off buttons with. Behind her are two or three baskets into which the different sorting are thrown. From the screen girls the rags go for a more careful sorting to the table girls. The heavily loaded baskets filled by them are slid into a little side room where the cutter is at work. A single girl feeds the rags into the low, rattling, grinding, jarring machine [cutting machine] that, six feet from the starting point, delivers them all cut in one to three-inch squares at the rate of a ton an hour. Until within a short time this work was done by hand on the knives in the room adjoining. The center drops the rags on a revolving strip of canvas which carries them down stairs, and lets them fall into another dusting machine [rag duster]. This machine is a huge drum of wire netting inside a box or

Paper cutters at Whiting Mill No. 1 circa 1890. Note the early electric lights dangling above each work station (from promotional literature circa 1875).

small room. Lift a door and look in and you see the rags rolling about within the drum, and below a thick deposit of linty "dust." The "dust" is nearly white and looks quite good enough for fine paper, but it is all sold to mills which manufacture the coarse paper used under carpets. The rags pass through three of these dusters one after another and then are caught on a strip of canvas which hurries them up to the ceiling, there to toss them down a steep incline of iron slats. Through this such stray buttons as have passed the sorter are supposed to fall. A vigilant watch is kept for buttons up to the last moment and there are many little devices for detaining them, so that there is a small danger of a customer's finding any in his paper.

After the dusting, the rags are pushed down through a convenient hole in the floor, which lets them fall into an immense cylinder tank or iron which holds fully three tons [rotary bleach boiler]. A mixture of lime water and soda ash is put in with the rags and the steam is turned on. In this slowly revolving tank the rags are boiled for twelve hours. They look to be well-cooked by the time they come out, for they have turned to a rusty brown. Numerous big boxes mounted on trucks are being trundled by the workmen to the sloppy floors of the boiler room to the apartment adjoining, where they are thrown into great oval vats known as "engines" [rag beater]. Each engine is furnished with a heavy, revolving iron wheel slatted with knives, which keeps the contents of the vat in motion and tears them to bits. A stream of water is turned on, and this water, though it came originally from the near canal, has been filtered to a purity that, if one tries it, will be found in taste and quality fully up to spring water. Chloride of lime is added for a bleach, and after six hours working over in this tank the contents are very white and pulpy.... It comes from the bleach of a delicious whiteness and looks good enough to eat. A paper pulp pudding, if it tasted as good as it looked, would be a rare dish. Now the pulp is washed in one of the engines [rag washer] and freed from the chloride of lime, and is kept grinding (in a hollander) for ten hours more before it is ready for the paper machine. Meanwhile, color has been added and alum to fix it, and now it goes down to the basement to the "chest" [stuff chest]. The chest is a big vat, with long arms revolving within to keep the contents stirring, and from this the pulp is pumped up to the machines.

The machine room is a noisy place and I can remember as a little fellow the half fright and awe with which I passed through this room, the [moving-wire] machine is so big and complicated and so loud in its clattering. The floors are wet and steamy vapors are rising from the damp paper, and the air here is warm and moist, whatever the outside weather. A machine is about a hundred feet long. At one end a little stream of pulp is allowed to flow in, varying in volume with the weight of paper desired. It is mixed with a stream of water that reduces it to a thin, milky fluid, that seems to be so nearly water that one doubts the possibility of ever getting such stuff into solid sheets of paper. All this end of the machines is in a violent, jarring motion to distribute the paper fiber evenly. Presently the fluid flows on to a long strip of revolving wire cloth and begins to lose its water. Fifteen feet distant, almost as if by a miracle, the substance flows away a solid, broad sheet of paper.

The moist sheet now passes beneath a revolving roll of wire [dandy roller], which bears on its surface raised letters and, perhaps, certain designs, and these stamp what is known as the watermark on the paper. You can see it if you look through a sheet held against the light, but it is not usually apparent otherwise. After a little further progress, you find the paper has become dry and hard. Lastly it passes through a vat of gluey animal sizing [tub-sizing attachment] and under a revolving cutter [rotary cutter] and is delivered at the end in fast falling sheets of the size desired. The paper then goes into the loft to be dried. The sheets are hung over poles in folds of ten or fifteen in a bunch, until a room is filled from floor to ceiling. Then steam is turned on and it is allowed to dry for forty-eight hours. It does not look very handsome as it comes from the lofts, it is so wrinkled and rough that you begin to think it's a failure after all, when the calender girls get it and run it through their machines [super-calenders] and it comes up as fair and as smooth as you please. Next the sorters take it. They sit at their table and keep the sheets swiftly turning — the perfect sheets being put in one pile, the slightly imperfect in another [seconds], and the more defective [retree or broken], which must be ground over, in a third. A counter takes the sheets and runs over them rapidly and lays them off in reams to be trimmed and wrapped for the market. The last work is that of the sealer who, with a stick of wax and a lighted gas jet, accumulates the neatly packed packages about him or her with astonishing rapidity. If the paper is to be put up in boxes, it has to be cut into small sheets and perhaps passed through a ruling machine. Then it is folded, pressed and banded. Envelopes have to be made to go with it, and it is interesting to watch the [envelope] machine, which takes the queerly outlined sheets cut for it and folds and gums and shapes them into envelopes. The girl at the machine counts and bands them, and another takes them away and puts them and the paper up in the pretty boxes awaiting. And now they are ready for the final packing and shipment to stationers the world over.

Part Three
The Wood Pulp Era

12

The Curtisville Exponent

The manufacture of commercial grade wood-pulp paper got its start in Germany (then the German States) during the 1850s. The process was slow to take root in other western European countries largely due to the lack of natural resources. Tidbits of news reached American ears, but few could afford risking large amounts of venture capital on a process that had yet to be proven anywhere west of the Rhine. The only papermaker in the U.S. to make the attempt happened to be the largest manufacturer in the state with the most paper mills, Platner & Smith of Lee, Mass.

Elizur Smith was born in a tiny village in Berkshire County, Mass., in 1812. When sixteen years old he cut his foot and was bedridden for several months. During this time he studied and read a number of books, eventually deciding to further his education at Westfield Academy. In 1830 Elizur took a job in Lee, Mass., clerking for the firm of John Nye & Co. at a salary of $20 per year including room and board. John Nye was a perennial selectman in Lee, and the young clerk made the acquaintance of many paper mill owners and investors during his time at the firm. Over 1834-35 Smith purchased half interest in the Turkey Mill located on Hop Brook in Tyringham, then owned by George Platner. As the partnership evolved Elizur took over management of the mill, leaving the firm's finances in the hands of Mr. Platner. It was about this time Elizur Smith made the rounds in New York, perhaps for the first time, and met with Charles Owen, who had given him a letter to be hand-carried back to Lee. Meanwhile, George Platner served on the vigilance (police) committee in Lee with Thomas Hurlbut, and such contacts would naturally lead to business transactions between the two firms.[1]

Like other successful papermakers in Berkshire County at the time, Platner & Smith (P&S) focused its efforts on fine letter and ledger papers. The firm had weathered the Panic of 1837 fairly well and in 1838 they purchased both the Union Mill and the Enterprise Mill in Lee Town Center from Joseph and Leonard Church. The two mills were renovated and converted to machine production, with the latter facility renamed the Eagle Mill. Between 1839 and 1841 P&S also leased the Baker Mill in East Lee, and the pair were so successful they even tried their hand at woolen manufacture. Platner & Smith's profits soared during the heady days of the 1840s, and the partners also perused independent side lines, such as George Platner's investment in his brother's paper mill in Ancram, N.Y., and Elizur Smith's interest in a mill in Russell with his brother, J.R. Smith, and Cyrus W. Field.

After the Western Railroad reached the town of Dalton, Platner & Smith leased the Defiance Mill, and by 1847 the firm was making fully one-third of its sales in Boston. To maintain their various operations Platner & Smith set up a machine shop in Lee. The shop superintendent, Robert McAlpine, was encouraged to innovate, and he began by developing a special grinder for the seven-foot long stop-cutter. The guillotine-style blades were patented by A. & W. Palmer and very expensive to replace, so the sharpener was a great money saver. The superintendent went on to develop a modification to the ruling machine called a striker. The striker automated the raising and lowering of the pens, eliminating the pen attendant, and allowing the paper machine to run faster.

Platner & Smith's collaboration with Owen & Hurlbut began with the rag trade, as in 1848 the former sent the South Lee Mill 50 bales of No.1 Leghorn rags from a shared consignment. Platner and Smith also ordered iron shafting from the South Lee Manufacturing Co., another enterprise of Owen & Hurlbut's. However, if there was one thing Elizur Smith did not wish to share, it was his good employees. This apparently was the case in 1849 when Smith sent a blistering complaint to Owen & Hurlbut: "[I] make it a note in our business not to let any manufacturers take away our men.... If your wish to peruse this game, you will find it one that two can play at."[2]

The Housatonic Mill was the original machine-mill built in Lee by Laflin & Loomis in 1827, and in 1850 it came up for sale. Platner & Smith owned all the other mills in Lee, and had been waiting for just this opportunity. After acquiring this key property the firm went in for waterpower improvements, extending the head-race as far as seven hundred yards above the mill. The rebuilding project took eight months and cumulated in the installation of two new Smith-Winchester moving-wire machines. In acquiring the Housatonic Mill, Platner & Smith also inherited contracts with James Gordon Bennett's *New York Herald*, which required a thousand dollars of newsprint a day. P&S had no experience making newsprint, but the company quickly learned the business and in looking for ways to improve was among the first to employ the Kingsland refiner in 1855. That same year the twenty-year partnership came to an abrupt end when George Platner passed away, leaving his affairs to his wife, Adeline.

As business continued on the rise, Platner & Smith converted a satinet factory and clothier's shop on Laurel Lake to papermaking, calling it the Laurel Mill. The firm now made great quantities of book paper, and according to an 1858 article in the *Valley Gleaner*, a Platner & Smith workmen developed a new paper trimmer that increased output by one-third. The firm suffered its first serious setback that same year when the Housatonic Mill was destroyed by fire. Losses were reported at $150,000 with just $30,000 covered by insurance. The following year the Union Mill experienced a devastating boiler explosion, and the combination of events put the firm into receivership. Adeline Platner sold her interest in the firm to Elizur Smith in 1860.

In 1861 Elizur Smith became one of the founding members of the Writing Paper Manufacturers of America, and his enterprise eventually emerged from bankruptcy in the high-flying climate of the Civil War. Both the Union and Housatonic mills were rebuilt, and to keep up with demand Smith also rented the Pleasant Valley Mill in Lenox. Owing to a shortage of rags during the war, the Eagle Mill was converted to manila, otherwise the rest of Smith's mills were running at capacity, these including the Housatonic Mill at 10,500 lbs. a day, the Columbia Mill at 12,500 lbs. per day, and the Pleasant Valley at 9,000 lbs. a day. With seven mills under his control and capitol assets topping $250,000, Elizur Smith was arguably the leading manufacturer in the state.

As reported in the *Valley Gleaner*, on the last Saturday in 1864 every employee of the Smith Paper Co. received a Christmas goose from Smith, which cost the entrepreneur $500. About a month later Elizur Smith announced his retirement, and for the occasion mill work-

The front and back of a picture postcard depicting the Columbia Mill sent by August R. Smith in 1907.

ers threw an oyster supper complete with music and dancing. The illustrious papermaker was presented with a forty-five-piece tea service, and he took to retirement on a 600-acre estate just west of Laurel Lake. Much of this land was low-lying and swampy, so Smith had some forty miles of drain tiles installed to bring it into production. Elizur Smith worked at raising thoroughbreds and was taken by all the latest farm machinery and gadgets.

Steinway, Pagenstecher et al., and the Smith Paper Co.

Theodore Steinway, of piano manufacturing fame, was buying sheet music from Germany on paper made from wood-pulp, and given the inordinately high price of paper in New York during the Civil War, Steinway discussed this latest development with an associate of his named Albrecht Pagenstecher. Albrecht then wrote his brother, Rudolph, in Wiesbaden, Germany, to investigate the source of this paper, and to see about bringing it to the United States.

Jacob Christian Schaffer invented the wood-pulp process in Germany in 1765. Schaffer had successfully made paper from hemp, bark, straw, cabbage, cattails, thistles, cornhusks, and wasps' nests, which demonstrated that most any fibrous material could be macerated and formed into sheets. Handmade rag paper was far superior to any alternative source at the time, but following the proliferation of the paper machine during the 1830s Schaffer's process was given a second look.

A weaver named Friedrich Keller patented a wood-grinding machine in 1840, and the first tests at a paper mill in Alt-Chemnitz produced a roll of newsprint made from a mixture of forty percent rag and sixty percent wood-pulp. In 1847 Henrich Voelter, a papermaker from Saxony, purchased Keller's patent, and went on to perfect the grinding process over a twenty-year period. Voelter installed his grinder at a mill in Wurttemberg. By this time the apparatus consisted of a revolving grindstone with three metal pockets arranged so the length of the log rested against the stone. Upon closing the pocket door hydraulic presses kept the logs pressed against the stone while a stream of cold water showered the rotating stone to keep it cool. Tiny bits of wood torn off the logs were carried by the water through a rake, where they dropped onto a slowly rotating wire cage (squirrel cage). Within this cage rotated two more cages of successively finer mesh such that wood particles eventually fell into one of three grades. Particles too coarse to make it through to the inner screen were sent to a refiner to be disintegrated into smaller pieces before being returned to the wire cage through a system of plumbing. After five years of development Voelter was finally ready to make paper, and in 1852 ground wood paper fortified with rag fiber was successfully made at Voelter's plant along with another mill in Silesia belonging to his son. The refined Voelter grinder was first exhibited at Paris in 1867.

The report coming back to Albrecht Pagenstecher was that the Voelter grinder could be obtained through an agent in New York by the name of Louis Prang. Pagenstecher then wrote his brother Alberto, who had accumulated a fortune as a railway engineer in South America, to assist with the financing. With technical assistance from Professor Ferdinand Hoffman, Pagenstecher now formed a company to obtain the U.S. rights to the wood-grinder, giving it an exclusive franchise for twenty years. At that time Lee, Mass., had the largest concentration of mills in the nation, so Pagenstecher & Co. chose to locate its first commercial wood-pulp mill in Berkshire County.

Pagenstecher & Co. put in an order for two Voelter grinders, and the inventor sent along six skilled grinder operators to ensure success. The location chosen was the small industrial village of Curtisville, a northern suburb of Stockbridge about four miles from Lee. The village had sufficient waterpower; the Lake Mahkeenac stream supported light manufacturing such as a foundry and several machine shops. Here Gaston Burgharth and the Curtisville Furnace Co. made wheels, shafts, brass castings, and a boiler door for the South Lee Mill during installation of their new cylinder-wire machine in 1839. The Curtisville mechanics also performed a variety of paper mill related tasks including balancing waterwheels, fitting and turning calender rollers, installing steam pipe, and rebuilding piston pumps. Curtisville fairly welcomed Pagenstecher & Co.'s new venture, and on March 17, 1867, the Curtisville Pulp Mill sent its first load of wood-pulp to the Smith Paper Company for the price of eight cents a pound.[3]

Elizur Smith's interest in alternative fibers began after he started making newsprint for the New York dailies. From 1850 to 1857 rag prices advanced from 8 cents a pound to 20 cents a pound. To reduce rag consumption Platner & Smith converted some of their mills to manila, and also made the first tentative steps to create ground wood-pulp. Following the methods of Keller and Voelter, Platner & Smith's workshop in Lee made a wood grinder in 1857, but this device didn't grind fast enough nor was the pulp fine enough for papermaking. Financial considerations forced the firm to abandon the project, but a few years later a group of Boston investors approached Elizur Smith with yet another alternative-fiber proposal. In 1862 Smith's National Mill on Lake May Stream in East Lee made paper from the life-everlasting plant, also known as the cudweed. Reporting in the January 14, 1863, edition of the *Boston Weekly Journal*, the editor claims: "The entire edition of the Journal today will be printed on paper made from wood, by a new process.... This paper is not a fair test of what the manufacturers propose to do when their arrangements are fully perfected, but it will certainly prove there are other materials than rags, which can be used successfully in the manufacture of white paper. We have been experimenting with this paper on portions of our editions for several days past, and the results have been highly satisfactory." However, despite these claims, subsequent research reveals that *Journal*s printed on vegetable paper were considerably inferior.

Following Elizur Smith's retirement his two nephews, Wellington and Dewitt, took over the Smith Paper Co. empire. Even after rag prices retreated to 14.5 cents per pound, Smith Paper continued the stringent cost accounting measures and modern managerial accounting techniques developed during the war. The Smiths were already familiar with the manufacture of straw and manila papers, and now working with Pagenstecher & Co., they found that that wood pulp actually had higher fiber content than straw. Wood-pulp also drained more rapidly than either straw or manila, meaning it could be made on the moving-wire machine. One problem they did uncover was that if not enough rags were used the wood fibers tended to rise to the surface of the wire, causing the paper to become very rough on one side. Nonetheless, the ability to make wood-pulp paper on the moving-wire machine proved the deciding factor, so Smith Paper contracted the Curtisville Pulp Mill to buy its entire output for the next year.

Despite their commitment to Pagenstecher & Co., Smith Paper was not entirely wedded to the outcome. An 1869 article in the *Berkshire Courier* reports that Smith Paper Co. had been conducting experiments with esparto paper. Esparto paper was invented in Great

Britain in 1857 by Thomas Routledge. Routledge made a fine white paper from common esparto grass that was readily available in Spain and North Africa. British papermakers subsequently adopted the process and imported large quantities of the grass, greatly relieving the rag market in that country. Still, Smith Paper found the added cost of beating and bleaching esparto pulp made it nearly as expensive as manila.

The Growth of Wood Pulp

When Henrich Voelter sent his first two wood grinders to the U.S., he assigned Fredrich Wurtzbach, a 31-year-old mechanic from Magdesprung, Germany, to manage the operations. One of the early challenges for Wurtzbach was to transport the wood pulp the five miles from Curtisville to Lee. Wood pulp contains sixty percent water, and when carried over any distance evaporation sets in and it hardens into brick. Since the pulp would dry out in any event, the manager tried shipping it in the form of "cheeses." This process consisted of shoveling pulp into small bags that were formed in a hand press into compressed blocks measuring 12" × 10" × 2". When the cheeses arrived at the paper mill they were dropped into the hollander to dissolve, a process that took about an hour.

The Curtisville Pulp Mill had a capacity of one-half ton a day, and when the contract with Smith Paper ran its course Wurtzbach began making pulp for other mills in Massachusetts such as those in Lawrence and Fitchburg. Wurtzbach erected a second mill in 1869, but he was no longer alone in the business, as a third pulp mill in Curtisville was built by Benjamin Burghardt, and Lewis Beach and James H. Royce converted an old textile mill in Lee to pulp production. Following these developments the Montague Paper Co. set up the first large pulp mill near their plant in Turners Falls, Mass. Here a 1000 horsepower turbine-wheel powered twenty-four grinders, half of these making No. 1 pulp, while the remainder produced No. 2 pulp for boards. The papermakers soon learned that aspen and lime were the best for white stock, while poplar was better for darker grades of paper. Pine and fir had fibers of superior length that worked best for strong papers such as wrapping or construction. As local stands of wood became depleted, William A. Russell of Lawrence, Mass., ordered Voelter grinders for out-of-state pulp mills being erected in Franklin, N.H., and Bellows Falls, Vt.

To ensure its own supply of pulp, Smith Paper began building pulp mills of its own, converting a former glass factory at Lenox and erecting a second plant in Lee from scratch. The firm owned all the waterpower on the Housatonic River from Lee to Columbia Village, and in 1868 installed some ten Leffel turbines to supply its needs. The firm's pulp mills ground about 30 to 40 cords of wood a week making up to twenty-five tons of pulp.

Beginning in January of 1868 the *New Yorker Staats-Zeitung* became the first New York daily to use wood pulp paper. From its initial price of eight cents a pound, wood pulp quickly dropped to between four and five cents a pound and newspaper publishers found it to be the most economical paper on the market. In 1870 the *New York World* became the first English-language publication printed on wood pulp paper. The *Providence Journal* and *Brooklyn Eagle* began using wood pulp newsprint in 1871, the *New York Evening Express* in 1872, and the *New York Times* and *Albany Argus* followed the next year. A host of publications made the switch over 1874–1875, including the *New York Weekly Times*, *Vebote* [Chicago],

New York Sun, New York Tribune, New York Herald, New York Journal of Commerce, New York Evening Post, and *Cincinnati Daily Gazette*. The rush to wood pulp rapidly depleted available stocks, so the above could not entirely avoid the use of rag paper until around 1881.

To meet the ever-increasing demand the wood pulp industry quickly moved into heavily forested regions of New York and New England. By the early 1880s the Curtisville pulp mills were all closed and Frederick Wurtzbach was hired by Smith Paper Co. to manage its pulp operations. As Curtisville quietly reverted back to light manufacturing, Pagenstecher & Co. contracted with the Curtisville Furnace Co. of Erastus Burghardt to build Voelter grinders. The scale of the work was such that Burghardt now employed 55 skilled workers, drawn principally from the ironworks of Troy. Benjamin Barker, who owned the other pulp mill in Curtisville, also went into the business of building grinders. Between these two concerns seven new patents were registered between 1869 and 1877. Burghardt made refinements to the sifting of flints to ensure longer fibers, while Barker substituted an emery wheel for a grindstone to reduce the need for sharpening.

Despite its proven success most American papermakers rejected wood pulp paper on grounds that it was of inferior stock. One obvious problem was that ground wood contained lignin, which, like an epoxy, significantly hardened any fibers to which it bonded. The shortness of wood fibers and the lignin problem made ground-wood paper even less durable than that made of straw. Furthermore, wood-pulp paper rapidly yellowed when exposed to sunlight (UV radiation). Despite these drawbacks wood pulp was destined for use as newsprint, which now fell from its high of twenty-five cents a pound in the early 1860s to just 8 or 9 cents per pound.

The problems inherent in wood pulp paper would be solved by Henry Lowe of Baltimore, who obtained the rights to caustic soda from its inventor, Hugh Burgess. Caustic soda was distilled from boiling lime, ash, and water in a kiln, then leaching or diffusing it through a charcoal filter. Hugh Burgess first patented the process in Great Britain in 1852, and in the U.S. in 1854, and had originally designed it to decompose straw, reeds, and other low cellulose materials. Lowe was the first to apply the soda process to wood pulp. During the late 1850s Lowe made tests with spruce, fir, and hemlock at the Warren Mill in Maylandville, Pa. There was still the problem of producing wood pulp in sufficient quantities to sustain a commercial enterprise, so Lowe continued his experiments with crops such as flax, straw, hemp, corn, bamboo, pineapple leaf, and coconuts. In 1861 Lowe began experimenting with Southern cane in the swamps surrounding Wilmington, N.C., but this work would be curtailed by the war. In 1868 a papermaker from Herkimer, N.Y., named Warner Miller finally put it all together, buying a soda digester from Lowe, then leasing grinders from Pagenstecher & Co. Miller established a pulp mill in the nearby town of Luzerne, N.Y., and before the wood pulp arrived at the mill it was processed in the digester. A digester was essentially a double boiler; the outer boiler held water and steam, while pulpwood chips cooked in an alkali solution within the inner boiler. Fibers came out softer and easier to process, and were used by the Herkimer Mill to make the first wood pulp book paper. Following this success other firms such as the Megargee Brothers and Jessup & Moore adopted the soda process.[4]

Warner Miller was the first to make use of soda pulp, but in the course of production he was troubled by the length of time it took the cheeses to dissolve in the hollander. As an alternative, Miller tried processing wood pulp into sheets with a cylinder-wire machine. The goal was not to make paper, but rather to make quick forming sheets about 3x5 feet square. Miller configured a wet machine to make wood pulp sections called "pulp lap." Pulp lap was less dense than cheeses, and readily dissolved in the beater. This application proved so beneficial the practice was rapidly adopted by Pagenstecher & Co., and quickly spread to the industry beyond.

Maine Ascendancy

The first paper mill in Maine was built on the Presumpscot River in Falmouth, Cumberland County. Local farmers Thomas Westbrook and Samuel Waldo built the mill in 1730, but having no skilled workmen, the latter traveled to England to find an experienced papermaker. Richard Fry, a master papermaker, took a lease on the mill — but on arrival he discovered a lack of local rags. The subsequent lack of progress caused Fry to be thrown in jail in Boston in 1737 for non-payment of the lease. The papermaker made numerous appeals to magistrates, the public, and the governor before finally securing his release.[5]

The first successful paper mill in the state was started around 1812 in Gardiner, Kennebeck County, by John Savels, who learned the trade at a mill in Milton, Mass. Another plant was set up in North Yarmouth, Cumberland County, in 1816 by Henry Cox, who also got his start working for James Boies of Milton. In 1828 a two-vat mill went up in Camden, Maine, this built for $5,000 by Ebenezer Barrett and John Swann. The plant manufactured about twenty reams a day, and continued in operation until 1841 when it was destroyed by fire.

The paper machine came to Maine in the early 1840s. According to the *Farmington Chronicle* of May 7, 1846, Maine had 9 paper mills employing 1,369 workers and produced $1,750,200 worth of paper a year. Another wave of mill building occurred during the 1850s after the arrival of the Boston & Portland Railroad. Beginning in 1852 the Dennison Mfg. Co. built six mills in and around Mechanic Falls in Androscoggin County. The first moving-wire machine in the state was erected in 1851 at the Copesecook Mill in Gardiner. This plant was purchased three years later by the Boston warehouse of Grant, Warren & Co. During the Civil War, Grant, Warren & Co. added a second machine to this mill, bringing its daily production to 11,000 pounds.

The abundance of poplar trees made the state a magnet for pulp mills; the first of these was established in the town of Norway around 1869. About this time, Samuel Dennis Warren, of the aforementioned Grant, Warren & Co., took over the Copesecook Mill and converted it to wood pulp paper. Over 1875-76 Warren adopted the soda process for the production of book and writing paper, putting in two new direct-fired stationary digesters at the Copesecook Mill. The direct-fire digester was essentially a double boiler heated by a coal fire at its base. Although the process added to the cost of the final product, soda-pulp paper had the strength and softness desired by printers, and became readily adopted in the publishing of books and periodicals. By 1880, the S.D. Warren Mill was one of the largest soda plants in the country, producing some three and one-half tons of paper a day.

Surprisingly, the first wood pulp paper made in Maine was at a mill in Waterville in 1855. Charles H. Hall of Portland successfully experimented with barks of trees and produced a wood pulp wrapping paper at the Messalonskee Mill. Wood bark, however, was not a sustainable source — only with advent of the wood-grinder could mills depend on a continuing supply of raw materials. Another advantage of wood-pulp was that papermakers could put on as many paper machines as they had grinders, and thus were only limited by the availability of waterpower. Existing rag mills in the state now became hot properties; for example, a single-machine mill at Bowdoin was converted to wood pulp, and preceded to run on four moving-wire machines, and the Yarmouth Mill, oldest in the state, was converted to wood pulp by the Forest Paper Co.[6]

By the early 1880s, new wood pulp mills were established in Brunswick, Livermore Falls, North Gorham, Skowhegan, and Snow Falls. In Jay, Hugh J. Chisholm erected the third largest paper mill in the country. Chisholm faced stiff competition on the river for waterpower, and he began by buying the Falmouth Mill, a former sawmill converted to papermaking in 1844 by Alvin Record. Chisholm then built a dam above the old mill and converted it to a wood pulp plant called the Umbagog Mill. Chisholm completed work at the Umbagog Mill in 1885 by installing two new moving-wire machines. Alvin Record responded by buying several saw and grist mills to further his water rights and after Chisholm erected the massive Otis Falls Mill, Record built the two-machine Jay Bridge Mill. However, by 1896 Record had fallen on hard times and sold the Falmouth and Jay Bridge Mills to Chisholm, who proceeded to remove the old mill dam and build a new pulp plant with thirteen grinders called the Riley Mill. Hugh Chisholm later founded the International Paper Co. with headquarters in Portland.

The manufacture of soda was improved by the Solvay process, first patented in Europe in 1861. A key component was soda recovery, and the way it worked was the mill burned up used liquor to make soda ash that, in turn, was sold back to the manufacturer to make more soda liquor. Roland Hazard and William B. Cogswell erected the first Solvay plant in the United States in 1881 at Syracuse, N.Y., however, a circulating questioner of 1883 revealed that many companies didn't understand the need for soda recovery, and so lost money on the process.

The soda process was just the first of several chemical cocktails developed during the 19th century. Each of these were similar in that pulp was boiled in a digester, the only difference being the chemistry of the cooking liquor. While chemists found numerous compounds to remove lignin, only a few distilling processes were cost effective to make cooking liquor in bulk. The sulphite-pulp process was an outgrowth of straw processing, and Benjamin Tilghman of Philadelphia first tried it on wood fibers in 1866. Tilghman cooked wood bark in a solution of sulfuric acid, but dealing with strong acids was no minor undertaking. He lined the interior of the digester with lead to withstand the attack, but then found the acid was so strong it damaged the wood fibers. With so many technical problems to overcome Tilghman traveled to Europe to obtain help from the burgeoning chemical industry. In Germany, Tilghman narrowly survived a patent dispute, and in Austria, chemists finally helped him develop a liquor that was weak enough to use without lead linings.

The Pictet-Brelaz process of 1883 consisted of heating to 105 degrees Celsius in a solution of 7 to 8 percent sulphurous acid, but still the digester had to have removable iron linings that needed changing about every three months. Iron, however, was much preferable to lead in both cost and ease of use, and the Parthington digester became the first spherical rotary boiler with removable iron linings designed for the sulphite process. Bolstered by the new digester and liquor, Tilghman returned to the U.S. to introduce his new method to American manufacturers. Back in Maine, the Penobscot Chemical Fiber Company converted a sawmill in Old Town to soda-pulp in 1882, and Tilghman now appeared on the scene to show them how the sulphite process could work with any kind of softwood, not just poplar. As a result, several new digesters were ordered for the Old Town Mill, and production of wood pulp at the plant grew to eighteen tons per day.

In 1889 President Grover Cleveland led a team of investors to build a new sulphite-pulp mill in central Maine called the Madison Mill. The state was already host to thirteen ground wood mills rated at 157 tons per day, and twelve chemical-pulp mills producing 182 tons of pulp per day. This activity eventually drove the price of wood pulp down to four cents a pound, and by the end of the century Maine had the third largest paper industry in the nation.

Papermaking in New Hampshire and Vermont

Founded in 1623, Dover, New Hampshire, was among the first half-dozen communities established in the New World. Dover possessed an excellent port facility, and the hills beyond provided a bountiful source of waterpower. Dover was also among the first communities to develop a textile industry, and in 1827 the region produced the first calico manufactured in the U.S. Over the years the worker villages at Salmon Falls and Great Falls were joined in 1849 to form the town of Rollinsford. While the railroad was slow to arrive,

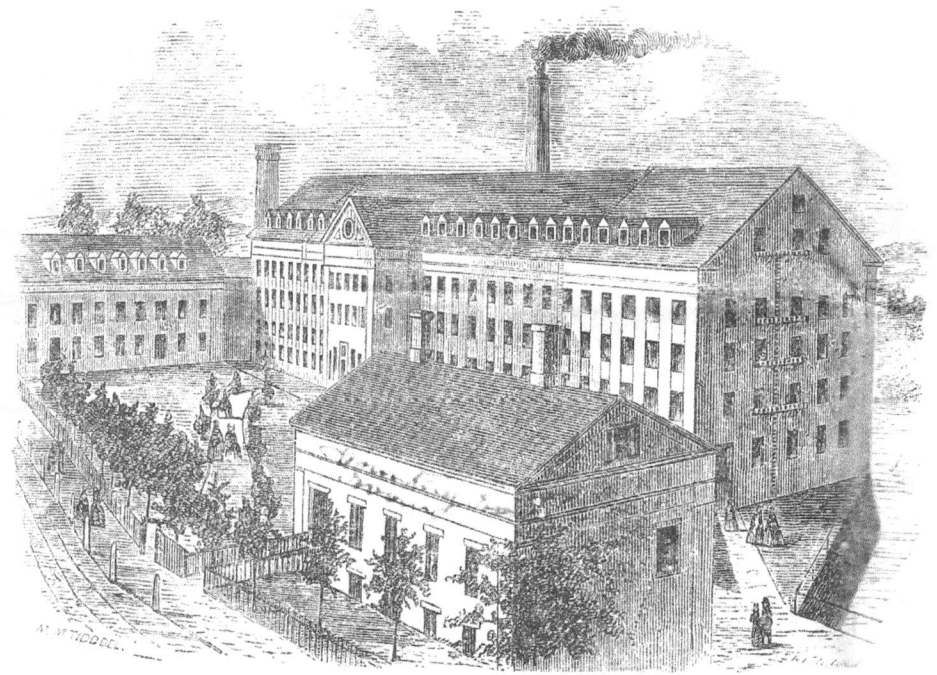

Stationery of Haley & Holden Mill, Salmon Falls, N.H., dated 1859. The counting room is identified by the correspondent as the building in the foreground.

the daily packet to Boston proved more than sufficient for Haley & Holden, who built an extensive paper mill at Salmon Falls in the early 1850s. This mill made stationery and envelopes for the Boston and New York markets in an imposing factory building four and a half stories high with a separate wing for paper machines.[7]

The first attempt at papermaking in New Hampshire began in 1777 when Richard Jordan, papermaker from Milton, Mass., petitioned the state legislature for a loan of two hundred pounds to construct a mill in Exeter. Jordan went back before the legislature the following year to petition for a rag monopoly; however, according to Dard Hunter, there is no physical evidence the Exeter Mill produced any paper at this early date. The first confirmed mill was erected in nearby Northumberland by Thomas Wiswall & Co. in 1823. Wiswall leased waterpower from Moses Wiggins' dam, and when Wiggins later died Wiswall purchased both mill and dam. About 1830 the papermaker added a D&J Ames paper machine equipped with a fire dryer, potentially the first paper machine in the state. Wiswall's son, Augustus Curtis, decided he'd rather go to sea; however, he later landed a job at a paper mill in Newton Lower Falls. About 1865 the Northumberland Mill came into possession of Chase, Robert & Co., which manufactured manila and straw paper on a double-cylinder machine.

In 1849 the Boston, Concord, and Montreal Railroad reached New Hampshire and new paper mills went up along the line at Claremont and Manchester. In Manchester the Martin Manufacturing Company erected a moving-wire machine in their facility on Canal Street next door to the Blodgett Edge Tool Company. In 1862 the firm of John Hoyt & Co. was contracted to build a new wheelhouse and boiler house for the paper mill, and Hoyt eventually came into possession of the mill in 1874. The company now became the Amoskeag Paper Mills Co., and in 1881 it purchased the edge tool factory, creating a complex of paper mill and machine shops.

The first straw mill in New Hampshire was established in Center Ossipee in 1857. A paucity of local rags kept conventional machine mills to a minimum, but the abundant waterpower made the region popular for manila mills such as those in Petersborough and Lancaster. The town of Ashland in Grafton County was only incorporated in 1868, but by this time it already supported one straw mill, two manila mills, and a leather board mill. In the early 1870s manila mills went up at Haverhill, Hinsdale, Bennington, Webster, West Heniker, Davisville, and Franklin. Remarkably, the Franklin Fire Department maintains a web page where it is reported that on July 7, 1872, fire struck the Winnipesaukee Mill. The main building was a two story 40 x 108 foot structure with connecting bleach house of 33 x 28 feet. Evidently, the fire started around the bleach boiler and quickly spread to overhead flooring. Franklin's Fire Engine No. 3 arrived quickly, but the building was already completely involved. Engine No. 1 followed shortly thereafter. At the center of the blaze were two rotary bleachers, one five feet in diameter and 22 feet long, weighing 13 tons, and the other six feet in diameter and weighing 10 tons. Both had been filled with steam and boiling stock, and when the cold stream hit a superheated bleacher it exploded, throwing debris over 200 feet across the river and knocking five firefighters plying a two and a half inch hose to the ground. Before the fire came under control a second bleach boiler exploded, sending sections of iron and debris over thirty feet in the opposite direction. Firefighters worked the fire for over three hours and were so exhausted that bystanders were pressed into service. Damage to the facility and loss of stock came to over $50,000.[8]

The wood pulp industry came to New Hampshire in 1883 when the White Mountain Pulp and Paper Co. built the first ground wood plant on the Dead River in Berlin (formerly Maynesborough). Berlin had offered ten years of tax exemption to pulp mills, and many took up the offer such as the Glen Manufacturing Co., who constructed a pulp and paper mill on the east side of the river. In 1887 Benjamin Tilghman introduced the sulphite pulping process at the Berlin mill of W.W. Brown. In 1888 the Berlin Mills Co. built the Riverside Groundwood Mill, followed by the Riverside Paper Mill in 1891.

The upper Connecticut River separating Vermont from New Hampshire supported a number of early paper mills. Revolutionary War General Ebenezer Walbridge and Ensign Joseph Hinsdill erected a paper mill in Brattleboro. By the turn of the century the Walbridge family assumed full control of the mill and continued in the business for many years thereafter. In 1811 William Fessenden, editor of the *Vermont Intelligencer*, and five others dismantled a sawmill and erected a large paper mill on Whetstone Creek at the other side of town. The Whetstone Creek Mill paid $32 for a pair of moulds from Nathan Sellers, who charged an additional dollar for sewing on letters of the watermark. Patty Fessenden inherited the estate in 1815, and she subsequently placed the printing and paper-making enterprises under the auspices of her father, John Holbrook. Holbrook was a publisher and bookseller in Brattleboro, as well as the town postmaster from 1794 to 1804. Holbrook & Fessenden sold handmade paper of the following styles: foolscap, pot, letter, demy, demy-blue, post, royal, copperplate, hanging, bonnet, binder, and wrapping. Holbrook & Fessenden later bought a paper mill in Watertown, N.Y., run for a time by an in-law.

Moving up the Connecticut River, the Guilford Mill in Green River Village was erected by Jonah Cutting in 1810, but sold soon thereafter to brothers Samuel and William Gregory. The mill made cap for $4 a ream, pot for $3 per ream, printing at $3.50 a ream, and wrapping at $1 a ream. Sometime around 1820 papermakers William and Richard Gookin started a mill near Center Rutland at Mead's Falls on the Otter Creek. In Putney Village, Solomon Stimpson, Lawrence H. Green, and Ebenezer W. Fairbanks established the Owl Mill on Sacketts Brook in 1819. The building budget was $4000, but expenses ran to $6442, leading the firm to mortgage the plant. Business was so slow to start, so to keep out of debtor's prison the firm successfully petitioned the General Assembly for relief from civil suits. In addition to the usual paper offerings, the Owl Mill reportedly made a green-colored stock. The Owl Mill was acquired in 1825 by William Robertson & Sons, papermakers newly arrived from Scotland. About 1830 the Robertsons added a cylinder-wire machine, reportedly the first in the state.

A little east of Bellows Falls is the town of Alsted, where in 1793 Major Elisha Kingsbury erected a paper mill on the Cold River. Alstead was one of those towns whose allegiance wavered during the Revolutionary War. In April of 1781 citizens decided to join Vermont, but returned to New Hampshire the next year. The worker community at the Cold River Mill grew into a separate community called Paper Mill Village, and business was so good that Kingsbury established another mill in Bellows Falls. Thomas Prentiss, a merchant from Langdon, N.H., came to Alston in 1834, and established the firm of Prentiss, Bemis & Co. Prentiss later acquired the Cold River Mill and set up a paper machine. The mill burned in 1868, but was immediately rebuilt by Fredrick Prentiss, Thomas's son. Fire struck the mill again in 1880, taking four businesses and a residence along with it. The plant was not rebuilt, and Alstead eventually incorporated Paper Mill Village.

The canal and locks at Bellows Falls were completed around 1801. Here the Connecticut River took an abrupt eastward bow before cascading 52 feet down a rocky gorge. The canal cut the bow off from the shoreline, and the area since was called the "Island." Bellows Falls now entered a period of growth that saw a number of new buildings constructed in the Italianate, Second Empire, Romanesque, Queen Anne, Federal, and Greek revival styles. In 1802 Elisha Kingsbury of Alsted constructed a paper mill at the base of the falls called the Rockingham Mill. This plant was placed under the management of his vatman, Bill Blake. After the mill burned down in 1812 it was replaced by a three-story masonry structure measuring 140x32 feet. Blake eventually acquired the mill and branched into printing and bookselling. River traffic declined significantly after the arrival of the railroad in 1849, but the canal continued to supply waterpower to the mills along its banks. Today, a small group of mostly brick industrial buildings stand on a lower slope near the river that encompasses the Rockingham Mill and the old Adams Grist Mill, which serves as a historical museum.

About 1810, Bill Blake & Co. erected another mill at Newbury, Vt., on the Wells River. Blake hired Stephen Reed of Londonderry, New Hampshire, to run the new plant, although by 1817 it was leased to Ira White, a printer from Bellows Falls. White opened a bookstore in Newbury, where he published schoolbooks on paper made at the mill. Munsell reports in 1827 that the firm of White & Gale of Vermont obtained a patent for a new mode of finishing paper.

Abijah Burbank, long-time papermaker of Worcester, Mass., retired to Vermont in 1788 and settled in Royalton, Windsor County. About 1801 his son, Abijah Jr., purchased a sawmill on the White River near Sharon, Vt., and converted it to papermaking. The mill dam was destroyed on two occasions by flooding, but the White River Mill remained profitable and became the model for future mills in the region. In the census of 1820 Abijah Jr. reported the mill had one water wheel, one roll and plate, one screw and press, one vat, ten pairs of moulds, and two sets of felts (about 250 felts). In 1806 younger brother Silas Burbank established a mill below the falls of the Winooski River in Montpelier, two years before it became the state's capital. In 1821 Samuel Goss, a former printer, built a mill on the opposite bank of the river in what was then the town of Berlin. Abijah's White River Mill was closed in 1840 after flooding washed away the dam for a final time.

Associates of the Burbank's, James and Stephen Billings, built a second paper mill in Montpelier on a site below the second falls of the Waits River. The brothers erected a mill, dam, sizing house, and drying house. Amazingly the lease called for five quires of good quality writing paper (about 125 sheets) to be supplied to the landlord, retired sea captain William Trotter of Bradford. After James Billings left the company in 1812 Captain Trotter renegotiated the lease to include ten quires of paper a year. Stephen Billings continued in the business with Daniel Kimball of Plainfield, New Hampshire, who later founded the Kimball Union Academy in Plainfield, N.H. Kimball became sole owner in 1817, and quit-claimed the mill to the academy, which ended up selling the facility to Elisha Hammond of Brookline, Mass., for $950.

In 1800 Joshua Henshaw of Albany, N.Y., and Joel Eaton of Amherst, Mass., erected a mill in Weybridge, Addison County, on the upper falls of Otter Creek (about 25 miles south of Montpelier). The mill won a contract to furnish the state with 400 reams of printing paper for the 1808 edition of the *Laws of the State of Vermont* for which Henshaw charged $2,162, including transport to and from the nearby town of Randolph. In the federal census of 1810 Henshaw reported employing just nine persons while his one-vat mill consumed some eighteen tons of rags per year.

In 1783 Bennington was a town of 2,000 on the border of New York about 30 miles northeast of Albany. Here David Russell ran the local post office out of the back of his print shop, where he published the *Vermont Gazette*. To supply the newspaper, Russell and the editor, Anthony Haswell, erected a paper mill on the Walloomsac River below Bennington Falls. This mill was later sold to Joseph Hinsdill and Ebenezer Walbridge, who operated the plant until 1797. The town got a second paper mill in 1812 when Joseph Hinsdill and Lucius Gibbs set up a plant at the confluence of the Paran Creek and the Walloomisac River.

Vermont's first writing paper mill was erected in Fair Haven in 1794. The *Rutland Farmer's Library* reports the building was a two-story affair measuring 70 x 43 feet. Colonel Matthew Lyon, the state's congressman from 1797 to 1801, who owned the mill, was conducting experiments with basswood bark as early as 1794. The Fair Haven Mill was later sold to investors from Castleton and burned down in 1806. The plant eventually came into possession of George Warren, a papermaker from Millbury, Mass., who rebuilt the mill and continued in the business thorough the 1820s.

The first straw mill in Vermont was established in Winchester in 1870, while a number of manila mills went up in Bellows Falls, Middlebury, North Bennington, West Derby, and Putney. In 1875 the Middlebury Mill was wiped out in a tremendous blaze that also burned twenty-nine businesses, two halls, six dwellings, and the highway bridge. The first wood pulp mill in Vermont was the Moosalamoo Mill in Salisbury, erected by the Kingsley Brothers in 1879. During the 1880s, the wood pulp industry literally invaded the state with pulp and paper mills established in Barnet, Bellows Falls, Lyndon, Middlebury, Morrisville, Passumpsic, Pittsford, Readsboro, and South Wallingford.

13

A Changing Industry

Following the Hudson River to its source one would eventually arrive at a portage road cut through the forest around Hadley Falls. Timberman Jeremiah Palmer owned a 700-acre parcel of land on both sides of the falls, and perhaps seeking to duplicate the success at Holyoke, formed the Palmer Water Power Company in 1858. The company put together a site plan for "Palmer's Falls," an industrial community with three canals running parallel to the river. The town would be serviced by a railroad, and featured a grand hotel and a river palisade along with numerous businesses, houses, schools, and parks. The company claimed the falls were capable of 15,000 horsepower, but as the potential investors soon discovered, the falls were reduced to a mere trickle in summer months. Palmer built the first canal, but when investors failed to materialize, plans for further development were quietly shelved.[1]

Following the Civil War a trustee of the waterpower company named Thomas Brown established a woolen mill and edge tool factory on one of the canal sites in what then was called the town of Corinth. The quiet life in Corinth was shattered when Thomas Brown was accidentally shot by his night watchman. Following his death the tool factory went up for sale and Pagenstecher & Co. came to town to investigate the possibility of establishing a new pulp mill. The abundant water power combined with the region's five million acres of forestland created ideal conditions for a pulp mill, so in 1869 Pagenstecher & Co. built a pulp mill in Corinth under the auspices of the Hudson River Pulp Co.

The Hudson River Company purchased the former edge tool factory in 1870 and converted it to papermaking with a 68" moving-wire machine. Ready access to the New York market and the lower cost of wood pulp paper was a boon to business, and in 1873 a new mill was built. This facility featured two new moving-wires producing 50 tons of news and printing paper a week. The power-hungry wood grinders were idled during periods of low flow, so to keep pace with the paper mill the firm began making extra pulp in the spring. In 1880 the Hudson River Co. made further waterpower improvements, constructing a crib dam and new stone raceway that increased the head from Palmer Falls to eighty-four feet. This improvement paved the way for more wood grinders and a new 112" moving-wire machine, the widest machine in the world at the time. As the forest receded a railroad spur was built from Corinth Station to alleviate the need to drag logs over two miles to the pulp mill. Corinth eventually got a nice hotel, as in 1874 the Palmer's Falls Hotel was built next to the entrance to the paper mill. Otherwise, mill workers chose to live in nearby Jessup's Landing, which came to be incorporated by Corinth in 1888.

Beginning in the early 1890s the Hudson River Pulp and Paper Mill began using steam engines to power its grinders during the summer months. The first boilers were fueled with wood, but the plant later converted to oil. In 1891 the first of five sulphite digesters was installed. By 1896 the mill was so reliant on steam power they replaced two of their moving-wire machines with models specifically designed to be run by a steam engine. Steam engines were first used in conjunction with the cylinder-wire machine at the Apsley Mill between 1810 and 1817. Attempts to do the same with the moving-wire machine proved less than successful, as early non-condensing engines vibrated excessively and were subject to surges, as the governor ran off the speed of the drive shaft. Such conditions created havoc on the moving-wire machine whose fast-moving web required the smooth and steady application of power. It would not be until the more sophisticated condensing engine came out in the 1860s that Donkin & Co. finally achieved a measure of success with the Fourdrinier. Even at that, the industry in general continued to relegate the steam engine to less critical functions such as driving hollanders or wood grinders. In 1898 the Hudson River Pulp & Paper Mill was sold to the International Paper Company, who as a result moved its headquarters from Portland to Cornith.

Water Pollution

The Lima Mills were established in Lima, Allen County, Ohio, in 1870 for the manufacture of straw board. In 1876 the owners were indicted for dumping refuse into Hog Creek that allegedly killed fish and made the water undrinkable. More than one hundred people gave testimony at the trial in Columbus, which resulted in a hung jury. This was not the first occasion that papermakers were taken to task for water pollution, nor would it be the last.

In 1859 there appeared a series of articles in the *Valley Gleaner* that closely followed the digging of a well by papermakers in South Lee. The well had been dug in vain, as the paper later reported, "They would succeed in reaching China without finding a supply." The Hurlbuts' search for fresh water appears to be in reaction to a chemical-laden Housatonic River. South Lee was immediately downstream of newsprint, straw, and manila mills in Lee and East Lee, and this was ten years before the advent of wood pulp. The Hurlbut brothers eventually resorted to the use of a filter reservoir built in the woods about 600 feet to the south of the plant. Filter reservoirs use successive layers of gravel and sand to purify the water on its way up to the surface.[2]

Over in Holyoke, the Newton brothers erected a soda-pulp mill in 1876. Named the Connecticut River Pulp Mill, it occupied a two-story brick building on Race Street between Appleton and Dwight. The plant was powered by two water turbines and had two direct-fired digesters, two 4000-pound rotary boilers, and a wet machine for pulp lap. The old digesters were replaced in 1877 with two Dixon digesters, signaling a switch to sulphite pulp. The new digesters were horizontal double boilers heated by steam pipes coiled around the interior. It seems the one drawback to the sulphite process was that moisture in the pulp reduced the effectiveness of the digester, leaving too much lignin in the fibers. In 1883 Martin L. Griffin, a consulting chemist, recalled a conversation one morning with the president, Daniel Newton, who called him to his office to say: "Mr. Griffin, we are having many difficulties with our customers over moisture in our pulp. I want you to come to the mill every morning and evening and sample the pulp. Take fifteen or twenty pounds each sampling, no homeopathic samples for me!"

13 — A Changing Industry

Connecticut River Pulp Mill circa 1875. About this time the mill was involved in a lawsuit over pollution in the Holyoke canal.

In 1878 the Holyoke Paper Company began digging a well on their property. Their facilities were located directly downstream from the Connecticut River Pulp Mill, and acid effluents from the sulphite process were apparently having an undesirable effect on paper manufacture. The Holyoke Mill was not the only plant impacted by the pollution, as on June 29 of that year the firms of Beebe & Holbrook and Massasoit Paper Company built a 100x30x15 foot filter reservoir upstream of their plants on Race Street. The cost of construction was reportedly paid for by the Newton brothers, since their plant was deemed responsible for fouling the canal water.

The Strange Case of Smith Gardner

Owen & Hurlbut (O&H) were among the more conservative papermakers. They tended to follow industry trends rather than set them, and for this reason it seems most unusual to find them in close communications with an inventor by the name of Smith Gardner. Reading between the lines it seems the inventor was from Boston, but other than that Gardner was tight-lipped about the nature of his business. It may not be that unusual for an inventor to play his cards close to the vest, but it's still difficult to figure what Owen & Hurlbut hoped to gain from the association. One thing for sure, Gardner was planning a trip to London, as in July of 1842 he writes the papermakers from Howard's Hotel in New York to say, "I should draw on you at 90 days time for $60—the amount agreed upon for you to pay during my stay in England."[3]

The first inkling of Gardner's line of work comes up in a letter from a clerk in London named John L. Osborn, who reports: "[I] regret that I am not more able to inform you of Mr. Gardner being actually at work. The machinery necessary for his operations is quite complicated and has been waiting some days for the movement of the steam engine in the building that we now occupy for the purpose of Mr. Gardner's experiments.... The owner having been disappointed in the completion of some works that are to be driven with the same power ... is not willing to start for us who require only the force of 1 horse [power] while the engine is calculated for 100. We should have removed entirely from the premises, but the situation being more appropriate than any other we can find ... Mr. G[ardner] ... thought it best to submit to the temporary concern, and besides great expense would be incurred in the removal of the machinery which is now fixed in ready for the application of the steam."[4]

From the previous letter it would appear Gardner had been working with some kind of water-powered machinery and had requested it be hooked up to a steam engine. The steam engine in this case could be moved about the mill, but evidently the owner had other plans for it and Gardner would simply have to make do. Whatever the work Gardner was engaged in, it seems he couldn't simply pick up and move to another facility. A single ray of light finally showed through when Osborn concluded his report: "By the next conveyance I hope to communicate to you more satisfactory intelligence, and if possible the results of different tests of the white lead for which several parties are now waiting. I have at hand analysis by respective chemicals a few small pieces that Mr. G. happened to bring with him, and they all concur in the same opinion of its superiority to any other they have here."

The nature of Gardner's relationship with Owen & Hurlbut may have remained a mystery if not for a Sept. 8, 1843, letter from an old friend, Asa Butler of Suffield, Conn. Thomas Hurlbut had managed Butler's mill twenty years before, and it seems the two still had mutual interests. The retired papermaker wrote: "Rec'd. a line from Mr. Thomas Hurlbut dated New York Sep 6th saying Mr. Lathrop wanted to rent the buildings belonging to the White Lead Company at Fishkill [N.Y.].... You may do as you think proper in meeting the same without prejudice to our former contracts giving us at all times the use of water-power and room to exhibit the manufacturing of white lead, with two [mixing] pans, and other machinery connected with it." In parting, Butler remarked: "P.S. I wish you would inform me if you got anything from Mr. Gardner by the last boat, and if so what is the story? I have been thinking as the Commitment River Company has a great water power monopoly whether we could not make some arrangement with them to take hold of this business. Let me have your thoughts on the subject."[5]

Sure enough, in subsequent letters from Gardner it is learned that a barrel of white lead had been shipped from London to the South Lee Mill. Thus, it appears Gardner was working for, or in conjunction with, the White Lead Company. Afterwards, Smith Gardner wrote from London: "Such a length of time has elapsed since I last heard from you that I am almost induced to believe you wish to discontinue our correspondence. I informed you in my last that we had obtained three patents on the continent and that I hoped soon to effect a sale. We have since that time made about two tons of white lead from litharge (lead monoxide) which has been used by some of the best judges and pronounced 'good' ... The gentleman who has joined us in taking out the patents ... offers to invest a few thousand pounds in the manufacture of the lead if anyone will join him and purchase the patents."[6]

Owen & Hurlbut provided funds to assist Gardner, but otherwise the papermakers don't seem to be very interested in the venture. The South Lee Mill was evidently making samples of wall paper for White Lead Company, but only as a favor to Asa Butler, who had sold his paper mill years before. White lead is the base component of what would become lead-based paints. Smith Gardner seems to have been working on improving the formula and obtaining a patent. Prior to the Civil War very few U.S. mills were engaged in the manufacture of wall paper. The industry evidently grew up around Troy, N.Y., before spreading north to Sandy Hill, N.Y., and North Benington, Vt., and this being the case, the paint factory at Fishkill was in a very strategic location. Following the war a number of mills in Pennsylvania got into the business, which really took off with the advent of inexpensive wood pulp paper. By the 1890s companies like the Maryland Wall Paper Co. in Baltimore and the Potter Wall Paper Mill in Boston established traveling sales forces to comb the countryside and take orders from small town merchants. The Syracuse Pulp & Paper Company, with offices in New York and Boston, claimed to be the largest wallpaper manufacturer in the country. By the end of the century the firm changed its name to the Syracuse Paint and Paper Co.

Color in Paper

A number of specialty mills came to Holyoke during in the 1880s. The Whitmore Manufacturing Co. occupied an immense building on Cabot Street where it produced surface coated lithograph paper and thirty-five colors of glazed and enameled papers, tag bristol board, and box board. The Nonotuck Paper Co.'s two-machine mill made twelve tons of paper a day, including flat and ruled writing, envelope, map, book, enameled book, coated label, and lithograph paper. The Hampden Glazed Paper Co. was established in an abandoned mill on the east side of the Connecticut River, and true to its name the firm made coated card, lithograph, high-finish glazed, and embossed papers. John C. Newton and brother Moses Newton founded the Chemical Paper Co. in 1880, and this grew to be one of the largest paper mills in the nation. The plant's real estate covered twenty-one acres, and the facility itself had 160,000 square feet of floor space. The daily output of the Chemical Paper Co. mill was between twenty-eight and forty tons of manila writing, envelope, lithograph, as well as embossed, glazed, and colored wrapping paper — the first of its kind in the U.S.

For the longest time color was something papermakers only did on an occasional basis. From about 1815 subtle shades of pink, blue, and green could be found in American-made papers, but since the vatman was constantly up to his elbows in pulp, colorants were used very sparingly. Color could also be achieved in other ways, such as the liberal application of blue, red, or green silk threads. At the South Lee Mill, Charles Owen used a fine sheet of Phoenix cap in an 1838 letter to his partner. This sheet is found to have been double-dipped; that is, the mould was dipped successive vats, one containing the standard white pulp, and the other containing pulp with blue silk fibers. The effect is stunning, which in part explains its popularity despite the high price.[7]

The earliest reference to blue paper within the Owen & Hurlbut correspondence comes in an 1829 letter from Washington, D.C., paper dealer William A. Davis. The South Lee Mill was said to have filled an order for blue foolscap and blue letter paper in 1830 for R. Farnham of Washington, D.C. These references are fairly confirmed by records of Coleman Sellers, who advised a papermaker in 1829 that indigo dye was unsatisfactory because it faded rapidly. For machine-made paper Coleman recommended "smultz," a powder made from cobalt (blue glass) that was mixed in hydrochloric acid before being added to the vat. Sellers wrote William Whiteman of Knoxville in 1834 regarding current recipes for colored paper. Evidently red color was achieved from madder root, while rose came from a combination of powdered cochineal and alum. Pink was made from Brazil wood treated with of tin muriate oxide and tin nitrate, and yellow ochre (a.k.a. straw yellow) came from quercitron bark (black or dryer's oak) combined with alum and sugar of lead (lead monoxide).[8]

By the 1840s there is a noticeable shift toward bolder and deeper colors. Similar changes are found in papers emanating from France, Germany, and England, where it is believed many of the trends originated. According to an 1850 letter from Platner & Smith to Owen & Hurlbut, the two firms were experimenting with a green textile dye from England called copperas. Generally speaking, American papermakers mixed their own colors until the late 1840s. For example, Cyrus Field's letterhead of 1845 makes no mention whatsoever of color, but his circular of late 1851 reports: We have this day reduced the price of French Ultramarine.... We guarantee it to be the finest, best, and the same quality as sold by Messrs. White & Sheffield as we receive our supply through them."[9]

Ultramarine was a compound of mineral blue tinged with green. The pigment was

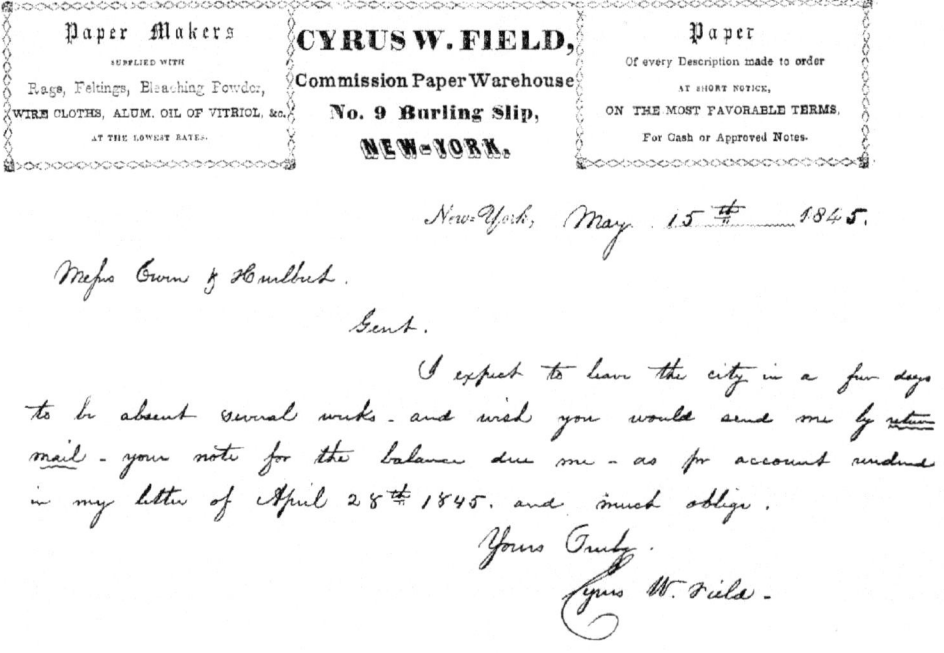

Cyrus W. Field letterhead of 1845. No mention is made of ultramarine at this early date.

added to the beater, but reportedly developed a serious molding problem with writing papers when used in conjunction with tub-sizing. Once this problem became known many papermakers avoided its use. Even after a new and improved formula came on the market, it still had few takers. The New York warehouse of C.H. Johneiden tried to get Owen & Hurlbut to use the new ultramarine with an offer to buy the paper once it was made. In a letter dated November 12, 1853, they wrote the South Lee Mill to say: "I received ... your favor of Sept 3 [noting] that you were not at that time in a position to execute my order for 100 reams letter paper. The said order has had to be executed ... by another house, and I beg to address you on the paper which was delivered to me, and to enclose a free sheet as another sample thereof. The same has been blued with a new kind of Ultramarine No.10, the best & most perfect product of the Nuremberg Ultramarine Works, which besides being fully as rich and as fine in color as the best French Ultramarine No.1 has the advantage over the latter of lasting stronger in tubs, and of adapting itself consequently more to a better sizing of the paper. The said Ultramarine No.10 is now used exclusively and permanently by your neighbors [Platner & Smith], who like it very much and I do not hesitate recommending to you also."

To get over past perceptions certain New York warehouses employed traveling salesman to help papermakers avoid molding problems. A February 1854 letter from Hereckenrath, Schneider & Co. to the South Lee Mill attests to the practice: "According to your order transmitted to us by our traveling agent Mr. Palmer, we have this day forwarded to you [by] Railroad one case No.10 Ultramarine as per enclosed receipt.... We have no doubt but that you will find the Ultramarine superior to any other brand you have been using, and we trust that you will continue to favor us with your order." Owen & Hurlbut were thus satisfied, and Hereckenrath, Schneider & Co. informed the South Lee Mill of further developments, writing: "We beg to enclose sample of No.10 Ultramarine which we offer you @35 6 mo. deliverable and payable here, and which lot has been used for the last 6 months and is still used in preference to any other kind of ultramarine by Messrs Platner & Smith. We also join sample of similar description of Ultramarine No.11, which we will sell you as low as 30c — same conditions, and which is also now in use by the above named gentlemen." Owen & Hurlbut continued buying ultramarine from Hereckenrath, Schneider & Co. through June of 1856.[10]

The color white is often taken for granted. Most American papers of the 18th and early 19th centuries were "natural white," a euphemism for pale yellowish or tan colored paper resulting from the use of hard water. During the 1820s ground lime served as an effective alkaline cleaner for rags and significantly improved the appearance of the final product, yielding more of an off-white color. The old method of bleaching in trays eventually gave way in the 1840s to chlorine-based powder or clay used in conjunction with the rag washer. These products generally came from Europe, for example in 1851 O&H received an offer from a New York warehouse for M. Von der Beck's bleaching clay, saying: "The cost of the bleaching clay in Hamburg $16.$1/2$ per ton.... The freight from Hamburg is about $4 to $4 $1/2$ ton and duty here 20%." In 1852 a Cyrus Field & Co. circular announced to all their customers, "We are sole agents in the United States for Mosphett's Superior Bleaching Powder and can supply all orders at the very lowest prices."[11]

The *Paper Trade Journal* reported that the first use of aniline colors in the paper industry began in 1867 with the color magenta, and magenta-colored envelopes do appear to have made the rounds in the U.S. about this time. The advent of aniline colors opened the door for specialized papermakers as the Nonotuck's, Whitmore's, and Chemical Paper Companies. Using manila paper for the most part, engineers at these mills developed techniques for color mixing, as described by Charles Thomas Davis in his 1886 *Manufacture of Paper*:

> Acid mordants are employed with aniline colors. Green vitriol (ferrous sulphate) is a mordant for black, gray, and violet, and also used in Berlin Blue. Lead nitrate forms white crystals that dissolve in water. Plumbic acetate (sugar of lead) made from dissolving litharge (lead monoxide) in vinegar and evaporating. Both of the above are poisonous. Yellow prussiate of potash is mordant for fixing other colors, and in combination with salts of iron make Prussian Blue or Berlin Blue. Potassium ferric cyanide red prussiate of potash produces salts with blue, like Turnbull Blue.
>
> Red shades come from wood extracts as Brazil woods, Jamaica wood, Nicaragua wood, Santa Martha wood, Pernambuco wood, or Sapian wood. The extract, called Braziline, is used in the making amaranth, crimson, purple rose, and similar shades.
>
> Yellow shades are produced with various vegetable, mineral, and aniline coloring matters. Barberry yellow comes from boiling with barberry root and alum. Quercitron-yellow comes from boiling with quercitron bark. Rust yellow from boiling with annotto and potash. Annotto comes from East India. Yellow grays come from fustic, alder bark, bablah or babool, sumac, saw-wort, turmeric, dyers broom, and American golden rod. Lemon yellow comes from turmerica and ordinary spirit of wine. Chrome yellow comes from bichromate of potash or bicarbonate of soda and water with sugar. Acetate of lead makes oranges, but is poisonous. Yellow ocher is another mineral pigment. Aniline-yellow used for chrome yellow, is soluble in water. Chrysaniline is a yellow power dissolved in alcohol gives a beautiful yellow color. Golden yellow comes from Aurin, or hydrochlorate of chrysaniline, [and] is somewhat water-soluble. Reddish yellow made from zinaline in form of cinnabar colored powder.
>
> Blue shades come from ferric salts and ammonium oxalate mixed in pulp, and treated with yellow prussiate of potash, and adding a weak acid solution. Adding red color can neutralize greenish tint. Ultramarine should not be added to the pulp until after resin size and alum have been added. Cobalt blue and a mixture of sulphite of copper and red extract can blue paper pulp. Sky blue uses yellow prussiate of potash and solution of acetate of iron. Aniline blues are more readily soluble in water, including bleu de Parme, bleu de Lyon, and bleu de Paris. Phenol blue, or azuline, is a product of creosote of coal tar and looks like ultramarine. Blue may also be added to yellowish pulp to improve its appearance.
>
> Green shades are made in a variety of ways. Sap green is made with buckthorn sap and wine spirits. Olive green from quercitron bark, Hungarian fustic, and dogwood berries. Picric green comes from Berlin or Prussian blue with the addition of picrid acid in water, or with chromate of lead. Aniline green produces brilliant green, Victoria green, or Russian green [and] is applied to pulp with alum.
>
> Brown shades primarily come from Venetian red (ferric oxide). Aniline red is greenish granular crystals or ground into red powder used to make magenta, mauve, rubine, and rose. Dark brown made with quercitron, logwood, sandalwood, Brazil wood, and Hungarian fustic. Light brown made as above with green vitriol or bichromate of potash. Catechu brown comes from catechu and green vitriol. Olive brown is made from logwood, quercitron, and Hungarian fustic. Coffee brown is an acetate of copper in water treated after with yellow prussiate of potash. Light leather brown is made from logwood liquor, Brazil wood dye liquor, and fustic dye liquor. Mifonce brown comes from logwood, gustic dye, and Brazil wood. Aniline browns are insoluble in water but soluble in sprit of wine or acetic acid. Bismarck brown comes from tarry black brown. Havana brown and phenol brown are made from creosote of coal tar.
>
> Violet shades are made from Prussian blue or Berlin blue or ultramarine, combined with dry shavings of logwood in spirits of wine or alcohol, then combined with alum. Violet can

also be made with extract of campeachy wood and alum. Aniline violets such as dahlia come from aniline red. Hoffman violet is reddish or blue only moderately soluble in water. Perkins violet is soluble in hot water. Parisian violet is not soluble in water. Rosaniline violet is difficult to dissolve in water.

Grey shades come from an extract of logwood, Indian fustic, tan liquor, green vitriol, and alum. Iron gray comes from logwood, tan liquor, and green vitriol. Dark gray comes from lampblack treated in solution of soda or potash. Aniline gray a.k.a. murine is soluble in boiling water. Black is made from lampblack or logwood and acetate of iron and alum. Aniline black produces dark aniline green. Deep indelible black comes from a combination of blue, yellow, red aniline and naphthalene, and alcohol.

Bronze shades come from red aniline, aniline purple, and alcohol. Brilliant bronze comes from blue, violet, or purple aniline with red aniline and acetic acid.[12]

A Self-Guided Tour of the South Lee Mill

In 1854 Owen & Hurlbut purchased the Potter Mill in nearby Housatonic Village in what may best be characterized as a fire sale. Housatonic Village is about 5 miles from South Lee by county road, and here the former owner established waterpower rights and built the mill, but the facility caught fire and burned down before the paper machine and related equipment were installed. After Owen & Hurlbut acquired the property they rebuilt the plant and installed a 62" moving-wire machine. This work proceeded fairly quickly, as an 1854 receipt from Bunking & Van Norstrand's New York warehouse confirms the papermakers took possession of several Fourdrinier machine belts and wires.[13]

Owen & Hurlbut's ream label for the Irving Mill (a.k.a. South Lee Mill) featured an illustration of a stand-alone tub-sizing machine. Tub sizing equipment is mentioned throughout O&H's day books from the 1840s, and it seems only fitting this image graced a ream of their blue laid letter. This particular machine had a reel of paper positioned such that the web passes under a copper cylinder partially submerged in a size bath. A pair of copper rollers squeeze the excess sizing, draining it back into the tub while a set of doctor plates constantly scrape the copper rollers to keep them clean. The sizing dries as the web passes around three steam-heated skeleton drums fitted with rotating fans to maintain a constant temperature. The web then emerges through a set of glazing rollers that give the paper a final finish before it is slit, cut, and stacked at the end of the line. The handle of the stop-cutter is clearly visible at the back; this device was used to halt the web momentarily as it is cut. Traditional loft-drying papermakers such as the Crane's decried these mechanical techniques, but this seems the only practical way to keep up with the rapidly increasing speed of the moving-wire machine.

When Owen & Hurlbut dissolved in 1862 Charles Owen and his son, Edward, formed the Owen Paper Company to manage the Potter Mill, while the brothers Thomas Otis and Henry Clay Hurlbut took over the aging South Lee Mill. O&H's longstanding bookkeeper, Henry D. Cone, transferred to the Owen Mill in 1857, and when Edward Owen died in 1865, Cone married his widow. Cone became sole proprietor of the Owen Mill in 1873, and immediately began building a new mill just downstream of the existing plant. Due to financial difficulties work on the new facility was abandoned in 1878, and in 1899 the Rising Paper Company purchased both the old Owen Mill and adjoining property and built an expansive paper-making facility.

In October of 1871 the *Scientific American* magazine ran an article about the Owen Paper Company, "An Example to Manufacturers." Here it is said: "The present owners of the property have purchased all of the land on both sides of the river for several miles.... The road is kept in admirable order, the fences are neatly painted, shade tree are judiciously planted, and little parks laid out.... All of the persons employed in the mills are provided

Ream label of Owen & Hurlbut's Irving Mill circa 1870, with a depiction of a sizing machine with a lone operator.

13 — *A Changing Industry* 177

A postcard depicting the Old Owen Paper Mill in Housatonic Village just before it was demolished in 1899.

Stationery advertising of A.G. Elliot & Co. paper warehouse, Philadelphia, circa 1890. Many of the high quality paper companies are among those listed. Note the Hurlbut Paper Co. is listed under Irving Mills, which burned down in 1876.

A Hurlbut Paper Co. advertisement showing its new brick mill in South Lee. The Housatonic Railroad steam engine is at the far left (from Beer's *County Atlas of Berkshire*, 1876).

with homes. Comfortable cottages, surrounded by gardens and flowers dot the hill sides [and] adorn the banks of the river.... There is a fine circulating library and reading room attached to the mill."[14]

The Hurlbut brothers continued with the South Lee Mill until 1871, when they rebuilt the dam on the Housatonic River and broke ground on a new brick factory on the opposite bank. The new plant, called the Hurlbut Mill, was outfitted with one 66" and one 80" moving-wire machine, four waterwheels (later replaced by three turbine wheels totaling 350 horsepower), and one steam engine of 250 horsepower. Several buildings of the Irving Mill on the northern side of the river continued to be used for tub sizing and finishing until they burned down on the Fourth of July, 1879. Financing for the new mill came from S.D. Warren & Co. to the tune of the $175,000. Evidently part of the payment was made in finished goods, and as one veteran worker recalls, five carloads of superfine paper were sent to Portland as part of the deal. The plant otherwise ran on high quality book and writing papers for the Prang Educational Co. of Boston, and mill correspondence reveals a wide range of sales, including bond paper for David M. Farlowe, paper dealers of Montreal, writing paper for the Albany News Co., drawing paper for the George Stephens warehouse in Cincinnati, and folio paper for the Buffalo Card & Envelope Co.

Financial pressures brought on by the wood pulp industry sent the Hurlbut Paper Co.

A photograph of Hurlbut Mill taken in 1997 shows the front door facing the South Lee bridge. Company offices are on the second floor above the basement. Even at this late date the sign painted on the side of building reads, "Hurlbut Paper Company."

into receivership in 1888, and the Hurlbut Mill was sold to Arthur W. Eaton, a stationer from East Hartford. Eaton initially founded the Eaton, Hurlbut Paper Co., then in 1893 partnered with Crane & Co. under the name of the Eaton, Crane & Pike Paper Co. The Pittsfield-based concern distributed boxed sets of stationery and envelopes supplied by the Hurlbut and Bay State mills.

In 1899 the Hurlbut Mill was acquired by the American Writing Paper Company. American was a large conglomerate of twenty-three mills incorporated in New Jersey. Most acquisitions were seemingly the last of the great rag mills that were now in dire need of inexpensive pulp lap to stay in business. Among the plants were the Oakland Mill in Manchester, Platner & Porter Mill in Unionville, Agawam Mill in Mittineague, G.K. Baird Mill in Lee, and thirteen mills in Holyoke, including the Albion, Beebe & Holbrook, Crocker, Dickinson, Holyoke, Nonotuck, Norman, Parsons No. 1, Riverside, and Wauregan mills. Moreover, American was not the only conglomerate — among the largest were the International Paper Co. with twenty-eight mills, the United Box Board and Paper Co. with twenty-five mills, and the Union Bag and Paper Co. with fourteen mills. It was the end of an era. Rag mills that didn't convert to manila or wood pulp simply went out of business, such that by the end of the century the number remaining could be counted on one hand.

In 1820 the Factory Insurance Co. of Hartford, Conn., had schematic drawings made of the Hurlbut Mill and outlying facilities. The four-story brick structure was primarily occupied with finishing equipment and offices. The third floor held the drying loft, while the loft area itself was used for paper storage. The second floor had a tablet room for binding,

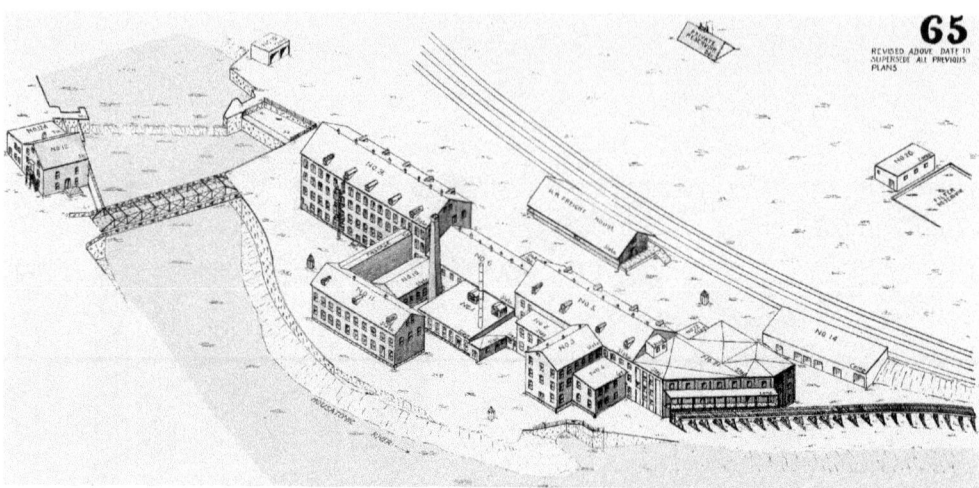

A drawing of Hurlbut Mill circa 1920. Note addition of stock house and railroad spur at right for delivery and storage of wood pulp.

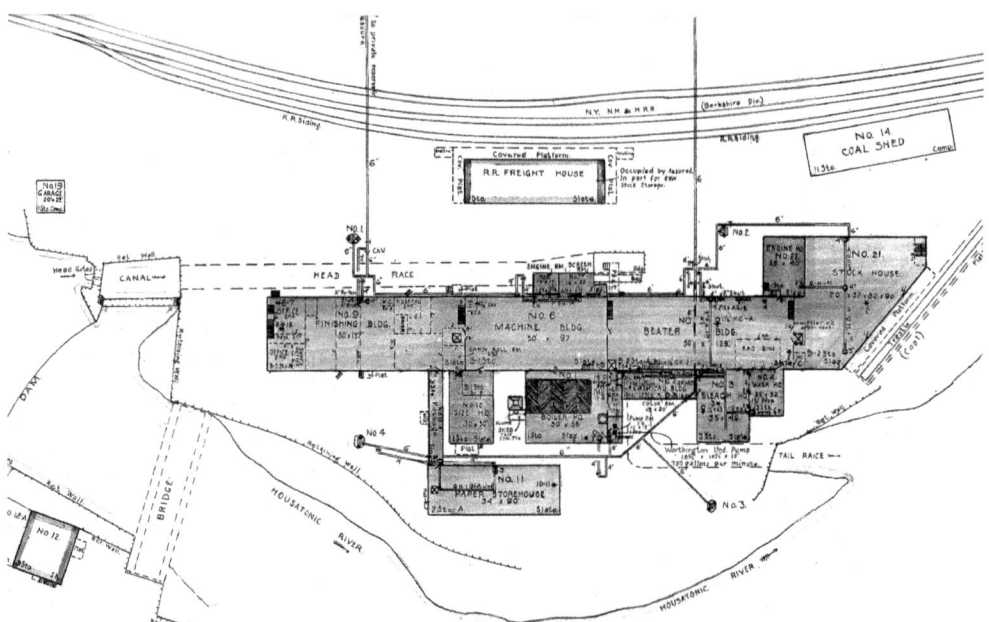

Top down schematic of Hurlbut Mill. The paper storehouse added at a later date is on the wrong side of the facility from the railroad tracks.

while the first floor held the finishing room with calenders and a trimming cutter. The daylight basement below had a machine shop and carpenter shop. The middle building was a two-story affair housing paper machines on both floors. The basement below held size tanks and the stuff chests, with a pipeline running about 600' to the filter reservoir in the woods south of the mill. The three-story building at the other end of the facility held the rag warehouse and engine rooms. Rag storage was on the top floor along with the rag duster, while

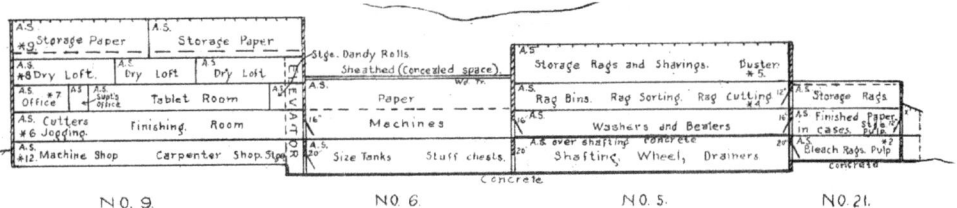

Schematic details of Hurlbut Mill interior. Note special closet for dandy roll storage in third floor tablet room.

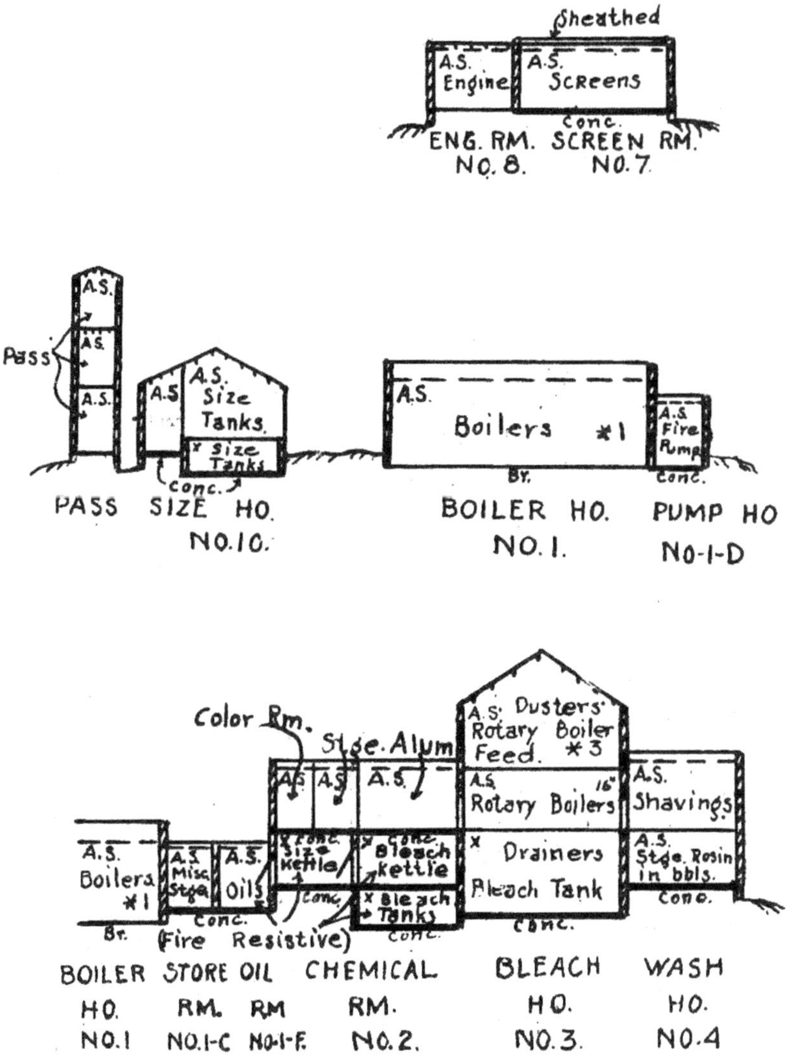

Schematic details of out buildings at Hurlbut Mill. The rotary boiler "feed" is on the top level.

the second story held rag bins, sorting tables, and rag cutting machines. Hollanders and washers occupied the main floor, while the basement below had turbine wheels and associated shafting. Two rotary bleach boilers occupied an attached building that also had a bleach retaining tank in the basement. The auxiliary boilers are housed in a separate building that includes the sizing room with storage tanks for the same. The main flume occupied a small shed to one side of the boiler house. The head-race originated on the Housatonic River just above the mill dam and ran directly under the factory basement, where the turbine wheels were mounted.

The original woodcut engraving from the 1870s used in the Hurlbut brothers' advertising may be compared to the insurance drawings to see just how the facilities were converted to the use of wood pulp. It seems the mill gained a new stock house for storage of pulp lap at the end of the facility adjoining the spur to the New York & New Hampshire Railroad. Another subsequent development was the addition of a chemical house for storing alum and bleach, and a room for the mixing and processing of colors.

14

West and Northwest

The first paper mill in Wisconsin was established in Milwaukee in 1848 to supply the *Milwaukee Sentinel and Gazette*. This machine-mill made ninety reams a week on a 32" cylinder-wire, but experienced flooding on occasion and always seemed to be in financial trouble. A second mill went up in Milwaukee during the Civil War, but this plant experienced a debilitating boiler explosion in 1867 and was never rebuilt. This was not a severe problem, as by this time much of the city's supply came from the Wisconsin's Fox River Valley.[1]

The Fox River Valley begins at the upper tip of Lake Winnebago about 40 miles west of Lake Michigan. At the upper end of the valley is Lake Winnebago, which drains north to Green Bay about 35 miles distant. The area is heavily timbered with oak, maple, hickory, ash, butternut, and basswood, and is also one of the best wheat producing regions of the country. As an inducement for settlement in the area the territorial government offered $10,000 in matching funds to any group or society that erected a school of higher learning. Amos Lawrence, a Boston Methodist, donated money for a "university in the wilderness" in honor of his wife's family, the Appletons. Lawrence University was thus founded in 1848 in the new town of Appleton, this occurring about three months before President James K. Polk signed legislation making Wisconsin a state. Outagamie County was created in 1851 with Grand Chute as the county seat, but within two years Grand Chute and neighboring settlements were incorporated by Appleton.

The Green Bay and Mississippi Canal was established in 1854 to provide for the transportation needs of the Fox River Valley. The first paper mill in the region went up in Appleton that same year, and as the story goes, the Richmond Brothers purchased the steam boiler and beaters out of the old East Side Mill in Chicago, transporting the machinery along the canal by steamboat all the way to the dockside doors of the mill. Here G. & N. Richmond made book, printing, and straw wrapping paper, and when the plant burned in 1859 it was quickly rebuilt by architects Clark & Barker, who installed a fire suppression system and a fireproof roof. The new factory was a four-story building 40x60 feet with a wing of 80x60 feet.

The first white settlement in the region was at Kaukauna, and residents of the region argued that this should have been Outagamie's county seat. The Chicago & Northwestern Railway came to Kaukauna in 1872, and from here it was 318 miles to Chicago and just 30 miles to Green Bay. The falls of Kaukauna held an estimated 3000 horsepower, and the first paper mill was also erected in 1872 to make straw wrapping for the Chicago market. Built on the site of a burned-out flour mill by Henry Hewitt and John Stoveken, the Eagle Mill was a three story affair with three water wheels totaling 70 horsepower, four rotary boilers, and a cylinder-wire machine housed in a separate wing. When the Eagle Mill ran

A photograph of Richmond Mill, Appleton, Wis., circa 1850, showing a load of straw overflowing a trailer. A horse-drawn delivery wagon is tethered at the front office.

into financial difficulties Stoveken sought help from a family member named Henry Frambach. Frambach was then in the mercantile business in St. Louis, and so liked the prospects in Wisconsin he moved his entire store to Kaukauna the following year. Frambach had been something of a war hero — at the battle of Shiloh in 1862 he snatched the flag from the hands of a fatally wounded color bearer and rallied his gun battery until help arrived. The young private was rewarded for his efforts with a detached assignment as a scout. Frambach proved invaluable in his new role, and on one occasion he was wounded and taken prisoner, but miraculously escaped. Given his exemplary service record, as he was being mustered out of the service in the fall of 1865, Frambach was awarded the rank of colonel by General Steel, chief of the scouting service. When Kaukauna was incorporated in 1876, Colonel Frambach, as he was called, became the town's first mayor. In July of 1878 a worker at the Eagle Mill was killed when a grinding stone suddenly burst. Afterwards, Stovekin, who owned a dry goods store in town, became concerned about the well-being of the mill and its workers, and so asked Frambach to take over. Frambach's solution was to convert the plant to manila and do away with the grindstone altogether.

The heavy concentration of waterpower in the region combined with easy access to the Chicago market made Outagamie County a prime location for paper manufacture. In 1872 Kimberly & Clark Co. built a rag mill in Neenah called the Globe Mill, which ran on book and printing paper. Following this success the firm then built a manila mill at the same location called the Neenah Mill. With the coming of wood pulp the forests of the region proved most inviting. Waterpower of the Kaukauna canal was improved in 1875 by expanding it to 14 feet wide and 580 feet long. In 1880 the Kaukauna Water Power Co.

A wood grinder (from advertisement in *Paper Mills of the World*, 1885).

Paper machine at Fox River Mill in Appleton, Wis., showing spools of paper being collected at the end of the line.

constructed a new waterpower canal furnishing 5000 horsepower to Outagamie County, and this project proved so successful that another canal of 2500 horsepower was completed the following year. Yet another canal of 2500 horsepower was built by the Hewitt Water Power Co. in 1887. During this time Kimberly & Clark went on a building spree, starting up a manila mill in Kaukauna and two new rag mills in Appleton.

The Eagle Mill burned down in 1881 with losses of $45,000, including $24,000 in machinery and $11,000 in stock. That same year the Meade-Edwards waterpower company completed a canal, bringing an additional 5000 horsepower to Kaukauna, and so Col. Frambach formed the Frambach Paper Co. with the intention of building a new and larger paper mill. Frambach engaged architect Oscar Burns of Appleton to construct what would be called the Badger Mill. Completed in 1885, the mill featured a 30' wide and 7' deep head race supplying twenty-one Elmore turbine-wheels delivering an estimated 500 horsepower. The pulp mill featured six wood grinders and a wet machine for the manufacture of pulp lap. The paper mill had four 1200-pound beating engines, three rag washers, one 84" Harper for the manufacture of book and writing paper, and one 64" double cylinder for making wrapping paper. A separate digester building was later constructed for the manufacture of soda pulp. About this time a third machine was installed and run on manila and red express (brown wrapping paper), and during periods of low demand the mill made wall paper for the National Wall Paper Assoc. In 1890 the Frambach installed three Perkins vertical digesters and converted the plant to sulphite pulp.

Colonel Frambach exhibited something of an aversion to royalties. At the Eagle Mill, he built his own bleach boilers, and as reported by the *Kaukauna Times* in April 1881: "Messrs Frambach and Stovekin ... by reason of a superior process peculiar to this institution ... are enabled to sell it at prices that compete sharply with ... other manufacturers. The process is covered by seven distinct patents ... and no infringement will be tolerated." In advance of building the Badger Mill Frambach began experimenting with wood grinders. His first patent in 1879 shows an invention that held the logs under water as the grinding wheel chewed off bits of wood. This apparently didn't work as well as thought, so Frambach submitted another invention consisting of a series of crushing rollers to tear away at planks of wood. Upon further review, Frambach submitted a third device similar to Voelker grinder except the logs were pressed lengthwise against the stone instead of widthwise. This minor difference was accepted, and while Frambach's version produced shorter wood fibers, this didn't seem to make much difference in the final product.

The Badger Mill would later have one turbine wheel dedicated to an electric generator supplying 250 Edison lamps strung throughout the plant. However, it seems Appleton upstaged Kaukauna once again when Henry J. Rogers' hydroelectric plant went on-line in September of 1882 —-just two years after the invention of the light bulb, and a mere twenty-four days after Edison started up the Pearl Street Station in New York. Appleton's electric plant supplied lights at two paper mills and one residence. Initial service was from dawn to dusk, although a second dynamo soon came on-line and operations extended to 24 hours. In those days there was no such thing as a voltage regulator, so the lights sometimes grew very dim. The system also lacked fuses, so on occasions when lightning struck the lines the lamps grew so bright they burst. On the other hand there were no power meters either, so customers simply paid a flat monthly fee.

In 1882 the firm of Gilbert & Whiting erected the Menasha Mill in Kaukauna. This plant had one 82" moving-wire machine and produced book and newsprint. In August of 1888 a fire began in a pile of wood shavings in the boiler room, and by the time the fire department arrived the building was fully involved. The fire engine laid down a stream of cold water that hit one of the superheated bleach boilers, causing an explosion and sending ten tons of iron and debris flying out of the building and across a sidetrack where a throng of spectators had gathered. The blow came so suddenly it stunned everyone; many people were injured and several were killed. Presence of mind soon prevailed and people hastened to the relief of the wounded and care for the dead. Body after body was found under collapsing timbers and masonry, some crushed beyond recognition. The injured were cared for in neighboring residences while the dead were removed to City Hall.

John Stovekin happened to be traveling in the northern part of the state along the Menominee River that divides Wisconsin from Michigan's upper peninsula and came across the 80 foot high falls at Quinnesec. The area was thick with timber, so in 1885 he formed the Northern Pulp Co., which purchased all the land for miles on both sides of the falls. The Menominee Pulp Mill had four Frambach grinders manufactured by the Atlas Iron and Brass works of Kaukauna powered by several Appleton Machine Co. horizontal turbine-wheels like those used at the Badger Mill. The site was still a little remote, as pulp lap had to be hauled five miles to the railhead in Quinnesec, so in 1892 when the Frambach-owned Qinnesec Water Power Co. bought the plant, it built a paper mill nearby.

In 1883 Oscar Thilmany, a German-American, established the Thilmany Pulp Co. in Kaukauna. Thilmany was taken by German moving-wire machines called the "Flying

Dutchmen" and the "Yankee." The Yankee was designed to make duplex paper with a smooth side and a rough side. The two-finish effect was obtained by replacing the tandem dryer with the more primitive multi-cylinder dryer. Thilmany installed his first German-made machine in 1889, and it proved so successful that four more machines were installed over 1891–1897.

In 1887 the Hewitt Water Power Co. completed another 2500 horsepower canal in the Fox River Valley, with two-thirds of the power dedicated to the new Otagauma Pulp and Paper Mill. That year the Kaukauna Water Power Company secured the rights to the Meade & Edward Canal for $75,000. This action spurred litigation by a formidable group including the Green Bay & Mississippi Canal Company, the Union Pulp Company, and the Fox River Paper and Pulp Company to prevent the Kaukauna Water from taking any more water from Fox River for hydraulic purposes. The litigation had a chilling effect on future waterpower projects in the region, and this particular water war was not settled until 1900 when the U.S. Supreme Court ruled in favor of the plaintiff.

By the census of 1890, Wisconsin ranked fourth in paper production in the U.S., and still had room to grow. In 1891 the Combined Locks Paper and Pulp Mills erected a new plant that consumed over 4,000 horsepower supplied by the Hewitt Water Power Co. This new mill boasted an astounding fifty-one turbine wheels ranging from 25 to 48 inches. The paper mill had four moving-wires in two machine rooms, and nine bleach boilers housed in a three-story boiler building next door. The pulp plant alone had sixteen grinders making sulphite pulp in three large digesters supplying five wet machines.

The Sherman Anti-trust Act of 1890 authorized the federal government to institute proceedings against trusts, but the vagueness of several provisions led to Supreme Court rulings that prevented federal authorities from fully implementing the act for several years to come. Following the collapse of paper prices in 1893, Col. Frambach organized the paper mills of the Fox River Valley and Wisconsin River into the Wisconsin Papermakers Association. Meeting in Neenah, J.A. Kimberly of Kimberly & Clark was elected president, Col Frambach vice president, and William Gilbert of Gilbert Paper Co. treasurer and secretary. The members agreed to cease all production for three days beginning in July to coincide with similar actions taken by trade associations in Cincinnati, Pittsfield, and New York. The Wisconsin papermakers, however, were dealt a more serious blow in 1895 when the federal government took control over the waters of Lake Winnebago. Evidently the town of Oshkosh had been complaining that diversion of water to canals in the Lower Fox River made the lake unfit for boating in the summer. So, the War Department was ordered to regulate lake levels, allowing just enough water for steam navigation on the Green Bay & Mississippi Canal along with just four additional inches (or 40,000 cubic feet of water) for all the mills in the Fox River Valley between Neehan and De Pere. This action caused all the paper mills in Neehan, Menasha, Appleton, Kaukauna, and Kimberly to reduce to half power during summer months.

Despite being equipped with a Grinnell automated sprinkler system, the Badger Paper Mill caught fire and burned down in 1897 with losses of $150,000. Due to generally depressed paper prices Col. Frambach chose not to rebuild, and the following year sold the Qinnesec Water Power Co. to Kimberly & Clark for $250,000 before moving to Colorado to speculate on mining interests. Frambach did eventually return to Wisconsin in 1902 to run the Cheboygan Paper Mill, where he continued in the business until his retirement in 1916.

Sugar House Mill

Owen & Hurlbut had long done business with a bookstore in Rochester, N.Y., run by Aling & Cory.[2] This small publishing house had been started in 1819 by E. F. Marshall, a cousin of Chief Justice John Marshall. William Aling later joined the firm as a clerk, and eventually bought the business in 1834. Here Joseph Smith, founder of Mormonism, was a frequent customer, coming from his home in Palmyra to read and exchange views on theology with the proprietors and their customers. In 1838 Smith and brother Hyrum were jailed in Missouri for defending themselves against a mob that had reacted angrily to Smith's teachings. The scene repeated itself in Carthage, Illinois, in June of 1844, but this time Smith and his brother were killed. In 1846 Brigham Young organized a western exodus of the Mormon faithful, first to Winter Quarters, Nebraska, and then to the Salt Lake Valley the following year. A printer in Council Bluffs, Iowa, originally served the Mormon community in Salt Lake, but the high cost of freight, as well as the uncertainties of transportation to the remote western region, prompted the church to look into in the printing and papermaking business for themselves.

Upon joining the Mormon church Thomas Howard was fired from his job at a paper mill in Woburn, Buckinghamshire, England, where he had worked for twenty-two years. Howard proceeded to Salt Lake City in 1851 and immediately offered to set up a paper mill. A facility was set aside six miles south of Salt Lake City, but owing to a lack of equipment it never opened. The following summer forty wagons of sugar-milling machinery arrived in town, and the equipment was set up in a 40x100 foot adobe building. However, operations stalled for a lack of raw materials, and after a year had passed Charles Howard approached Brigham Young about the use of some of the machinery. The papermaker converted the beet grinder to a rag engine and made good use of the hydraulic press. He had a vat and waterwheel built locally, and piece felts made at Matthew Guant's woolen mill on the Jordan River. The hand mill was set up in the public work shops in Temple Square powered by the City Creek, and here Howard produced newsprint for the June 1854 issue of the *Deseret News*. The church's four year rag collection effort yielded rags, scrap paper, carpet shavings, etc., but amounted only to 150 pounds, so without a steady supply Howard immediately shifted to making wrapping and box board. Six months later the sugar mill was scheduled for reopening and all the machinery had to be returned. The papermaker later wrote: "[As] much as the need for paper to provide the medium of information to knit and unite the people of the Valley ... residents in those parlous years were wearing their rags, and were not of a mind to go unclothed to provide the feed stock for a paper mill."[3]

The church never gave up on plans for a paper mill, and in 1860 Brigham Young bought a 36" cylinder-wire from the Nelson Gavitt Co. in Philadelphia for $8500. The paper machine arrived in Salt Lake City later that year and was erected in the adobe building formerly occupied by the sugar mill, which by this time had failed yet again. Thomas Howard found himself back in business, and with the aid of Z. Derrick he began making rag pulp fortified with waste paper in two 150 pound rag engines. The Sugar House Mill now made newsprint, wrapper, book, writing, binder board, and envelopes, and Howard later commented: "We had no materials or facilities for any bleaching or sizing. I managed to secure an indifferent bleach by using quicklime. My sizing ... I made with a little rosin I was fortunate enough to get hold of. My envelopes were tinted with copper, and as a mordant I obtained a wagonload of crude alum from the southern part of Utah. It could not properly be called alum, it was, rather, dirt, strongly impregnated with alum salt. It served my purpose very well, however."[4]

The task of organizing rag collection in the Utah Territory would be assigned to George Goddard, who ranged as far as Ogden in the north and Provo to the south, collecting over a ton of rags. The following year Goddard did even better, netting over five tons, and while these efforts were laudable, it was still less than a half the material needed to keep the Sugar House Mill operating for a month. The *Deseret News* took over operations of the paper mill in 1867, went through some very lean years at first, producing just two tons of newsprint a year (the equivalent of a single day's production at most eastern mills). But success finally came with the arrival of the railroad in 1869, as the Deseret Mill was finally able to obtain enough raw materials on the New York Market to stay open permanently. In 1883 the mill machinery was removed to a wonderful three-story facility called Granite Mill, which was three years in the making. A good deal of thought went into the mill, for example effluent from the rag washers and bleach boilers were directed to a settling basin instead of simply dumping it into the Big Cottonwook Creek. The owners now installed a new 62" Smith-Winchester moving-wire machine to boost production to 1500 pounds of news and print paper a day, while the old cylinder-wire machine continued in service making manila and straw wrapping. The mill was eventually closed in the 1890s with the advent of wood pulp, but during its time it was one of the most remarkable achievements in the West.

San Francisco's Supply

Samuel Penfeld Taylor earned just $20 crewing on an old schooner traveling around the horn, and arrived in San Francisco in 1849. The following year he left for the gold fields, accumulated $5,692 of gold dust, and returned to the city to open a lumberyard on California Street. In the search for timber he took a small boat to the fishing village of Sausalito and traveled north through Marin County where he found a one hundred acre parcel of redwood forest for sale on the north slope of Mt. Tamalpais. The property had a strong creek running down the mountain to the shore that would provide sufficient waterpower for a saw mill. Following his success Taylor returned to Fall River, Mass., in 1854 to marry his hometown sweetheart, Sarah Washington. Certain members of the family attending the wedding came from Saugerties, where they worked at the local paper mill, and upon learning of Taylor's dwindling lumber supply they convinced him to open a paper mill. Taylor then traveled to South Windham, Conn., to look into buying a used cylinder-wire machine, and after returning to San Francisco found a willing partner in V.B. Post's paper warehouse.[5]

Taylor's mill would be powered by two large overshot water wheels, and as construction neared completion the paper machine was shipped round the horn to the port of Bolinas. Newsprint was in high demand in San Francisco, so much to the delight of the local press, the Pioneer Mill officially opened in November of 1856. A rag market had been established in San Francisco's Chinatown district, and from here bales of rags were shipped by schooner to Tomales Bay. From here the bales were be loaded on a scow and floated on the tide to Taylor's warehouse, where a team of oxen transported them to the mill. Henry Russell, a papermaker from Massachusetts, was manager of the plant, which made newsprint for the

San Francisco Bulletin, the *Alta Californian*, and the *California Farmer*. Local newspapers became so dependent on this mill that messengers were often sent to Telegraph Hill, where the arrival and departure of ships were recorded, to ask, "Has the sloop from Bolinas yet hove in sight?"[6]

In 1883 S.P. Taylor & Co. spent $165,000 on a new paper-making facility in what was now called the town of Paperville. The San Geronimo Mill would be powered by a 30" Leffell turbine wheel along with a 150 horsepower Corliss steam engine. The steam engine was necessary since the county water department began diverting flow from the paper mill creek to city reservoirs. The Corliss engine would be used to power the wood-grinders, and in later years another twenty horsepower steam engine was added just to run the super-calenders. The San Geronimo Mill made newsprint, lithographic, and book paper, while the old plant was converted to manila for wrapping paper, sack, and fruit wrap. Taylor & Co. obtained shavings from a jute mill at San Quentin Prison, and occasionally imported jute from as far away as Calcutta.

Francis Blake and Charles F. Robbins were former printers who had been associated with the paper trade in San Francisco since the early 1850s at 516 Sacramento St., where they sold book, news, wrapping, card stock, and book binders board. In 1874 Blake, Robbins & Co. amalgamated with the Lick Mills Paper Mfg. Co., contributing $28,000 and $7,725 worth of wrapping paper in exchange for stock to acquire the former Lick flour mill. This was no ordinary flour mill, rather it had belonged to millionaire James Lick, the famous Bay Area real estate magnate.

James Lick sold a successful piano manufacturing business in Lima, Peru, before moving to San Francisco in January of 1848. Carrying $30,000 in Peruvian doubloons, he arrived in the city a mere seventeen days before gold was discovered in Sacramento. Lick's intention was to start another business, but when news of the gold strike reached the city in May, landholders scrambled to sell their properties to rush off to the gold fields, so Lick stayed behind and bought as much property as he could. By the end of the year his holdings amounted to over one million dollars against an investment of just $10,000. Lick purchased an old flour mill located on an artesian basin in Alviso, a small port town that served San Jose, the territorial capital. In 1850 he expanded his holdings in the area to 200 acres and built a new flour mill. Given the millionaire's network of contacts in South America, he had the interior walls and trim of the plant made from solid mahogany and other exotic hardwoods, spending a staggering $200,000 in the process. He went on to establish the fabulous Lick Hotel in San Francisco, and constructed a 24-room mansion on the grounds of his beloved flour mill. Believing the Santa Clara Valley to be the ideal growing area, Lick concentrated on fruit tree business, and in 1852 he wrote his son in Fredericksburg, Pa., "If you cannot sell your store, give it away. Come at once. I have enough for both of us." John Lick arrived in California in 1855 and was put in charge of the flour mill, but citing health reasons, he returned to Pennsylvania after just three years.[7]

In 1865 Lick offered to sell the flour mill with water rights for $500,000, but found no buyers. In 1872 he donated the mill to the Thomas Paine Society of Boston, which brokered a deal for $18,000 with Adolph Pfister who with help from Blake & Robbins planned to turn it into a paper mill. The paper market in San Francisco was somewhat underdeveloped, and even with the coming of the railroad many eastern manufacturers refused credit to warehouses in the west because of the great distances involved. Col. Frambach of Wisconsin seems to have been one of the few manufacturers to serve the San Francisco market. A cylinder-wire machine was set up in the mahogany mill for the manufacture of book paper and newsprint. Due to the high cost of raw materials the owners struggled to earn a profit. Still, the mahogany mill won a certain acclaim within the industry, and in 1877 the plight of the mill came to the attention of Alfred Denison Remington, a papermaker from Watertown, N.Y. Alfred and his two brothers, Hiram and C.R. Remington, had entered the papermaking business in Watertown during the Civil War, and the Remington Paper Co. was one of the first in that state to convert to sulphite pulp in the early 1880s. It seems Alfred owned stock in several gold mines in the West, and during his travels he stopped in San Jose to visit the famous mill. Remington was so impressed he bought the facility in 1878, while simultaneously establishing a paper warehouse in San Francisco. Within a few years the Lick Mill was converted to sulphite pulp and began supplying newsprint to the *San Jose Mercury*, *San Francisco Examiner*, *San Francisco Chronicle*, and *San Francisco Call Bulletin*. The plant also became the sole supplier of paper bags in the west, producing a million sacks a year valued at $180,000.

The Lick Mill was unfortunately lost to fire in 1882, but Remington built in its place a new three-story triangular shaped mill with its own blacksmith and machine shops. The new plant featured a 68" moving-wire machine, four beaters, one rotary bleach boiler, three Babcock tubular boilers, one Corliss engine, one small steam engine, and a 36" Acme self-clamping paper cutter. A newspaper article from 1895 reports the Remington Mill made pulp from poplar, spruce, or hemlock, hauling in the logs via the S&P Railroad. The mill also made fruit wrap from wood pulp mixed with cotton rags, red express (brown wrapping paper) from wood pulp and burlap, and druggists paper from bleached wood pulp and rag. After newsprint dropped to a mere 5.5 cents a pound in 1898, the mill was closed and all the machinery sold off.

The Oregon Trail

William Buck bought a 54" cylinder-wire machine from Smith Winchester & Co., and in 1866 installed it at a mill on the Willamette River in Oregon City, Oregon. The *Portland Oregonian* reported in January of 1867: "We have on our table some samples of Oregon made paper, the first ever manufactured in this state; they are from the Oregon city paper mills and at the hour of writing are not more than twenty-four hours old.... The quality is especially remarkable. We feel very much like boasting and jubilating a little over the first effort of Oregon in paper manufacture." The celebration proved short lived. The local rag market never developed and the Oregon City Mill was soon converted to straw.

H.L. Pittock, publisher of the *Portland Oregonian*, bought the Oregon City Mill in 1884 and formed the Columbia River Paper Co. Pittock and company erected a pulp mill at Young's River near the port of Astoria, and for the paper mill the company purchased six hundred acres of land on the Washington side of the border near Portland. This real estate encompassed the entire La Camas Creek, more than a mile in length, as well as several square miles of mountain lakes that fed into it. A series of three dams was constructed at the south end of the lakes to provide waterpower via a 2000 foot raceway running down to the new mill located near the banks of the Columbia River. Turbine wheels totaling 300 horsepower and a new 78" moving-wire machine were barged in from the railhead in Portland and set up at the LeCamas mill, which began manufacturing four tons of cottonwood-pulp newsprint a day. The old cylinder-wire machine from the Oregon City Mill now made butcher's and grocer's paper from straw, and when straw was out of season it made manila paper. The *Vancouver Independent* reported in 1885 that fifty tons of jute were shipped directly from Calcutta.

The mill at LeCamas was supplied entirely by barges on the Columbia River, and for the next twenty years pulp lap arriving at riverside docks was transported to the plant by horse-drawn railroad cars. Shipments of paper went out by a ferry run by the Hosford brothers to service the needs of farmers taking their produce into Portland. Disaster struck the mill after only a year of operation when a fire started in the bleach room, but on this occasion onlookers were kept well away by firefighters. Fortunately much of the machinery was saved, and losses of $95,000 were primarily in finished paper and pulp lap. With insurance of $45,000 a new plant building was immediately built using fireproof brick. Vice President J.K. Gill then proceeded on a whirlwind trip to Appleton, Holyoke, Worcester, Boston, and New York to take in the latest the paper industry had to offer. Gill brought back a new Rice, Barton & Fales 86" moving-wire machine and two sulphite digesters. The new bleach boilers, however, were made locally by the Willamette Iron Works.[8]

After the fire a new supervisor named Charles Smith West took charge of the mill. West came from a family of English papermakers; his father, George West, immigrated to this country to work for Zenas Crane. Charles was born in Lee, Mass., in 1847, and began his career at the Bay State Mill. He was later tutored by his uncle, Charles Cornelius West,

A postcard depiction of Columbia Mill circa 1900. Horse-drawn rail cars delivered pulp lap from docks on Columbia River.

A postcard depicting O'Neill's straw mill in Lebanon, Oregon, circa 1900.

who owned the Bancroft Mill in Becket, before landing his first job as a machine tender at a mill in Watertown, N.Y. In 1877 West began work at the new Turners Falls Mill in Franklin County, Mass., making nine hundred pound rolls of newsprint for the *New York Times*. The young mechanic got the urge to move west, and so took a job at the Van Nortwick Mills in Batavia, Illinois, before accepting the position at the LeCamas Mill.

Extending the railroad to Portland was no easy task. The first tracks were laid down along the Columbia River to ease portage around the Cascade Rapids, and eventually the railroad traveled the length of the Columbia River from Walla Walla to Portland. In 1881 an agreement was made to connect the system to the Union Pacific by way of Granger Wyo., this followed by the consolidation of lines in western Oregon in 1887 finally brought the trans-continental railroad to the port of Portland. The vast timber and waterpower of the region combined with the railroad made it an attractive location for more paper mills. In 1889 the Willamette River Pulp and Paper Company established a new mill in Oregon City in 1889, and bought out the struggling Columbia River Paper Co. the following year. Also about this time the Crown Paper Company erected a large pulp and paper mill on the Willamette River in the town of West Linn.

In 1883 the O'Neill brothers of Massachusetts traveled to California and started up a straw mill in Soquel, a small town in Monterey Bay located between Santa Clara and Watsonville. The South Coast Mill was the vision of Edward O'Neill, who had learned the trade in Lee, Mass. A papermaker from Ireland named Charlie Callaghan later bought the mill, and O'Neill then traveled five hundred miles north to Oregon, where he founded another mill in the town of Lebanon. Lebanon was in the middle of straw county in the southern Willamette Valley about fifty miles south of Portland. Here the Albany-Santiam Canal, built in 1872, provided ample waterpower for O'Neill's new plant situated on the

South Santiam River. A. Zellerbach & Sons, stationers of San Francisco, built a sulphite mill Lebanon in 1899. Water power for this mill came from the Lebanon–Santiam Canal, and the plant ran on pulp lap shipped by rail from Oregon City or West Linn. The San Francisco based company had to ship finished goods by way of Wyoming, but this was still more convenient than hiring a schooner.[9]

Appendix I: Directory of 19th Century Paper Mills

The graphs presented earlier in this work were based on a census taken during research into this book. Prior works have generalized mills as belonging to one class or another, but the reality is quite revealing, as various strategies were used to fit the situation and the times. The basic categories inventoried were the kinds of materials used, selected machinery, and types of paper made.

This database is restricted to the commercial classes of paper: rag, straw, manila, wood-pulp, leather, etc. Dates are provided when a mill changed status or added a major new line. Composite papers such as manila could be made with either jute, hemp, straw, and other materials, or combinations thereof, so to be clear, the primary intent is to record the kinds of materials coming in the door without regard to the types of paper made. In some cases the records don't provide that detail, but depending on other factors one can usually make an educated guess.

Equipment lists were limited to twenty different classes, including the mill's power supply (waterwheels, turbines, steam engines, etc.). For surviving eighteenth-century mills, the census tries to identify whether a mill was pre-industrial or not. For early industrialized mills the census also tries to record the type of waterwheel employed; essentially, overshot wheels are powered by water falling over the top, while undershot wheels are propelled by water flowing underneath. Dates are provided for major mill conversions, expansions, or consolidations, while the comment line identifies calamities or other event of importance.

The end product for each mill was determined by market environment and exposure. Here the paper market is divided into some thirty classes, most of them (e.g., book paper) are self-explanatory. The most basic categories from the beginning of the century were printing and writing. Actually, both were the same product, only writing came with external sizing and printing did not. The old generic names didn't reveal a paper's use; for instance, foolscap came in both writing and printing versions. The class of printing paper here includes everything such as leaflet, handbill, plate, map, and music paper. The wrapping class extends to sheathing, express (heavy paper for shipping), butcher, butter, flour, candy, tobacco, and fruit wrappers. The makers of bond and banknote paper were also largely responsible for special papers as parchment, bank check, tax stamp, and safety papers, so these types are all lumped together. Cartridge paper, revolver paper, shot shells, nitrated paper, and anything related to firearms is another class. Cover stock encompasses cards, card middles, bristol board, and leather board, while board stock is generally thicker and includes things like binder board, press board, shield board, trunk board, and the like. Building paper includes tar, roofing, and sand paper, while hardware refers to utility items such as pharmaceutical cups, pails, bowls, foot baths, and the like. Folio paper includes art paper and specialty items for invitations and weddings. Stationery is something of a catch-all covering all writing papers except ledger (and tablet), which are special types of book paper with their own category.

Accord Mills (est. 1852) in Accord, Ulster Co., NY, was powered by two waterwheels making straw wrapping paper on a cylinder-wire machine. Owner code: JLS

Acushnet Mill (est. 1883) in North New Bedford, Bristol Co., MA, made wrapping paper and felt lining on a cylinder-wire machine and experimented with cedar bark paper. This plant also used a steam engine to power auxiliary equipment. Owner code: JAB

Adam Mill (est. 1783) in Salisbury, Litchfield Co., CT, was a hand mill that closed in 1842. Owner code: JAC

Adams & Co. Mill (est. 1858) in Chagrin Falls, Cuyahoga Co., OH, was powered by two waterwheels and three steam engines. This mill made manila paper on a double cylinder machine. Owner code: JAE

Adriatic Mill (est. 1830) in South Manchester, Hartford Co., CT, was powered by one waterwheel and two steam engines. This mill made rag printing paper on a cylinder-wire machine. Owner code: KLQ

Advertiser Mill (est. 1836) in Louisville, Jefferson Co., KY, made news paper on a cylinder-wire machine.

Africa Mill (circa 1880) in Valley Forge, Chester Co., PA, used a moving-wire machine and a Jordan refiner in the manufacture of book, stationery, and manila-colored mediums. Owner code: KMK

Agawam Mill (est. 1851) in Mittineague, Hampden Co., MA, was powered by two waterwheels and one steam engine. This mill made card, stationery, and wall paper on a moving-wire machine. Owner codes: JAI, KWG

Ager Mill (est. 1848) in Lyonsdale, Lewis Co., NY, had two waterwheels and made manila wrapper and filter paper on a cylinder-wire machine. Owner code: JAJ

Akron Mill (est. 1872) in Akron, Summit Co., OH, made wrapping paper on a cylinder-wire machine. The plant was powered by two steam engines. Owner code: KIK

Akron Straw Mill (est. 1872) in Akron, Summit Co., OH, made straw board on two cylinder-wire machines. This plant was powered by two steam engines. Owner code: JAK

Albany City Mills (circa 1880) was located in Albany, Albany Co., NY, and made wrapping paper on a cylinder-wire machine. Owner code: JSN

Albion Mill (est. 1869) at Holyoke, Hampden Co., MA, was powered by five waterwheels and three steam engines. The plant used Jordans and three moving-wire machines in the manufacture of book and news paper. It was later renamed Crocker Mill No. 2. Owner code: JAL

Allegan Mills (est. 1882) at Allegan, Allegan Co., MI, made straw wrapper on a cylinder wire machine. This mill was powered by two waterwheels. Owner code: KHV

Allen Brothers Mill (est. 1857) at Sandy Hill, Washington Co., NY, was a wall paper factory with a cylinder-wire machine and two moving-wire machines. This mill was powered by nine water turbines and one steam engine. Owner code: JAP

Allen Mill (est. 1818) at Leominster, Worcester Co., CT, was a hand mill that continued until around 1841. Owner code: JAR

Alling Mill (est. 1883) at Bangall, Duchess Co., NY, had one waterwheel powering a cylinder-wire machine for the manufacture of straw wrapping paper. Owner code: JAU

Amoskeag Mill (est. 1850) at Manchester, Hillsborough Co., NH, had one moving-wire machine powered by a waterwheel for the manufacture of book paper. Owner codes: KCB, JVD

Anchor Steam Mill (est. 1824) at Pittsburgh, Allegheny Co., PA, was a handmade mill that used a steam engine to power the rag engines. Owner code: JUM

Ancram Mill (est. 1842) at Ancram, Columbia Co., NY, produced straw wrapping paper on a cylinder-wire machine. Owner codes: KIY, KHS

Andover Mill (est. 1880) at Freeland, Baltimore Co., MD, was powered by two waterwheels.

It made straw wrapping paper on a cylinder-wire machine. Owner code: JRB

Androscoggin Mill (est. 1852) at Mechanic Falls, Androscoggin Co., ME, was powered by three waterwheels and a steam engine. This mill made book and news paper on a moving-wire machine. Owner code: JMG

Antelope Mill (est. 1881) at 28-34 Reade Street in New York City produced manila wrapping paper on a cylinder-wire machine. Owner codes: JEW, KFM, JOM

Antietam Mill (est. 1856) at Hagerstown, Washington Co., MD, had three waterwheels and two steam engines. The mill made book and news on a moving-wire machine. Owner code: JBL

Appleton Paper & Pulp Mills (est. 1873) at Appleton, Outagamie Co., WI, had two moving-wire machines and used a Jordan in the manufacture of news paper. Owner code: JBM

Appomattox Paper Mill (est. 1876) at Petersburg, VA, manufactured straw wrapping paper on a cylinder-wire machine.

Aqueduct Mills (3) (est. 1885) at Dayton, Montgomery Co., OH, made straw board and carpet lining felt on a double-cylinder and nine wet machines. The plant was powered by four steam engines. Owner code: JTJ

Arlington Mill (est. 1864) at Salisbury Mills, Orange Co., NY, was powered by three water turbines and two steam engines. This mill made book and news paper using the Kingston refiner. Owner code: KJT

Arlington Mill (est. 1883) at S. Manchester, Hartford Co., CT, was powered by two waterwheels. This mill made leather board on a cylinder-wire machine. Owner code: JHJ

Armour Bros. Mill in Stanley, Morris Co., NJ, was established about 1880. The plant used a cylinder-wire and double cylinder machine in the manufacture of manila building and roofing paper. Owner codes: JBO, KFL

Aroostook Mill (est. 1840) at Gardiner, Kennebec Co., ME, made rag paper on a moving-wire machine. Owner code: JUP

Ashland Mill (est. 1865) at Manayank, Philadelphia Co., PA, made news and printing paper on two moving-wire and one cylinder-wire machine. The plant had a Jordan refiner, and was powered by four steam engines. Owner code: KLW

Athens Mill (est. 1847) at Athens, GA, made news and wrapper paper on a moving-wire machine. The plant was rebuilt following a fire in 1856, but burned again in 1861. Owner code: JHW

Athern Lower Mill (est. 1844) at Lockland, Hamilton Co., OH, made straw wrapper paper on a cylinder-wire machine. Owner codes: KLX, JIL, KHE, JPN, JPQ, JYE. The Athern Upper Mill was established two years later. This facility made straw wrapper and building supplies on a cylinder-wire machine. Owner codes: JBP, JAT, JPO, JRV, JRU

Atlantic Mill (est. 1830) at S. Manchester, Hartford Co., CT, made printing paper on a cylinder-wire machine. This plant was powered by two waterwheels and a steam engine. Owner code: KLQ

Atlantic Mill (est. 1860) at Weymouth, Atlantic Co., NJ, made manila twine or rope on a cylinder-wire machine. Owner code: KVO

Atlas Mill (est. 1878) at Appleton, Outagamie Co., WI, was powered by two waterwheels. This mill made newsprint on two cylinder-wire machines. Owner code: JBR

Atlas Mill in Quincy, Adams Co., IL, was established around 1880. This mill had four water turbines and two steam engines and made straw board on a cylinder-wire machine. Owner code: KJQ

Auburn Mill (est. 1831) at Watertown, Jefferson Co., NY, made book, news, and stationery paper on a cylinder-wire machine. This plant was later the Knowlton Brothers Mill. Owner codes: JYO, JYN

Auburn Mill at Chesterville, Chester Co., PA, was established around 1883; it made manila tissue on a cylinder-wire machine powered by a waterwheel. It also had a steam engine to power auxiliary equipment. Owner code: JMX

Augustine Mill (est. 1843) in Rockland New Castle Co., DE, made book paper on three moving-wire machines. The mill converted to wood pulp in 1870. Owner code: JWN

Ausable Chasm Mill (est. 1884) in Ausable Chasm, Essex Co., NY, made wrapping paper and bogus manila on a cylinder-wire machine. This mill was powered by two waterwheels. Owner code: KKP

Bacon Mill (est. 1852) at Lawrence, Essex Co., MA, produced book and stationery paper on three moving-wire machines. The plant had a Jordan refiner and power was provided by two water turbines and two steam engines. Rebuilt in 1857. Owner code: JBW

Badger Mill (est. 1883) at Kaukauna, Outagamie Co., WI, made book paper from wood pulp on a Harper machine. It also made wrapping paper from both manila and straw (when in season) on a double cylinder machine. This facility was powered by twenty-one water turbines. Owner code: JPE

Bailey's Leather Board Mill (est. 1867) at Ashland, Grafton Co., NH, made leather board on two cylinder-wire machines powered by two waterwheels. Owner code: JBY

Baker Mill (est. 1807) at Dorchester, Suffolk Co., MA, was a hand mill that continued until 1825. Owner codes: JKB, JKA, JCB, JCC

Baker Mill (est. 1838) at East Lee, Berkshire Co., MA, made stationery paper on a cylinder-wire machine. This plant burned down in 1842. Owner codes: JCC, KIX, KVG

Baldwinsville Mill (est. 1875) at Baldwinsville, Worcester Co., MA, was powered by a waterwheel and steam engine. This facility made wrapping paper and carpet lining felt on a cylinder-wire machine. Owner code: KNW

Baldwinsville Mill at Baldwinsville, Onondaga Co., NY, was established around 1880 and made straw wrappers on a cylinder-wire machine. The plant was powered by four waterwheels and one steam engine. Owner code: JXX

Ballston Spa Mill (est. 1885) at Ballston Spa, Saratoga Co., NY, made bogus manila on a double-cylinder machine. Owner codes: KRA, JWT

Baltimore Mill at Millis, Middlesex Co., MA, was established around 1880; it made manila wrapper paper on a cylinder-wire machine. This mill was powered by two waterwheels. Owner code: KUO

Bancroft Mills (est. 1847) at Becket, Berkshire Co., MA, had a second plant added in 1857. The plant made newsprint on a moving-wire machine and manila on a cylinder-wire; it had a Jordan refiner and each facility was powered by a waterwheel and a steam engine. Owner codes: JFZ, JGB

Bannister Mill (est. 1807) at Petersburg, Dinwiddie Co., VA, was a hand mill that continued until about 1853. Owner code: JCG

Bates Mill was a manila mill established in 1876 at W. Cummington, Hampshire Co., MA. Owner code: JCR

Bath Island Mill (est. 1825) at Bath Island, Albany Co., NY, was a hand mill that made news, wrapping, and stationery paper. This plant was destroyed by fire in the 1850s. Owner code: KJC

Bath Mills (est. 1876) at Bath, Aiken Co., SC, made book, news, and manila-colored wrapping on a moving-wire machine. This mill was powered by six waterwheels and a steam engine. Owner code: JCM

Battenkill Mill (est. 1869) at Middle Falls, Washington Co., NY, made leather board on four board machines. This plant was powered by five waterwheels. Owner code: KPA

Bay State Mill, Old (est. 1851) at Dalton, Berkshire Co., MA, made ledger paper on a moving-wire machine. The plant failed in 1857 and was later rebuilt and reopened in 1878. Owner code: JKL. The new Bay State Mill made linen and linen-composite stationery paper and envelopes. This facility was powered by two steam engines and two waterwheels. Owner code: JKP

Beardsley Mill (est. 1850) at Elkhart, Elkhart Co., IN, made news and wrapping paper on a moving-wire machine. This plant had a Jordan refiner and was powered by two waterwheels and a steam engine. Owner code: JCZ

Beaver Brook Mill (est. 1840) at Dracut, Middlesex Co., MA, made paper on a cylinder-wire machine. The mill was powered by three waterwheels and a steam engine. Owner code: KHD

Beaver Brook Mill (est. 1868) at Danbury, Fairfield Co., CT, made manila wrapper and hardware paper on a double cylinder machine. The mill was powered by a waterwheel and a steam engine. Owner code: KCP

Beaver Brook Mill (est. 1880) at Shirley Village, Middlesex Co., MA, made leather board on a wet machine. The facility was powered by two waterwheels and one steam engine. Owner code: JIZ

Beaver Creek Mill (est. 1806) at Beaver Creek, Columbiana Co., OH, was a hand mill that made linen and linen-composite stationery and printing paper until around 1851. Owner code: JJY

Beaver Dam Mill (est. 1851) at Ercildoun, Chester Co., PA, was a manila mill that made hardware paper on a double-cylinder machine. The facility was powered by a waterwheel and a steam engine. Owner code: JPX

Beaver Falls Mill (est. 1883) at Beaver Falls, Beaver Co., PA, produced glazed paper on a cylinder-wire machine. Mill was powered by a steam engine. Owner code: JDA

Bechtel Mill (est. 1800) at Germantown, Philadelphia Co., PA, was a hand mill that closed about 1833. Owner code: JDB

Beckley Mill (est. 1800) at Beckleysville, Baltimore Co., MD, was a hand mill that made book and news paper. Sometime later a moving-wire machine was installed and operations continued until 1894. Owner code: JDF

Bedford Mill was established in 1898 at Big Island, VA. Owner code: JDH

Beebe & Holbrook Mill (est. 1871) at Holyoke, Hampden Co., MA, was the former Hampden Mill. This plant made stationery on a moving-wire machine. The facility was powered by three waterwheels. Owner code: JDI

Beehive Mill (est. 1799) at Upper Darby Township, Delaware Co., PA, was a hand mill that converted to a cotton factory in 1860. Owner codes: KCK, KAI

Bellows Falls Mill (est. 1872) at Bellows Falls, Windham Co., VT, was a manila tissue mill with one cylinder-wire machine and three waterwheels. Owner code: JOT

Beltonford Mill (est. 1883) at Lancaster, Lancaster Co., PA, made envelopes, folio paper, and manila-colored mediums on a moving-wire machine. This mill had a Jordan refiner, and was powered by two steam engines and two water wheels. Owner code: KQA

Berkshire Mill (est. 1801) at Dalton, Berkshire Co., MA, produced handmade paper until its conversion to machine manufacture in 1832. This mill made linen and linen-composite stationery and ledger paper on a cylinder-wire machine powered by a waterwheel. Owner Code: JKQ. The facilities were remodeled in 1866 and further outfitted with a moving-wire machine and a steam engine. Owner codes: JHH, KJD, JHD, JKR

Bicking Mill (est. 1769) at Lower Merion Township, Montgomery Co., PA, was a hand mill that continued until around 1830. Owner code: JDU

Bicking Mill (est. 1791) at Downington, Chester Co., PA, was a hand mill that added a cylinder-wire machine and began making board stock. The plant was powered by two waterwheels and two steam engines. Owner codes: JDX, JDV

Big Creek Mill (est. 1826) at Madison, Jefferson Co., IN, was a two-vat mill that continued until around 1836. Owner code: KEH

Billings Mill (est. 1810) at Montpelier, VT, was a hand mill that continued until about 1851. Owner codes: JEA, JYB

Bird's Upper Mill (est. 1830) at East Walpole, Norfolk Co., MA, made book paper on a cylinder-wire machine until operations were moved to a moving-wire machine in 1872. The cylinder machine then made manila hardware paper. The plant was powered by four waterwheels and four steam engines. Manufacturing proceeded with wood pulp in 1887. Owner code: JEE

Bishop Steam Mill, Pittsburgh, Allegheny Co., PA, was a handmade mill circa 1800 that employed steam engines to power its rag engine(s).

Black River Mill (est. 1808) at Watertown, Jefferson Co., NY, was a hand mill that continued until around 1828. Owner codes: JHM, JUN, JYN

Black River Mill (est. 1883) at Brownville, Jefferson Co., NY, made straw and manila paper on a double cylinder machine. The facility was powered by three water turbines and a steam engine. Owner code: KAW

Blackwater Mill (est. 1871) at Webster, Merrimack Co., NH, made leather board on two cylinder-wire machines powered by two waterwheels. Owner code: JUR

Blauvet & Gilmor Mill (est. 1861) at Lee, Berkshire Co., MA, employed a moving-wire machine in the manufacture of paper twine. Owner code: JEO

Bleything Mill (est. 1832) at Hanover, NJ, made paper on a cylinder-wire machine. Owner code: JEP

Bloomfield Mill at Bloomfield, Essex Co., NJ, was established around 1840.

Bloomingdale Paper Works (est. 1862) at Bloomingdale, Passaic Co., NJ, made manila tissue paper on a cylinder-wire machine powered by two waterwheels. Owner code: KVY

Bloomsburgh Mill (est. 1882) on Light Street in a town of the same name in Columbia Co., PA, was a manila mill that employed a double-cylinder machine. Owner code: JEQ

Blossvale Mill (est. 1872) at Blossvale, Oneida Co., NY, made straw wrapping on a cylinder-wire machine. This plant was powered by three waterwheels. Owner code: JRZ

Boies Family Mill (est. 1793) at Dorchester, Suffolk Co., MA, was a hand mill that burned down in 1805. Owner codes: JER, KRF

Boies Mill (est. 1769) at Milton, Norfolk Co., MA, was a hand mill that continued until 1830. Owner code: JER

Boonton Mill (est. 1880) at Boontown, Morris Co., NJ, made newsprint on a moving-wire machine and manila or straw board or card on a cylinder-wire machine. Owner code: JEV

Bowdoin Mills (est. 1868) at Topsham, Cumberland Co., ME, made newsprint on three moving-wire machines. The mill had a Jordan refiner. Owner code: JFA

Bowling Green Mill (est. 1805) at Bowling Green, Logan Co., KY, was a hand mill that operated until about 1849.

Boyce Mill at Franklin, Delaware Co., NY, was established around 1870. This mill made straw board on a cylinder-wire machine. Owner code: JFC

Bradford Mill in Bradford, Orange Co., VT, was established around 1880. This manila mill made tissue paper on a cylinder-wire machine powered by two waterwheels. Owner code: JFE

Bradley Mill was established in 1838 at Dansville, Livingston Co., NY; it burned down the same year and was not rebuilt. Owner code: JFF

Branch Mill (est. 1883) at Westfield, Union Co., NJ, was a manila mill that made card stock on a cylinder-wire machine. Owner code: KHH

Brandywine Mills (4) (est. 1787) at Wilmington, New Castle Co., DE, began as a hand-craft mill specializing in linen and linen-composite bond, printing, and stationery paper. It installed waterwheel(s) and rag engine(s) in 1808 and expanded to three 2-vat plants. A cylinder-wire machine was added in 1817 for the manufacture of book paper and newsprint. Following bankruptcy the facilities were closed by 1844. Owner code: JQM

Brick Mill (est. 1850) at Fitchburg, Worcester Co., MA, made newsprint and card stock using a Jordan refiner. The mill had a moving-wire machine, two waterwheels, and three steam engines. Owner code: JKZ

Brick Mill in Skaneateles Falls, Onondaga Co., NY, was established about 1880 and made newsprint on a moving-wire machine. This plant had two waterwheels and a steam engine. Owner code: KUX

Bridge Mill at Pawtucket, Providence Co., RI, was established around 1880. This plant made newsprint on a moving-wire machine. Owner code: JFL

Bridgeport Mill (est. 1883) at Bridgeport, Fairfield Co., CT, made newsprint on a moving-wire machine. The mill was powered by a waterwheel and two steam engines. Owner code: JFM

Bridgewater Mill in Bridgewater, Plymouth Co., MA, was established around 1870. This manila mill made hardware paper on a cylinder-wire machine. Owner code: JUQ

Brightwood Mill (est. 1873) at Hinsdale, Cheshire Co., NH, made manila tissue and toilet paper on a cylinder-wire machine. The plant featured six waterwheels. Owner code: JOQ

Bristol Mills (2), established in 1871 at Bristol, Grafton Co., NH, made straw board mill and bogus manila on three cylinder-wire machines. Owner code: KCG

Broadway Mill (est. 1847) at Cleveland, Cuyahoga Co., OH, made book paper on a moving-wire machine powered by a water wheel. A steam engine was used to drive auxiliary equipment. Owner code: JJA

Bronx Mill (est. 1800) at Bronx, Queens Co., NY, was a handmade mill powered by a mill dam now part of the grounds of the New York City Zoo. The mill burned down in 1822. Owner code: KBD

Brooklyn Steam Mill in Brooklyn, Kings Co., NY, was established around 1880. This plant had two moving-wire machines for the manufacture of newsprint and wall paper. Owner code: KIM

Brookside Mill in Mill River, Berkshire Co., MA, was established around 1880. This plant featured a moving-wire machine powered by a waterwheel. Owner code: JFO

Brookside Mill (est. 1884) at S. Manchester, Hartford Co., CT, made card and board stock on a wet machine powered by a waterwheel. A steam engine powered auxiliary equipment. Owner code: JHK

Brookville Mill (est. 1834) at Brookville, Franklin

Co., IN, made paper on a cylinder-wire machine. Owner code: KIH

Brown Leaf Mill in Chatham, Columbia Co., NY, was established around 1870. This mill made straw wrapper on a cylinder-wire machine. Owner code: KDO

Brownsville Mill (est. 1832) at Brownsville, Fayette Co., PA, made rag paper on a cylinder-wire machine. Owner code: JYV

Bryand Mill (est. 1898) at Kalamazoo, Kalamazoo Co., MI, featured two moving-wire machines. Owner code: JFX

Buck Run Mill, Old (est. 1795) at E. Fallowfield Township, Chester Co., PA, was a hand mill that closed in 1839. The facilities reopened in 1850 as a machine mill. Owner codes: JQH, JDQ, KEC, JNM, JXR

Buck Run Mill (est. 1864) at Ercildoun, Chester Co., PA, made board stock on a cylinder-wire machine powered by a waterwheel. Owner code: KYL

Buckeye Mill (est. 1843) at West Wheeling, Ohio Co., OH, made straw wrapping on a cylinder-wire machine. Auxiliary equipment was powered by two steam engines. Owner code: KTH

Buckeye Mill (est. 1859) at West Carrollton, Montgomery Co., OH, made book and news paper on a moving-wire machine. Owner code: JPQ

Buffalo Mills (est. 1849) at Shelby, Cleveland Co., NC, made paper on a cylinder-wire machine.

Bulletin Mill (est. 1847) at New Orleans, Orleans Co., LA, made news paper on a cylinder-wire machine. Owner code: KFK

Burbank Mill (est. 1776) at Sutton, Worcester Co., MA, was initially a one-vat mill that hand-crafted book and news paper. In 1789 it added rag engine(s) powered by a waterwheel and put on a second vat. A cylinder-wire machine was added in 1824, but the facility closed in 1836. Owner code: JGG

Burnside Mill (est. 1851) at Burnside, Hartford Co., CT, had four waterwheels and two steam engines. It made manila-colored envelope paper on two moving-wire machines, later adding a Jordan refiner when that became available. Owner code: JSK

Cadwallader Mill (est. 1800) at Birmingham, Huntingdon Co., PA, was a hand mill that operated continually until 1839. Owner code: JGO

Caldwell Mill (est. 1873) at Caldwell, Essex Co., NJ, made straw board on a single cylinder-wire machine powered by a waterwheel. Auxiliary equipment was powered by a steam engine. Owner code: KCV

Caledonia Mill (est. 1840) at Westminster, Carroll Co., MD, made straw wrappers on a cylinder-wire machine powered by two waterwheels. Owner code: JND

Caledonia Mill (circa 1870) at Whippany, Morris Co., NJ, made colored mediums on a moving-wire machine until the plant closed around 1883. Owner code: JJF

Calico Mill was at Millburn, Essex Co., NJ. Owner code: JTW

California Mill (est. 1880) at Dansville, Livingston Co., NY, made manila wrapping paper on a cylinder-wire machine powered by a waterwheel. Auxiliary equipment was powered by a steam engine. Owner code: KSW

California Mills (est. 1883) was at Stockton, San Joaquin Co., CA. This mill had one cylinder-wire machine and two moving-wire machines. It made news paper from rags using a Jordan refiner, and wrapping paper from manila or straw when that was in season. Auxiliary equipment was powered by two steam engines. Owner code: JGQ

Callicoon Mill (est. 1873) at Callicoon Depot, Sullivan Co., NY, made straw wrappers on a cylinder-wire machine powered by a waterwheel. Auxiliary equipment was powered by two steam engines. Owner code: JPT

Calumet Mill (est. 1880) at Clarksville, Pike Co., MS, made straw wrappers on a cylinder-wire machine. Auxiliary equipment was powered by two steam engines. Owner code: JIX

Camden Mill (est. 1828) at Camden, ME, was a hand mill that operated until 1841, when it burned down. Owner code: JCN

Camp Mill (est. 1838) at, Jefferson Co., NY, made rag paper on a cylinder-wire machine. It burned down in 1838 and was not rebuilt.

Campbell Mill (est. 1850) at Norfolk, Norfolk Co., MA, made manila carpet liners on a double-cylinder powered by two waterwheels. Owner code: JGS

Cannelton Mill (est. 1873) at Cannelton, Perry Co., IN, made wrapping paper on a single cylinder-wire machine. Auxiliary equipment was powered by three steam engines. Owner code: JGW

Canney Mill (est. 1857) at Center Ossipee, Carroll Co., NH, made straw board on a single cylinder-wire machine. Owner code: JGX

Canoe Creek Leather Mill (circa 1870) was at Millburn, Essex Co., NJ. Owner code: JOC

Canton Mill (est. 1863) at Canton, Stark Co., OH, made wrapping paper on two cylinder-wire machines. Auxiliary equipment was powered by a steam engine. Owner code: JGY

Carew Mills (est. 1848) at South Hadley, Hampshire Co., MA, made ledger and stationery paper on a single moving-wire machine. The plant burned down in 1871 and was rebuilt in brick. A second moving-wire machine was added in 1892. Owner code: JGZ

Carlyle Mill (est. 1898) was in Carlyle, IL. Owner code: JHA

Carroll Mills (est. 1877) at Mill River, Berkshire Co., MA, made book paper on a cylinder-wire machine and a moving-wire machine powered by four waterwheels. Auxiliary equipment was powered by four steam engines. Owner codes: JHC

Carrollton Mill (est. 1859) at West Carrollton, Montgomery Co., OH, made bogus manila on a double-cylinder machine. Owner code: JPQ

Cascade Mill (circa 1880) at Penn Yan, Yates Co., NY, made straw wrappers on two cylinder-wire machines. Owner code: JHP

Castle Mill (est. 1822) was in Lee, Berkshire Co., MA. This hand mill was closed in 1832 and later reopened as the Laurel Mill. Owner codes: JIJ, JKK

Cayudulta Mill (est. 1850) at Sammonsville, Fulton Co., NY, made straw board on a cylinder-wire machine powered by two waterwheels. Auxiliary equipment was powered by a steam engine. Owner code: JUB

Cecil County Mills (circa 1880) at Elkton, Cecil Co., MD, made book and news paper using a Jordan refiner on a single moving-wire machine. Owner code: JHQ

Cecil Mill (est. 1836) at Elkton, Cecil Co., MD, made book and news paper using a Jordan refiner on a moving-wire machine powered by a waterwheel. Auxiliary equipment was powered by two steam engines. This mill was later sold and reopened as the Rising Sun Mill. Owner code: JHQ

Cedar Falls Mills (est. 1882) at Cedar Falls, Black Hawk Co., IA, made straw wrappers on a cylinder-wire machine. The plant and auxiliary equipment was powered by five waterwheels. Owner code: JHR

Centennial Mill (est. 1855) at Dalton, Berkshire Co., MA, made ledger paper on two moving-wire machines powered by a waterwheel. Auxiliary equipment was powered by two steam engines. Owner code: KVI

Centennial Mill (est. 1872) at Valatie, Columbia Co., NY, made straw wrappers on two cylinder-wire machines, and was powered by four waterwheels and a steam engine. Owner code: JLU

Central Fall Mill (est. 1884) at Greenwich, Washington Co., NY, was a small wall paper mill that used a single cylinder-wire machine. Owner code: JBI

Central Mill (est. 1862) at Indianapolis, Marin Co., IN, made book and news paper on two moving-wire machines powered by a single waterwheel. Auxiliary equipment was powered by two steam engines. Owner code: KMD

Central New York Mill (circa 1880) at Chittenango, Madison Co., NY, made manila wrappers on a single cylinder-wire machine. Owner code: JCX

Champlain Straw Board Mill (est. 1870) at Champlain, Clinton Co., NY, made straw board on a board machine and was powered by two waterwheels. Owner code: KWD

Chapin and Gould Crescent Mill (est. 1858) at Huntington, Hampshire Co., MA, made wrapping paper on a moving wire machine and was powered by four water turbines. Owner code: JHV

Chaplin Mill (est. 1807) was at Chaplin, Windham Co., CT. Originally begun as a hand mill, this plant made manila and press board on a wet machine and a double cylinder machine, and was powered by two waterwheels. The mill closed in 1854. Owner codes: JHJ, JIP

Charter Oak Mill (est. 1865) at Burnside, Hartford Co., CT, had a moving wire machine powered by two waterwheels. It made folio paper and stationery. Auxiliary equipment was powered by two steam engines. Owner code: JNH

Chelsea Mill (est. 1834) at Greenville, New London Co., CT, had five moving-wire machines and made newsprint using the Kingston refiner. The plant was dismantled in the early 1890s. Owner code: JHY

Chemical Paper Co. Mill (est. 1880) at Holyoke, Hampden Co., MA, had two moving-wire machines and three Harper machines, and was powered by thirteen water turbines. This plant made stationery, envelopes, card stock, binder board, and colored stock, and used the Jordan refiner. Owner code: JHZ

Cheney Mill (circa 1882) at Manchester, Hillsborough Co., NH, had two moving-wire machines and was powered by three waterwheels and four steam engines. The mill produced colored paper and card stock. Owner code: JIC

Chester Creek Mill (est. 1826) was at Upper Providence Township, Delaware Co., PA. This hand mill operated until 1837. Owner code: JNB

Chester Mill (est. 1853) at Huntington, Hampshire Co., MA, made stationery and card stock on a moving-wire machine powered by three water turbines. A steam engine powered the auxiliary equipment. Owner code: JID

Chester Mill (circa 1880) at Modena, Chester Co., PA, made book and news on a moving-wire machine and a Gould machine. Owner code: JWN

Chestertown Straw Board Mill (est. 1882) at Chestertown, Kent Co., MD, made straw board on a single cylinder-wire machine. Auxiliary equipment was powered by three steam engines. Owner code: JIE

Childs Mill (est. 1868) was at Rockford, Kent Co., MI. This plant had a cylinder-wire machine and two board machines for the manufacture of straw board and wrapping paper. The plant was powered by three waterwheels and a steam engine. Owner code: JIF

Chillicothe Pulp and Paper Mill (est. 1885) at Chillicothe, Ross Co., OH, made card board from wood-pulp on a wet machine. Owner code: KDB

Church Mill (est. 1838) at Rochester, Monroe Co., NY, made rag paper on a cylinder-wire machine until it was destroyed by fire in 1841. Owner code: JIH

Cincinnati Steam Mill was established in 1835 at Cincinnati, Hamilton Co., OH. Owner code: KIJ

Circleville Mill (est. 1883) at Circleville, Pickaway Co., OH, made straw board on four cylinder-wire machines. Owner code: KJB

City Mill (est. 1844) at Manayank, Philadelphia Co., PA, had two moving-wire machines and made newsprint and bogus manila using the Jordan refiner. Auxiliary equipment was powered by four steam engines. Owner code: KGB

Clackamas Mill (est. 1867) at Clackamas, Clackamas Co., OR, made book and news paper on a moving-wire machine powered by two waterwheels. Owner code: JJK

Clark Mill (est. 1859) at Dayton, Montgomery Co., OH, made straw paper on a cylinder-wire machine. Owner code: JIW

Clark Mill (est. 1883) at Marseilles, La Salle Co., IL, made straw wrapper when in season, and bogus manila using a Jordan refiner on cylinder-wire machine. The plant was powered by three waterwheels and a steam engine. It was rebuilt 1884 following an unspecified disaster. Owner code: JIU

Clear Spring Mill (est. 1885) at New Hope, Bucks Co., PA, made bristol board. Owner code: KXV

Clifton Mill (circa 1820) at Upper Darby Township, Delaware Co., PA, was a hand mill that closed in 1867. Owner codes: JBC, KAC

Clifton Mill (est. 1846) at Clifton, Hamilton Co., OH, had a cylinder-wire machine and a moving-wire machine. It made book, news, and blotter paper as well as manila wrappers. The plant burned in 1860 and was rebuilt. Owner codes: KRV, KGA, JYD

Clifton Mill (est. 1866) at Clifton, Green Co., OH, made straw wrappers on a cylinder-wire machine. The plant was powered by two waterwheels and two steam engines. Owner code: JJB

Clinton Mill (circa 1880) at Clinton, Clinton Co., IA, made straw wrappers on a cylinder-wire machine. Owner code: JJC

Clinton Mill (est. 1883) at Clinton, Middlesex Co., CT, made manila paper on a cylinder-wire machine. The plant was powered by a steam engine and two waterwheels. Owner code: KHP

Clinton Mill at Steubenville, Jefferson Co., OH, made wrapping paper on a cylinder-wire and a moving-wire. The plant was powered by four steam engines. This was the former Clinton Steam Mill. Owner code: JTB

Clinton Steam Mill (est. 1813) at Steubenville, Jefferson Co., OH was a hand mill that added a cylinder-wire machine in 1830. The plant was powered by a steam engine, and it later became the Clinton Mill. Owner codes: KMR, JUM

Clipper Mills (est. 1775) at Hoffmanville, Baltimore Co., MD, was a hand mill that made book, news, and wrapping. A moving-wire machine was installed in 1848 for production of the same. Facilities were powered by two waterwheels and a steam engine. Owner code: JUH

Clyde River Mill (est. 1872) at West Derby, Orleans Co., VT, made manila tissue on a cylinder-wire machine. The plant was powered by three waterwheels. Owner code: JJD

Coates Mill (est. 1820) at Russellville, Chester Co., PA, was a hand mill that began making straw wrapping in 1853 and straw board in 1860 on a cylinder-wire machine. Owner codes: JJE, JDC

Coatsville Mill (est. 1806) made board stock on a cylinder-wire machine at Coatsville, Chester Co., PA. Owner code: JDW

Cobbossee River Mill (est. 1811) at Gardiner, Kennebec Co., ME, was a hand mill that continued until about 1850. Owner code: KML

Cobbossee Mill, New (est. 1860) at Gardiner, Kennebec Co., ME, made manila paper on two cylinder-wire machines. Owner code: JUP

Codorus Mill (est. 1846) at Cordorus Creek, York Co., PA, made manila tissue, stationery, and ledger paper on a cylinder-wire machine. Owner codes: JIV, JWO, JUU

Cold River Mill (est. 1793) at Paper Mill Village, Cheshire Co., NH, was a hand mill until installing a cylinder-wire in 1834. Plant was rebuilt following a fire in 1868, but again destroyed by a fire in 1880. Owner codes: JYG, KJH

Collins Mill (est. 1872) at N. Wilbraham, Hampton Co., MA, made stationery paper on two moving-wire machines. The plant was powered by two steam engines and six waterwheels. Owner code: JJH

Collomer Mill (est. 1880) at Russellville, Chester Co., PA, made manila wrapping on a double-cylinder. The mill was powered by a turbine wheel. Owner code: KCX

Colt Mill (est. 1847) was at Pittsfield, Berkshire Co., MA. Rebuilt in 1862 and failed in 1878. This facility was later sold and renamed the Government Mill. Owner codes: KXE, JJI, JCS

Columbia Mill (est. 1830) at Chatham Center, Columbia Co., NY, made straw wrapping on two cylinder-wire machines. This plant was powered by two steam engines and four waterwheels. Owner codes: JSA, JSJ

Columbia Mill (est. 1834) at Rock Bridge, Boone Co., MO, made news paper on a cylinder-wire machine. Owner code: JJJ, JZH

Columbia Mill (est. 1840) at Lee, Berkshire Co., MA, made book and news paper on a moving-wire machine. Owner codes: KID, KOB

Columbia Mill (est. 1884) at Lacamas, Clark Co., WA, made manila wrapping and wood pulp news paper on a cylinder-wire and two moving-wires. The plant was powered by two waterwheels. These facilities were rebuilt in 1885 following a fire. Owner codes: JJJ, JJK, KWT

Columbus Mill (est. 1867) at Columbus, Franklin Co., OH, made news paper on a moving-wire machine. This plant was powered by two steam engines. Owner code: JBG

Combined Locks Paper and Pulp Mills (est. 1891) at Outagamie Co., WI, made wood-pulp paper on four moving-wire machines. This plant was powered by five waterwheels. Owner code: JJL

Commonwealth Mill (est. 1872) at Plainwell, Kalamazoo Co., MI, made book and news paper on a cylinder-wire machine. This plant was powered by two waterwheels. Owner code: KBH

Comstock Mills (est. 1870) at Comstock Bridge, New London Co., CT, made card stock on a cylinder-wire machine. The plant had a Jordan refiner and was powered by three waterwheels. Owner code: JFR

Congress Mill (est. 1798) at York, York Co., PA, was a hand mill that made stationery paper; it was later named the York Mill. Owner code: JYF

Congress Mill (est. 1852) at East Lee, Berkshire Co., MA, made collar paper and paste board on a moving-wire machine. This plant was powered by a steam engine and waterwheel. Mill was rebuilt following a flood in 1854. Owner codes: JJW, JBZ

Conneaut River Mill (est. 1865) at Conneaut, Ashtabula Co., OH, made manila wrapping on a cylinder-wire machine. This plant was powered by two steam engines and a waterwheel. Owner code: KOK

Contoocook Valley Mill (est. 1871) at West Henniker, Merrimack Co., NH, made book, card, and news paper on a moving-wire machine. The plant had a Jordan refiner and was powered by three waterwheels. Owner code: JJP

Cook Mill (est. 1883) at Bridgeport, Fairfield Co., CT, made manila hardware on a cylinder-wire machine powered by a waterwheel.

Cooperstown Mill (est. 1806) at Toddsville, Otsego Co., NY, was a hand mill that closed around 1860. This facility later became the Otsego County Paper Works. Owner code: KIN

Copesecook Mill (est. 1851) at Gardiner, Kennebec Co., ME, made book and news paper on a moving-wire machine. The plant had a Jordan refiner and was powered by seven waterwheels. It later became the Warren Mill. Owner codes: KUF, JRI

Corinth Mill (est. 1869) at Corinth, Saratoga Co., NY, made news paper on a moving-wire machine. This plant had a Jordan refiner, and was powered by nine waterwheels. Owner codes: JVF, JWB

Corralitos Mill (circa 1880) at Watsonville, Santa Cruz Co., CA, made straw wrapping and board on two cylinder-wire machines and a board machine. This plant was powered by two steam engines and a waterwheel. Owner codes: JFS, JJT

Coshocton Mill (est. 1865) at Coshocton, Coshocton Co., OH, made straw wrapping on a cylinder-wire machine. This plant was powered by two steam engines. Owner code: JJV

Council Bluffs Mill (circa 1880) at Council Bluffs, Pottawattomie Co., IA, made straw wrapping on a cylinder-wire machine. Owner code: KJY

Courier Mill (est. 1817) at Louisville, Jefferson Co., KY, was a hand mill that made news paper at this location until about 1840.

Cowles Mill (est. 1866) at Unionville, Hartford Co., CT, made manila paper on a cylinder-wire machine. This plant was powered by a steam engine and two waterwheels. Owner code: JJZ

Cox Mill (est. 1806) at Chicopee, Hampden Co., MA, was a hand mill until installing a cylinder-wire in the 1830s. Owner codes: JKC, JBB, KSG

Cox Mill (est. 1819) at Zanesville, Muskingum Co., OH, was a hand mill until it burned in 1836 and was retrofitted with a cylinder-wire. The mill made news, wrapping, printing, construction, and telegraph paper, and the plant was powered by three steam engines. Owner codes: JQV, JQU, JKE, JKF, KKV

Coy Mill (est. 1884) at West Claremont, Sullivan Co., NH, made manila tissue on a cylinder-wire machine. This plant was powered by two waterwheels. Owner code: JKH

Craig Mill (est. 1793) at Georgetown, Scott Co., KY, was a hand mill that burned down in 1836. Owner code: JKI, JKJ

Crane Bros. Mill (est. 1845) at Westfield, Chantanqua Co., MA, made ledger, and writing paper on a moving-wire machine and collar paper on a double-cylinder. Owner codes: JKN, KXE, JKM

Crane Mill was established at Ballston Spa, Saratoga Co., NY, about 1840. Owner code: JKO

Crehore Mill (est. 1845) at Needham, Middlesex Co., MA, made printing and press board on a cylinder-wire machine. This plant the combined the former Grant and Ware Mills. Owner code: JKS, JKT, JKU, JKV

Crescent Mill (est. 1848) at Lockland (later Crescentville), Hamilton Co., OH, made book and wrapping paper on two cylinder-wire machines. This plant was powered by a steam engine and two waterwheels. Owner code: JPM, JPP

Crescent Mill (est. 1850) at Livingston, Columbia Co., NY, made straw wrapping on a cylinder-wire machine. This plant was powered by a steam engine and three waterwheels. Owner codes: JEB, JLY

Crichton Mills (2) (circa 1880) at Chittenango, Madison Co., NY, made manila wrapping paper on two cylinder-wire machines. Owner code: JKX

Crocker Mill No. 1 (est. 1871) at Holyoke, Hampden Co., MA, made card, collar, and board stock on three moving-wire machines. This plant was powered by a steam engine and a water turbine. Owner code: JKY

Crocker Mill No. 2 (est. 1878) at Holyoke, Hampden Co., MA, made book and news paper on three moving-wire machines. The plant had a Jordan refiner, and was powered by three steam engines and five waterwheels. It was the former Albion Mill. Owner code: JKY

Crouse Mill (est. 1812) at Kinnikinnick Creek, Ross Co., OH, was a hand mill that closed around 1857. Owner codes: JWA, JLC

Crow Hollow Mill (est. 1826) at Lee, Berkshire Co., MA, made stationery and news paper on a cylinder-wire machine powered by a waterwheel. The facilities were rebuilt following a fire in 1833. Owner code: JZB

Crown Mill was established in 1890 at West Linn, Clackamas Co., OR. Owner code: JLD

Crum Creek Mill (est. 1779) at Springfield Township, Delaware Co., PA, was a hand mill that became a textile factory in late 1830s. Owner code: KAJ

Crystal Palace Mill (est. 1882) at Troy, Rensselaer Co., NY, made wood-pulp paper on a moving-wire machine and manila paper on a double cylinder. This plant was powered by four waterwheels. Owner code: KBT

Cumberland Mills (circa 1870) at Brunswick, Cumberland Co., ME, made book paper on six moving-wire machines. The plant had a Jordan refiner. Owner codes: KUF, JRI

Cumberland River Mill was established in 1838 at Nashville, Davidson Co., TN. Owner code: KWA

Cummings Mill (est. 1819) at Schoharie River, Schenectady Co., NY, was a hand mill that continued until around 1848. Owner code: JLG

Cummington Mill (est. 1867) at W. Cummington, Hampshire Co., MA, made linen and linen composite bond paper and ledger on a moving-wire machine. This plant was powered by a steam engine and a waterwheel. Owner code: JFU

Curtis Mills (3) (est. 1789) at Needham, Middlesex Co., MA, was a hand mill that expanded in 1794 and 1814. The site was rebuilt in 1824, and received two moving-wire machines in 1828. This mill made bond, ledger, and stationery. Site rebuilt again in 1834, and mill closed in 1860. Owner codes: JNS, JLJ, JLI

Cushman Mills (est. 1835) at N. Amherst, Hampshire Co., MA, made leather and straw board on two cylinder-wire machines. The mill was powered by two steam engines and six waterwheels. Owner code: JLL

Cuyahoga Mill (est. 1850) at Cuyahoga Falls, Summit Co., OH, made cigarette paper on a moving-wire machine. Owner code: JLN

Darby Creek Mill at Upper Darby Township, Delaware Co., PA, was a hand mill and was later renamed the Beehive Mill. Owner code: KRT

Darby Mill (est. 1883) at Plain City, Madison Co., OH, made straw wrapping paper on a cylinder-wire machine. Auxiliary equipment was powered by two steam engines. Owner code: KIU

Darby Mill (circa 1883) at Darby, Delaware Co., PA, made manila tissue on a cylinder-wire machine. Owner code: JPY

Davenport Mill (est. 1885) at Hamburg, Sussex Co., NJ, had a one cylinder-wire machine. Owner code: JLP

Davey Mill (circa 1880) at Jersey City, Hudson Co., NJ, made binder board on a single cylinder-wire machine. Auxiliary equipment was powered by two steam engines. Owner code: JLR

Davey's Board Mill (est. 1841) at Bloomfield, Essex Co., NJ, made binder board on a moving-wire machine. The plant was powered by two waterwheels and one steam engine. Owner code: JLQ

Davidson Mill (est. 1811) at Trenton, NJ, was a hand mill that operated continuously until 1825. Owner code: JWP

Day Mill (est. 1813) was in Burlington, Chittendon Co., VT. This a hand mill operated continually until 1833. Owner code: JLX

Dayton Mill (est. 1876) at Dayton, La Salle Co., IL, made straw wrapping paper on a cylinder-wire machine. The plant was powered by two waterwheels. Owner code: KWY

Dayton Mills (3) (est. 1846) at Dayton, Montgomery Co., OH, made book and news paper using the Jordan and Kingsland refiners. The plant operated three moving-wire machines and was powered by two waterwheels and two steam engines. Owner code: KDB

Decker Creek Mill (est. 1825) at Cincinnati, Hamilton Co., OH, was a rag mill that made paper on a single cylinder-wire machine powered by a waterwheel. Owner code: JRF

Defiance Mill (est. 1823) at Dalton, Berkshire Co., MA, began as a hand mill that transitioned to machine-made paper in 1840. The mill made ledger paper on a single cylinder-wire machine powered by a waterwheel. Auxiliary equipment was powered by two steam engines. The facility was rebuilt in 1852 following a fire. Owner codes: JHE, KIX ('44), JHU, KVJ, KVI

Delaware Mills (est. 1793) was at Rockland, New Castle Co., DE. This hand mill made book and bond paper. The plant was rebuilt in 1814 following a fire and closed in 1825. Owner codes: KYO, JWN, KLO

Delaware Mills (2) (est. 1885) at Delaware, Delaware Co., OH, was a wood pulp mill that made book, news, and wrapper paper on two cylinder-wire machines and one moving-wire machine. The plant was powered by four water turbines and three steam engines. Owner code: JQO

Delaware Water Gap Pulp & Mill (est. 1884) at Stroudsburg, Monroe Co., PA, made manila blotter paper on one cylinder-wire machine. Owner codes: JMC, JIA

Delphi Mill (est. 1859) at Delphi, Cass Co., IN, operated a double-cylinder machine. The plant burned down in 1860. Owner code: JDD

Delphos Mill (est. 1878) at Delphos, Van Wert Co., OH, made straw wrappers on two cylinder-wire machines powered by a waterwheel. Auxiliary equipment was powered by a steam engine. Owner code: JMD

DeMeza Mill (circa 1880) at Chittenango, Madison Co., NY, made manila wrappers on a cylinder-wire machine powered by two waterwheels. Owner codes: JEB, JLY

Derby Mills (est. 1872) at Birmingham, New Haven Co., CT, made hardware, card, and board stock, and colored mediums on a cylinder-wire and a moving wire machine. The plant had a Jordan refiner and was powered by nine waterwheels. Owner code: KWR

Deseret Mill (est. 1865) at Salt Lake City, Salt Lake Co., UT, made news paper on a single cylinder-wire machine. This was the former Sugar House Mill that was later removed to the Granite Mill. Owner code: JMH

Devitt Mill (est. 1839) at St. Charles, Kane Co., IL, began as a hand mill and introduced a cylinder-wire machine in 1842. The plant was powered by a single waterwheel. The mill made news and wrapper paper until it was abandoned in 1866. Owner codes: JMI, JGK, JWF

Dexter Mills (est. 1846) at Windsor Locks, Hartford Co., CT, used a Jordan refiner and made colored mediums on a moving wire machine. It also made glazed paper as well as tissue paper and bogus manila on two cylinder-wire machines. The plant was powered by six water turbines. Owner code: JMK

Dhacutt Mill (circa 1880) at Skaneateles Falls, Onondaga Co., NY, made newsprint on a Gavitt machine. The plant was powered by two waterwheels and one steam engine. Owner code: KUX

Diamond Mill (est. 1852) at Mechanic Falls, Androscoggin Co., ME, made book and news paper on a moving-wire machine. The plant was powered by three waterwheels and a steam engine. Owner code: JMG

Diamond Mill (circa 1880) at Shortsville, Ontario Co., NY, made news and wrapper paper on a single cylinder-wire machine. Owner code: KPG

Diamond Mills (est. 1872) at Hart Lot, Onondaga Co., NY, made book and news paper on a single moving-wire machine, and was powered by two water turbines. Owner code: JML

Diamond Mills (circa 1880) at Chatham, Columbia Co., NY, made straw wrappers on a single cylinder wire machine, and was powered by two water wheels. Owner code: KEQ

Dickinson & Clark Mill (est. 1869) at Holyoke, Hampden Co., MA, made book and stationery paper using a Jordan refiner. The plant featured two moving-wire machines, and was powered by one water turbine and one steam engine. Later it was Dickinson Mill. Owner code: JMP

Dickinson Mill (est. 1883) at Holyoke, Hampden Co., MA, was the former Dickinson & Clark Mill, and it made stationery and book paper using a Jordan refiner. The plant had three moving-wire machines and was powered by one water turbine and one steam engine. Owner code: JMQ

Dodge Mill (circa 1880) at Constantine, St. Joseph Co., MI, made straw board and carpet lining. Owner code: JMT

Dominion Mill, Old (circa 1880) at Richmond, Henrico Co., VT, made blotter paper on a cylinder-wire machine. The plant was powered by two steam engines. Owner code: KTB

Dorchester Mill (est. 1813) was in Dorchester, Suffolk Co., MA. This hand mill operated continuously until 1830. Owner code: JKB

Dorlans Mill (est. 1834) was at Dorlans Mill, Chester Co., PA. This mill made wall paper on a single cylinder-wire machine. The plant was powered by two waterwheels and one steam engine. Owner code: JMW

Douglass Mill (circa 1880) at Pennsburgh, Montgomery Co., PA, made wrapper paper on a single cylinder-wire machine. Owner code: JZV

Dove Mill (est. 1794) at Lower Merion Township, Montgomery Co., PA, was the former Schutz Mill, and it operated continuously until 1829.

Dove Mill (est. 1840) at Modena, Chester Co., PA, made book, card, envelope, and print paper from rag stock until converting to wood pulp in 1870. The plant had a Brightman engine and a single moving wire machine powered by a water wheel. Auxiliary equipment was powered by a steam engine. Owner code: KDF

Dove Mills (est. 1750) at Lower Merion Township, Montgomery Co., PA, consisted of two plants that made hand-crafted book, news, and stationery paper. The mill was rebuilt in 1830 following a fire and continued in operation through the 1840s. Owner codes: JBD, JBF, KMP

Dover Mill (circa 1880) at Charles River Village, Norfolk Co., MA, made building paper on a single cylinder-wire machine and was powered by two waterwheels. Owner code: JTZ

Dubuque Mill (est. 1883) at Dubuque, Dubuque Co., IA, made straw wrapper on a cylinder-wire machine. The plant was powered by two steam engines. Owner code: JNA

Duckett Mill (est. 1798) at Aston, Delaware Co., PA, was a hand mill that operated continuously until 1830 (a.k.a. Sweedesberrg Mill on Chester Creek). Owner codes: KAZ, JNB

Dundee Mill (circa 1880) at Dundee, Monroe Co., MI, made straw wrappers on a cylinder-wire machine. Owner code: JWR

Eagle Mill was established in 1765 at Milton, Norfolk Co., MA. This hand mill was one of the original Neponset mills, and it made book, news, and print paper. Operations continued in 1840 with the installation of two moving-wire machines that produced colored mediums. Owner codes: KQB, KRF

Eagle Mill, Old (est. 1789), at Suffield, Hartford Co., CT, was a hand mill that made book paper. In 1833 operations continued with a cylinder-wire machine powered by a waterwheel. The plant was destroyed by fire in 1877. Owner codes: JGM, JBB, JGL, KSI, KHJ, JZK

Eagle Mill (est. 1833) at Downington, Chester Co., PA, made wrapper paper on a double-cylinder machine. The plant was powered by two waterwheels and two steam engines. Owner code: JRS

Eagle Mill (est. 1839) at Chatham, Columbia Co., NY, made straw wrappers on a single cylinder-wire machine. The plant was rebuilt in 1840 following a fire. Owner code: JAG

Eagle Mill (est. 1840) at Lee, Berkshire Co., MA, was the former Enterprise Mill, and it made manila book and news paper on a cylinder-wire and moving-wire machine. Owner codes: KIX, KOB

Eagle Mill (est. 1852) at Mechanic Falls, Androscoggin Co., ME, made book and news paper on a moving-wire machine. The plant was powered by three waterwheels and a steam engine. Owner code: JMG

Eagle Mill (est. 1859) at West Carrollton, Montgomery Co., OH, made manila paper on a double-cylinder machine. Owner code: JPQ

Eagle Mill (est. 1863) at Ballston Spa, Saratoga Co., NY, had a moving-wire machine. The plant was powered by three waterwheels and two steam engines. Owner code: KVG

Eagle Mill (est. 1872) at Kaukauna, Outagamie Co., WI, made straw paper on a cylinder-wire machine, transitioning to manila paper in 1878. The plant was powered by three waterwheels. The facility burned down in 1881. Owner code: KPT.2

Eagle Mill (circa 1880) at Marcellus Falls, Onondaga Co., NY, made straw wrapping paper on two cylinder-wire machines. Owner code: JZM

Eagle Mill (est. 1880) at Putney, Windham Co., VT, made manila tissue paper on a cylinder-wire machine. The plant was powered by three waterwheels and a steam engine. Owner code: JJG

Eagle Mill (circa 1880) at Burnside, Hartford Co., CT, made book and stationery paper on a moving wire machine. The plant was powered by two waterwheels and one steam engine. Owner code: KTQ

Eagle Mill (est. 1880) at Dansville, Livingston Co., NY, was a manila mill that made hardware paper and wrappers on a single cylinder-wire machine, and was powered by two waterwheels. Owner codes: JYQ, JYP

Eagle Mill was established in 1882 at Wellsville,

Litchfield Co., CT. This manila mill made hardware paper on a cylinder-wire machine. The plant was powered by a waterwheel and a steam engine. Owner code: KBL

Eagle Mill (est. 1883) at Franklin, Warren Co., OH, made book and news paper on two moving-wire machines. Owner code: JNG

Eagle Mills (2) (est. 1840) at Bentley Springs, Baltimore Co., MD, consisted of two facilities that made straw wrapping paper on cylinder-wire machines. The machines were powered by waterwheels and auxiliary equipment by two steam engines. One mill facility burned in 1851; the other closed in 1854. Owner codes: JND, JUH

East Chatham Mill (est. 1838) at East Chatham, Columbia Co., NY, made straw wrapping paper on a cylinder-wire machine. The plant was powered by two waterwheels and one steam engine. Owner codes: KRK, JVB

East Hartford Mill (est. 1779) at Burnside, Hartford Co., CT, was a hand mill and operated until 1837 before being sold and renamed the Burnside Mill. Owner code: JVE

East Walpole Mill (circa 1880) at East Walpole, Norfolk Co., MA, made manila paper on a double-cylinder machine. Owner code: JUO

Easthampton Mill (circa 1880) at Loudville, Hampshire Co., MA, made manila tissue paper on a cylinder-wire machine. The plant was powered by two waterwheels and two steam engines. Owner code: KFS

Eau Claire Pulp & Mill (est. 1882) at Eau Claire, Eau Claire Co., WI, made newsprint on a moving-wire machine. The plant was powered by three water-wheels. Owner code: JNK

Eclipse Straw Board Mill at Wooster, Wayne Co., OH, made straw board on a cylinder-wire machine. The plant was powered by two steam engines. Owner code: JVR

Eden Mill (circa 1880) at Whippany, Morris Co., NJ, made manila tissue on two moving-wire machines. Owner code: JML

Elk Grove Mill (est. 1863) at Hickory Hill, Chester Co., PA, made straw board and manila wrappers on a cylinder-wire machine powered by a waterwheel. The plant also used refuse cane from an attached sorghum mill. Owner code: KCT

Elk Horn Mill (est. 1860) was in Elk Horn, Allegheny Co., PA. This manila mill made wrapping paper on two cylinder-wire machines. Auxiliary equipment was powered by a steam engine. Rebuilt and converted to wood-pulp in 1884. Owner code: JQX

Elkhart Combination Board Mill (est. 1870) at Elkhart, Elkhart Co., IN, made board and card stock on a cylinder-wire machine. The plant was powered by three waterwheels. Owner codes: KOL, JNO

Elkhart Mill (est. 1876) at Elkhart, Elkhart Co., IN, made newsprint on a moving-wire machine. The plant was powered by five waterwheels. Owner code: JNP

Elkhart Tissue Mill (est. 1876) at Elkhart, Elkhart Co., IN, made tissue paper on a cylinder-wire machine. The plant was powered by three waterwheels. Owner code: JNQ

Elkridge Mill (est. 1818) was at Elkridge, Howard Co., MD. This hand mill continued in operation until 1843. Owner code: JMJ

Elliot Mill (est. 1884) was in City Mills, Norfolk Co., MA. This manila mill made wrapping paper on a cylinder-wire machine. Owner code: KTR

Ellis Mill (est. 1832) at Norwood, Norfolk Co., MA, was a manila mill that made carpet linings on a double-cylinder machine. The plant was powered by two waterwheels and two steam engines. Owner code: JNT

Elmwood Mill (est. 1867) at Elmwood, Peoria Co., IL, made straw wrappers on a cylinder-wire machine. The plant was powered by four steam engines. Owner code: KRO

Ellsworth Mills (est. 1883) at Ellsworth, Vigo Co., IN, made straw board and wrappers on two double-cylinder machines. The plant was powered by five steam engines. Owner code: KQT.3

Empire Mill (est. 1850) at Cuyahoga Falls, Summit Co., OH, made bogus manila on a cylinder-wire machine. Owner code: JLN

Empire Mill (est. 1863) at Ballston Spa, Saratoga Co., NY, produced colored mediums on a moving-wire machine. The plant was powered by three waterwheels and two steam engines. Owner code: KVG

Empire Mill (est. 1868) at Chatham, Columbia Co., NY, made bogus manila and straw wrapper on two cylinder-wire machines. The plant was powered by three waterwheels and two steam engines. Owner code: KDJ

Empire Mills (2) (circa 1880) at Schaghticoke,

Rensselaer Co., NY, made straw wrappers on two cylinder-wire machines. The plants were powered by six waterwheels. Owner code: KWO

Empire State Mill (est. 1867) at Richmondville, Schoharie Co., NY, made straw board on a cylinder-wire machine. The plant was powered by a waterwheel and a steam engine. Owner code: KVL

Enterprise Mill (est. 1808) at Lee, Berkshire Co., MA, was a hand mill, sold in 1830 and renamed the Eagle Mill. Owner code: JII

Enterprise Mill (est. 1873) at Honeoye Falls, Monroe Co., NY, made straw board on a cylinder-wire machine powered by two water turbines. Owner code: KFY

Enterprise Mill (est. 1880) at Rockton, Winnebago Co., IN, made printing paper using a Jordan refiner on a single cylinder-wire machine. The plant was powered by three waterwheels. Owner codes: KJU, JXU

Enterprise Straw Board Mill (est. 1882) at Wilmington, Will Co., IL, made straw board on a cylinder-wire machine. The plant was powered by three waterwheels and two steam engines. Owner code: JXL

Erie Mill (est. 1881) at Erie, Erie Co., PA, made manila building and hardware paper on three cylinder-wire machines. The plant was powered by five steam engines. Owner code: KUN

Essex Mill (circa 1880) at Bloomfield, Essex Co., NJ, made book paper using the Kingsland refiner on two moving-wire machines and a Gavitt machine. Owner code: JNZ

Essex Mill (circa 1880) at Essex Junction, Chittenden Co., VT, made wall paper on a cylinder-wire machine. The plant was powered by three waterwheels. Owner code: JVJ

Essex Mills (est. 1880) at Lawrence, Essex Co., MA, made leather board on two cylinder-wire machines powered by a single waterwheel. Owner code: JLW

Eureka Mill (est. 1850) at Livingston, Columbia Co., NY, made straw wrappers on a cylinder-wire machine. The plant was powered by three waterwheels. Owner codes: JEB, JLY

Eureka Mill (est. 1882) at Bridgeport, Montgomery Co., PA, made manila-colored mediums on a moving-wire machine. The plant was powered by two steam engines. The facilities were rebuilt in 1884 following a fire. Owner code: JLO

Eurica Mill (est. 1860) at Stockport, Columbia Co., NY, made straw wrappers on a single cylinder-wire machine powered by a waterwheel. Owner code: KSQ

Evansville Mill (est. 1858) at Evansville, Vanderborgh Co., IN, made straw wrappers on a double-cylinder machine. The plant was powered by three steam engines. Owner code: JPS

Excello Mill (est. 1872) at Franklin, Warren Co., OH, made stationery paper on a moving-wire machine. Owner code: JSO

Excelsior Mill (est. 1863) at Ballston Spa, Saratoga Co., NY, had a moving-wire machine. The plant was powered by three waterwheels and two steam engines. Plant was converted to wood-pulp around 1878. Facilities were rebuilt following a fire in 1882 and renamed the Remington Mill. Owner code: KVG

Excelsior Mill (est. 1867) at Southington, Hartford Co., CT, made straw board on a single wet machine powered by a waterwheel. Owner code: JDM

Excelsior Mill (est. 1833) at White Hall, Baltimore Co., MD, made straw wrappers on a single cylinder-wire machine. The plant was powered by one waterwheel and two steam engines. Owner codes: JGI, JYW

Excelsior Mill (est. 1873) at Holyoke, Hampden Co., MA, made book and printing paper with the Jordan refiner on a single moving-wire machine. The plant was powered by three water turbines and one steam engine. Owner code: JOA

Excelsior Mill (circa 1880) at Lockland, Hamilton Co., OH, made carpet lining on a cylinder-wire machine. Owner code: JPQ

Eyster Mill (est. 1869) at Halltown, Jefferson Co., WV, made straw board on a cylinder-wire machine. The plant was powered by two steam engines. Owner code: JOB

Fair Grove Mill (est. 1855) at Hamilton, Butler Co., OH, made book and news paper using the Kingsland refiner on one moving-wire machine. The plant was powered by waterwheels and steam engines. Owner code: KOP

Fair Haven Mill was established in 1794 at Fair Haven, VT. This hand mill burned in 1806 and was rebuilt. Owner code: KBI

Fairchild Mill (est. 1827) at Bridgeport, Fairfield Co., CT, made newsprint using the Jordan refiner on a moving-wire machine powered by two waterwheels Owner code: JOE

Fairfield Mill (est. 1848) at Montoursville, Lycoming Co., PA, made manila wrappers on a cylinder-wire machine. The plant was powered by a steam engine and four waterwheels. The plant was rebuilt in 1882. Owner code: KPJ

Fall Mountain Mills (7) (est. 1875) at Bellows Falls, Windham Co., VT, was a pulp and paper mill that made manila card on a cylinder-wire machine and newsprint on three moving-wire machines. The plant was powered by twenty-nine waterwheels. Owner code: JOF

Falls City Mill (est. 1863) at Louisville, Jefferson Co., KY, made book paper on two moving-wire machines. Auxiliary equipment was powered by two four steam engines. Owner code: JFK

Falls Creek Mill (est. 1838) at Ithaca, Tompkins Co., NY, made rag paper on a cylinder-wire machine powered by a waterwheel. The plant was destroyed by fire in 1842 and not rebuilt. Owner code: KBO

Falls Mill (est. 1829) at Upper Oxford Township, Chester Co., PA, was a hand mill that converted to manila wrapping in 1850 with the installation of a cylinder-wire machine powered by a waterwheel. A water turbine was installed in 1870. The plant burned down in 1872 and reopened as the Collomer Mill. Owner codes: JOG, JDR

Falls Mills (est. 1853) at Scotch Plains, Union Co., NJ, made card stock on a cylinder-wire machine. The plant was powered by two steam engines and two waterwheels. Owner code: KMW

Fallsburgh Mill (est. 1881) at Fallsburgh, Sullivan Co., NY, made straw wrapping paper on a cylinder-wire machine. The plant was powered by two waterwheels. Owner code: JZS

Falmouth Mill (est. 1844) was at Jay, ME. Owner codes: KJZ, KGQ

Fandango Mill (est. 1864) at Millburn, Essex Co., NJ, made binder board on two cylinder-wire machines. The plant was powered by two steam engines and a waterwheel. Owner codes: JWE, JTS

Fargo Mill (est. 1883) at Fargo, Cass Co., ND, made collar paper on a double-cylinder machine. The plant was powered by two steam engines. Owner code: JOH

Farley Mill (est. 1881) at Wendell Depot, Franklin Co., MA, made card stock and wrapping paper on a cylinder-wire machine. The plant was powered by four waterwheels. Owner code: JOI

Fayetteville Mill (circa 1880) at Fayetteville, Onondaga Co., NY, made manila filter paper on a cylinder-wire machine. Owner codes: JCX, JCW

Featherman Mill (est. 1816) was at West Nottingham, Chester Co., PA. This handmade mill was powered by a waterwheel and continued in business until 1832. Owner code: JOL

Fern Rock Mill (est. 1870) at Atglen, Chester Co., PA, made straw board on a cylinder-wire machine powered by a waterwheel. This was the former Sadsbury Mill, and later Lyndonette Mill. Owner code: KYN

Fibre Mill (est. 1863) at Ballston Spa, Saratoga Co., NY, made manila paper on a double-cylinder machine. The plant was powered by two steam engines and three waterwheels. Owner code: KVG

Fitchburg Mills (est. 1864) at W. Fitchburg, Worcester Co., MA, made bogus manila, card, and wall paper with the Jordan refiner on two moving-wire machines. The plant was powered by three water turbines and four steam engines. The facility was converted to wood pulp in 1867. Owner code: JOR

Five Miles Mill (est. 1775) at Manchester, Hartford Co., CT, was a hand mill that burned in 1778 and was rebuilt. It operated continuously until 1833. Owner codes: JQZ, JEC, JLZ

Flat Rock Mills (est. 1844) was at Manayank, Philadelphia Co., PA. This manila mill made newsprint using the Jordan refiner on two moving-wire machines. The plant was powered by five steam engines. Owner code: KGB

Flax Leather Board Mill (est. 1850) at South Natick, Middlesex Co., MA, made leather board on two wet machines. The plant was powered by three waterwheels. Owner code: JHB

Fond du Lac Mills (2) (est. 1876) at Fond du Lac, WI, made straw board.

Fonda Mill (circa 1880) at Fonda, Montgomery Co., NY. This manila mill made paper on a cylinder-wire machine. Owner codes: KRA, JWT

Fordham Mill (est. 1855) at Hamilton, Butler Co., OH, made straw board on a moving-wire machine and employed a Kingsland refiner. The plant was powered by two waterwheels and three steam engines. Owner code: KOP

Forest Mill, Old (est. 1819) at East Lee, Berkshire Co., MA, was a handmade mill that produced book and stationery paper. The plant was powered by a waterwheel. Owner codes: JII, JPU, KWI, JQM. It began making paper with a moving-wire machine in 1846 and thereafter was called the Upper Forest Mill. Owner codes: JDN, JBZ, JVU

Forest Mill (est. 1835) at East Lee, Berkshire Co., MA, made stationery paper on a cylinder-wire machine powered by a waterwheel. Auxiliary equipment was powered by two steam engines. Owner codes: JDN, JPU ('67)

Forest Mill (est. 1867) at Amsterdam, Montgomery Co., NY, made wall paper on a moving-wire machine. The plant was powered by two steam engines. Owner code: KPS

Forest Mill (est. 1873) at Cheney, Delaware Co., PA, made construction paper on a cylinder-wire machine. The plant was powered by a steam engine and a waterwheel. Owner code: JLE

Forest Mill (est. 1883) at S. Manchester, Hartford Co., CT, made board stock. The plant was powered by a steam engine. Owner code: JHJ

Forest Street Mill (est. 1847) was in Cleveland, Cuyahoga Co., OH. This manila mill made paper on two moving-wire machines. Auxiliary equipment was powered by a steam engine. Owner code: JJA

Fort Edward Mill (est. 1853) at Fort Edward, Washington Co., NY, made newsprint using the Jordan refiner on two moving-wire machines. Owner code: KRG

Fort Madison Mill (est. 1882) at Fort Madison, Lee Co., IA, made straw wrapper on a cylinder-wire machine. The plant was powered by two steam engines. Owner code: JOV

Fort Miller Mill (est. 1867) at Fort Miller, Washington Co., NY, made wall paper on a cylinder-wire machine. The plant was powered by six waterwheels. Owner code: KTI

Fort Ned Mills (circa 1870) at Norwich, New London Co., CT, made book and news paper using a Jordan refiner on a cylinder-wire and a moving-wire machine. Owner code: KKB

Fort Orange Mill (est. 1876) at Castleton, Rensselaer Co., NY, made book paper, binder board, postal cards, and bogus manila on two moving-wire machines. The plant was powered by two steam engines and a waterwheel. Owner code: JOW

Fort Wayne Mill (est. 1883) at Ft. Wayne, Allen Co., IN, made straw wrapping on a cylinder-wire machine. The mill was powered by a steam engine and a waterwheel. Owner code: KRR

Foster Mill (est. 1845) at Newton Lower Falls, Middlesex Co., MA, made board stock on a wet machine. This is the former Island Mills and later Wiswall Mill. Owner codes: JOX, KKN

Fox Mill (circa 1880) at Sauquiot, Oneida Co., NY, made newsprint on two cylinder-wire machines. Owner codes: JPC, JRG

Fox River Mill (est. 1882) at Yorkville, Kendall Co., IL, made straw wrapping on a cylinder-wire machine. The plant was powered by three waterwheels. Owner code: JPB

Fox River Straw Board Mill (est. 1882) at Appleton, Outagamie Co., WI, made straw wrapping and board on a board machine. The plant was powered by two waterwheels. Owner codes: JBN, JDZ

Franklin Mill (est. 1803) at Suffield, Hartford Co., CT, was a hand mill that made stationery and wrapping paper. It closed in 1848 and later was Suffield Mill. Owner code: JEH

Franklin Mill (est. 1810) was in Gwynns Falls, Baltimore Co., PA. This hand mill in Dickeysville closed in 1845. Owner code: JZX

Franklin Mill (est. 1821) at Elkhorn, Taylor Co., KY, was a hand mill that added a cylinder-wire machine in 1830. Owner codes: JXS, KPK

Franklin Mill (est. 1826) at Nether Providence Township, Delaware Co., PA, made book and news paper using a Jordan refiner on a moving-wire machine. The plant was powered by a waterwheel and two steam engines. The facility was rebuilt following a fire in 1883. Owner codes: KGZ, KAF, KAI

Franklin Mill (est. 1825) at Chesterville, Chester Co., PA, made book and news paper on a cylinder-wire machine. The plant was powered by two steam engines and a waterwheel. Owner code: KBN

Franklin Mill (est. 1845) at Richmond, Henrico Co., VA, made wrapping, stationery, news, printing, and envelope paper on two moving-wire machines. The plant on 8th St. was powered by a waterwheel. Owner codes: JPF, JDK

Franklin Mill (est. 1848) was at Newton Upper Falls, Middlesex Co., MA. This mill made wrapping and wall paper on a cylinder-wire machine. It was rebuilt as White Mill following a fire in 1862. Owner code: JPH

Franklin Mill (est. 1855) at Hamilton, Butler Co., OH, made newsprint using a Kingsland refiner on a moving-wire machine. The plant was powered by two steam engines and a waterwheel. Owner code: KOP

Franklin Mill (est. 1865) at Franklin, Warren Co., OH, made stationery on a moving-wire machine. The facilities were incorporated into the Franklin Mills. Owner code: JSO

Franklin Mill (est. 1866) at Holyoke, Hampden Co., MA, made stationery, envelope, and collar paper using a Jordan refiner on a single moving-wire machine. The plant was powered by a water turbine. Owner code: JPG

Franklin Mill (est. 1872) at Northeast, Erie Co., PA, was a manila mill that made newsprint on a cylinder-wire machine. The plant was powered by two steam engines and a waterwheel. Owner code: JPJ

Franklin Mills (2) (est. 1873) at Franklin, Warren Co., OH, made book and news paper using a Jordan refiner on two moving-wire machines. The plant was powered by three steam engines and a waterwheel. Owner code: JPI

Friend & Forgy Mill was established circa 1880 at Franklin, Warren Co., OH. This manila mill had a double-cylinder machine. Owner code: JPL

Friend & Fox Mill (est. 1848) at Lockland, Hamilton Co., OH, made book and wrapping paper on a moving-wire machine. The plant was powered by a steam engine and two waterwheels. Former Crescent Mill. Owner code: JPM

Friend & Tangeman Mill (est. 1851) at Lockland, Hamilton Co., OH, was rebuilt following fires in 1841 and 1862. Owner codes: JPO, KQN

Fuller Mill (circa 1880) at Milton, Norfolk Co., MA, made book, news, printing, and bogus manila on a moving-wire machine. Owner code: KRF

Fulton Mill (est. 1809) at Oxford, Chester Co., PA, was a handmade mill powered by a waterwheel. Reopened as the Coates Mill in 1820. Owner code: JWQ, JPQ

Fulton Mill (circa 1880) at Wheeling, Ohio Co., WV, made news and filter paper on a moving-wire machine. Owner code: JSL

Fulton Mill (est. 1883) at Atlanta, Fulton Co., GA, made manila paper on a cylinder-wire machine. The mill was powered by a steam engine and two waterwheels. Owner code: KMT

Gardiner Mills (est. 1884) at Gardiner, Kennebec Co., ME, made book and news paper on a moving-wire machine. Owner code: KKQ

Garlock Mill (est. 1880) at Chittenango, Madison Co., NY, made straw board on a cylinder-wire machine powered by two waterwheels. Owner code: JPV

Garrett Mill (est. 1789) at Red Clay Creek, New Castle Co., DE, was a hand mill that closed in 1812. Owner code: JQA

Gem City Mill (circa 1880) at Quincy, Adams Co., IL, made straw wrappers on two cylinder-wire machines. Owner code: KJQ

Genesee County Mill (est. 1868) at Leroy, Genesee Co., NY, made straw and rag wrapping paper on a cylinder-wire machine. The plant was powered by a water turbine and a waterwheel. Owner code: JWY

Genesee Mill (est. 1864) at Rochester, Monroe Co., NY, made news paper on two cylinder-wire machines and two moving-wire machines. The plant was powered by eleven waterwheels. Owner code: KLM

Genesee Mill (est. 1866) at South Avon, Livingston Co., NY, made straw board on a cylinder-wire machine powered by two waterwheels. Owner code: JST

George Mill (circa 1880) at Wellsburgh, Brooke Co., WV. This manila mill made paper on a cylinder-wire machine. Owner code: JQE

Gevel Mill (est. 1781) at Downington, Chester Co., PA. This hand mill closed in 1840. Owner code: JQF

Ghent Mill (est. 1874) at Ghent, Columbia Co., NY, made straw wrapping paper on a cylinder-wire machine. The plant was powered by a steam engine and three waterwheels. Owner code: JQG

Gillespie Mill at East Walpole, Norfolk Co., MA. This mill closed in 1800. Owner codes: JEE, KRF

Glen Mill (est. 1835) at Glen Mills, Delaware Co., PA, made linen and linen composite bond, book, ledger, and wrapping paper using a Jordan refiner on three moving-wire machines. The mill was powered by three steam engines, two water turbines, and a waterwheel.

The plant converted to manila paper in 1885. Owner codes: KWL, KWU, JQQ

Glen Mill (est. 1871) at Westfield, Hampden Co., MA, was a manila mill that made wrapping paper on a cylinder-wire machine. The plant was powered by a steam engine and three waterwheels. Owner code: KJK

Glen Mill (est. 1883) at Berlin, NH, made wood pulp paper. Owner code: JQP

Glen Union Mill was established in 1863 at Ballston Spa, Saratoga Co., NY. This manila mill made paper on a double-cylinder machine. The plant was powered by two steam engines and three waterwheels. Owner code: KVG

Glendale Mill (circa 1880) at Atlanta, Fulton Co., GA, made manila paper on a cylinder-wire machine. Owner code: JQR

Glendale Mill (est. 1835) at Glendale, Berkshire Co., MA, made manila paper on a cylinder-wire machine and newsprint on a moving-wire machine. The plant was powered by four waterwheels. Owner code: JHS

Glengarry Mill (est. 1882) at Lee, Berkshire Co., MA, made book, blotter, and bogus manila and used a Jordan refiner. The plant was powered by two steam engines and a water turbine. Owner code: JQS

Glenmount Mill (est. 1852) at Freeland, Baltimore Co., MD, made straw wrapper on a cylinder-wire machine powered by two waterwheels. Owner code: KNA

Glens Falls Mill at Glens Falls, Warren Co., NY, made newsprint on two moving-wire machines. The plant was powered by two steam engines and two waterwheels. The facilities were rebuilt in 1884 following a fire. Owner code: JQT

Glens Falls Pulp Co. (circa 1880) at Glens Falls, Warren Co., NY, made wood pulp. Owner code: JQT

Glenwood Mills (est. 1778) at Upper Darby Township, Delaware Co., PA, was a hand mill leased over 1828–1838. It continued as a machine mill (?) until it closed in 1862. Owner codes: KAA, KAB, KAC, JBF, JAZ

Globe Mill (circa 1827) at Middletown, Butler Co., OH, was a manila mill that made wrapping paper on a cylinder-wire machine. Owner code: JCL

Globe Mill (est. 1872) at Neenah, Winnebago Co., WI, made book and printing paper using a Jordan refiner on a moving-wire machine. Owner code: JYC

Globe Mills (est. 1873) at Rockford, Winnebago Co., IL, made straw board on a cylinder-wire machine. The plant was powered by five waterwheels. Owner code: JXJ

Globe Tissue Mill (est. 1883) at Elkhart, Elkhart Co., IN, made manila tissue on a cylinder-wire machine powered by two waterwheels. Owner code: JQW

Gold Leaf Mill (est. 1853) at Troy, Rensselaer Co., NY, made paper on a cylinder-wire machine. Owner code: KNY

Golden Mill (est. 1867) at Golden, Jefferson Co., CO made bogus manila, printing, hardware, and wrapping paper on a cylinder-wire machine powered by a waterwheel. The facilities were powered by a steam engine. Owner code: KUY

Gorham Mill (est. 1881) at North Gorham, ME, made wood pulp paper.

Government Mill (est. 1879) at Pittsfield, Berkshire Co., MA, made linen and linen composite printing and bond paper on a moving-wire machine. This was the former Colt Mill. Owner code: JSS

Grafton Mill (circa 1870) at Bristol, Grafton Co., NH, made manila on a cylinder-wire machine. It installed a moving-wire machine for production of wood-pulp paper around 1884. The plant was powered by a steam engine and three waterwheels. Owner codes: JAM, KFJ

Graham Mill (est. 1835) at Hamilton, Butler Co., OH, made stationery, binder board, and wrapping paper on a cylinder-wire machine powered by a waterwheel. The plant began making straw wrapping in 1857. The plant flooded in 1866 and was abandoned after 1868. Owner codes: JRF, JAW

Graham Mill (est. 1880) at South Rock Is., Winnebago Co., IL, made straw wrapping on a cylinder-wire machine. The plant was powered by four waterwheels and a steam engine. Owner code: JRE

Granite Mill (circa 1880) at Vernon Depot, Tolland Co., CT, made board stock. Owner code: KQL

Granite Mill (est. 1883) at Butlerville, Salt Lake Co., UT, made book and news paper on a moving-wire machine, and manila or straw wrapping on a cylinder-wire. The plant was powered by three waterwheels. Owner code: JMH

Grant Mill (est. 1809) at Needham, Middlesex Co., MA. This hand mill made card stock. It continued until 1836 and was incorporated into the Crehore Mill. Owner code: JRH

Grave Run Mill was established circa 1880 at Grave Run Mills, Baltimore Co., MD. Owner code: JSQ

Great Bend Paper Co. (est. 1868) at Great Bend, Jefferson Co., NY, made wall paper on a cylinder-wire machine. The plant was powered by three water turbines. Owner code: JRL

Great Spring Mill (est. 1800), Solebury Township, Berks Co., PA, was a hand mill that operated up to 1838. Owner codes: JVZ, JWA, JXR ('40)

Great Spring Mill (circa 1870) at Lambertville, Hunterdon Co., NJ, made manila news on a cylinder-wire machine. Owner code: JIY

Great Western Mill (est. 1809) was at Schoharie Bridge, Schenectady Co., NY. This hand mill operated up to 1839. Owner code: KXT

Green Mill (est. 1845) was at Lancaster, Hamilton Co., OH. Owner code: JRM

Greenville Mills (2) (est. 1846) at Greenville, Greenville Co., SC, made news paper on two cylinder-wire machines.

Greenwater Mill (est. 1843) at East Lee, Berkshire Co., MA, made rag paper on a cylinder-wire machine. Owner codes: KPX, JTO, KBE, JBZ

Greenwich Mill (est. 1863) at Greenwich, Washington Co., NY, made manila wall paper on a cylinder-wire machine powered by two waterwheels. Owner code: JBJ

Greenwood Mill (est. 1867) at East Lee, Berkshire Co., MA, made stationery paper on a moving-wire machine. The plant was powered by two steam engines and a waterwheel. This was the former Mountain Mill. Owner codes: JDO, KQQ

Gregory Bros. (est. 1810) at Green River Village, VT, was a hand mill that made printing and wrapping paper until 1854. Owner code: JRR

Greylock Mill (est. 1850) at Adams, Berkshire Co., MA, made ledger, stationery, and folio paper on three moving-wire machines. The plant was powered by three steam engines and three waterwheels. Owner code: JFU

Groton Center Mill (est. 1801) was at Groton Center, Middlesex Co., MA. This hand mill operated until 1840 when a moving-wire machine was installed. The plant made book and news paper using a Jordan refiner. Equipment was powered by two waterwheels and a steam engine. Owner code: KRF

Grove Mill (est. 1813) was at Lebanon, Warren Co., OH. This hand mill closed in 1832. Owner code: JNE

Grove Mill (circa 1880) at Newburgh, Orange Co., NY, made book paper using the Jordan refiner on a moving-wire machine. Owner code: JAD

Gunpowder Mill, Old (est. 1780) at Shamburg, Baltimore Co., MD, operated as a hand mill until 1830, when it added a cylinder-wire machine. Plant was closed in 1838. Owner code: JET

Gunpowder Mill (est. 1841) at Hoffmanville, Baltimore Co., MD, made book and news paper, and after 1833 manila or straw wrapping on a cylinder-wire machine. The plant was powered by a steam engine and two waterwheels. This mill closed sometime after 1864. Owner code: JUH

Haddam Neck Mill (est. 1883) at Haddam, Middlesex Co., CT, made manila wrapping paper on a cylinder-wire machine powered by two waterwheels. Owner code: JUW

Hagey Mills (2) (est. 1755) at Lower Merion, Montgomery Co., PA, was a hand-crafted mill until around 1764. It began making manila on a double-cylinder machine during the 1840s. Owner code: JRT

Haldeman Mills (2) (est. 1854) at Lockland, Hamilton Co., OH, made manila wrappers and carpet lining on a cylinder-wire machine and a double-cylinder machine. The plant was powered by four steam engines and two waterwheels. Owner code: JRU

Hallett Mill (est. 1868) at Riverhead, Suffolk Co., NY, made straw board on a cylinder-wire machine powered by a waterwheel. Owner code: JRY

Hamilton Mill (est. 1848) at Hamilton, Butler Co., OH, made stationery paper on a cylinder-wire machine brought from the Phoenix Mill in Ohio. Auxiliary equipment was powered by a single steam engine. It later became the Hamilton Tissue Mill. Owner code: JNW

Hamilton Straw Lumber Mill (circa 1880) at Lawrence, Douglas Co., KS, made straw hardware. Owner code: JSB

Hamilton Tissue Mill was established in 1878 at Hamilton, Butler Co., OH. This manila mill made tissue paper and bogus manila on a cylinder-wire machine. The plant was powered by two steam engines and two waterwheels. It was the former Hamilton Mill. Owner code: KNV

Hammanossette Mill (est. 1870) was in Madison, New Haven Co., CT. This manila mill made cartridge paper on two cylinder-wire machines. The plant was powered by two steam engines and two waterwheels. Owner code: JJR

Hammond Mill (est. 1882) at Jackson, Jackson Co., MI, made wrapping paper from rag, straw, and manila on a cylinder-wire machine. Owner code: JSE

Hampden Glazed Paper Co. (est. 1881) at South Hadley, Hampshire Co., MA, made manila card and board, bogus manila, stationery, and envelopes. Owner code: JSF

Hampden Mill (est. 1862) at Holyoke, Hampden Co., MA, had a moving-wire machine and three waterwheels. It later became the Beebe & Holbrook Mill. Owner code: JSG

Hampshire Mills (est. 1866) at South Hadley, Hampshire Co., MA, made bond and stationery on two moving-wire machines. The plant was powered by a steam engine and four waterwheels. Owner code: JSH

Hanna Mill (est. 1850) at Fitchburg, Worcester Co., MA, made news paper and card with a Jordan refiner and moving-wire machine. The plant was powered by two steam engines and two waterwheels. Owner code: JKZ

Harbottle Mill (circa 1880) at Brooklyn, Kings Co., NY, made manila roofing paper on a cylinder-wire machine. Owner code: JSM

Hardwick Mill (circa 1880) at Furnace, Worcester Co., MA, made book and news paper with a Jordan refiner and a moving-wire machine. The plant was powered by two steam engines and four water turbines. Owner code: KGX

Harris & Cox Mill (est. 1813) was in North Yarmouth, Cumberland Co., ME. This hand mill operated until 1829. Owner code: JSW

Harris Mill (est. 1849) at Harrisville, Burlington Co., NJ, made manila hardware on a cylinder-wire machine. Owner code: JSX

Hartford Mill (est. 1855) at N. Manchester, Hartford Co., CT, produced book, card, and wall paper with a Jordan refiner on a double-cylinder machine. The plant was powered by steam engine and waterwheel. Owner code: JEY, KVV

Hartland Mill (est. 1872) at Middleport, Niagara Co., NY, made straw wrappers on a cylinder-wire machine. The plant was powered by two waterwheels. Owner codes: JTC, KDP

Harvey Mill (est. 1853) at Wellsburgh, Brooke Co., WV, made straw wrapping on a cylinder-wire machine. The plant was powered by a steam engine. Owner code: JTD

Harwood Mill (est. 1868) at N. Leominster, Worcester Co., MA, made leather board and hardware on three wet machines. The plant was powered by a steam engine and three waterwheels. The facility was rebuilt in 1884. Owner codes: JTF, JTE

Hasting Mill at Springfield, Clarke Co., OH, made wrapping paper on two cylinder-wire machines. Owner code: JTG

Haverhill Mill (est. 1870) at Haverhill, Grafton Co., NH, made manila and straw wrapping paper on a cylinder-wire machine. The plant was powered by two waterwheels. Owner code: JTI

Haverhill Mill (est. 1884) at Bradford, Essex Co., MA, made newsprint on two moving-wire machines. The plant was powered by four steam engines. Owner code: JTI

Hawkins Mill (circa 1880) at Islip, Suffolk Co., NY, made straw board on a cylinder-wire machine. Owner code: JTL

Hawley Mill (est. 1808) at Moreau, Saratoga Co., NY, was a hand mill making stationery and book paper. Owner code: JTM

Hazen Mill at Shirley, Middlesex Co., MA, made manila paper.

Heilman Mill (est. 1793) at Swatara, Lebanon Co., PA, was a hand mill that operated until 1828. Owner code: JTP

Henshaw Mill (est. 1800) at Weybridge Upper Falls, Addison Co., VT, was a hand mill that operated until 1841. Owner code: JTT

Herkimer Mill (est. 1849) at Herkimer, Herkimer Co., NY, made news and colored mediums on two moving-wire machines. Following the plant's conversion to wood pulp in 1868 it made book and wrapping paper. Owner codes: JTU, JXV, KDV

Hibernian Mill (est. 1854) at Windsor Locks, Hartford Co., CT, made wrapping paper on a cylinder-wire machine powered by a water turbine. The plant was rebuilt in 1856 following a fire. Owner codes: KIA, KJM

High Bridge Mill (circa 1880) at Chatham, Columbia Co., NY, made straw wrappers on a cylinder-wire machine. The plant was powered by a steam engine and two waterwheels. Owner code: JBK

Highland Mill (circa 1880) at Newburgh, Orange Co., NY, made stationery with a moving-wire machine. The plant was powered by three waterwheels. Owner codes: KRN, JHG

Highland Mill (est. 1883) at S. Manchester, Hartford Co., CT, made board stock on a wet machine powered by a steam engine. Owner code: JHJ

Hill Mill (est. 1860) at Washington, DC, made book and news on a double-cylinder machine. The plant was equipped with a Jordan refiner, and was powered by the C&O Canal. Later named the Potomac Mill. Owner code: JTX, JUA

Hilldale Mill (circa 1880) at West Chester, Chester Co., PA, made wrapping paper on a double-cylinder machine. Owner code: JFB

Hinsdale Mills (est. 1842) at Hinsdale, Cheshire Co., NH, made manila tissue using a Jordan refiner. The plant had two cylinder-wire machines powered by four waterwheels. Owner code: KLH

Hinsdill Mill at Bennington, Bennington Co., VT, was a hand mill that closed in 1800 and became the Paran Creek Mill. Owner codes: JUD, JUC

Hoboken Mill (est. 1883) at Hoboken, Hudson Co., NJ, made building paper on a cylinder-wire machine powered by a steam engine. Owner code: KKG

Hoffman Mill at Houcksville, Carroll Co., MD, made straw wrappers on a cylinder-wire machine. The mill on east branch of Patapsco River was powered by a steam engine and two waterwheels. This plant later became the Pine Grove Mill. Owner code: JUH

Hollister Mill (est. 1855) at Milton, Saratoga Co., NY, made bond, newsprint, and wall paper on a Harper machine. The plant was powered by three steam engines and two waterwheels. Owner code: JMU

Hollywell Mill (est. 1790) at Chambersburg, Franklin Co., PA, produced handmade bond and book paper until 1851. Auxiliary equipment was powered by a waterwheel. Owner Code: KNM. The Hollywell Straw Mill began in 1828 with one cylinder-wire machine for the manufacture of straw book, news, card, and wrapping paper. The plant was powered by one steam engine and two waterwheels. Owner codes: KNO, JTV

Holyoke Manila Mill (est. 1875) at Holyoke, Hampden Co., MA, operated a single cylinder-wire machine. Owner code: KLK

Holyoke Mill (est. 1857) at Holyoke, Hampden Co., MA, made bond, ledger, and stationery on three moving-wire machines. The plant was powered by eight water turbines. Owner code: JUS

Hoosic Mill (est. 1853) at Ann Arbor, Washtenaw Co., MI, made straw wrapping paper on a cylinder-wire machine. The plant was powered by two waterwheels. Owner code: JJU

Hope Mill (est. 1813) at Catskill, Greene Co., NY, produced handmade paper until 1831. This plant next made board stock and collar paper on a cylinder-wire machine. The plant was powered by two waterwheels. Owner codes: JBS, KIT

Housatonic Mill, Old (est. 1827) at Lee, Berkshire Co., MA, made news paper on a cylinder-wire machine powered by a waterwheel. Owner Code: JZB. The plant sold in 1850 and then made bogus manila book and news paper on two moving-wire machines. A steam engine powered auxiliary equipment. Owner codes: KIX, KOB

House Mill (circa 1880) at Rainbow, Hartford Co., CT, made board stock on a wet machine. The plant was powered by a steam engine and a waterwheel. Owner code: JUV

Howard & Lathrup Mill (est. 1828) at Canal Village, Hampshire Co., MA, made rag paper on a cylinder-wire machine powered by a waterwheel. The plant was dismantled in 1848. Owner code: JUX

Howell Mill (est. 1814) at Lockport, Williams Co., OH, was a hand mill that operated continuously until 1834. Owner code: JUZ

Howland Mill was established in 1831 at Troy, Rensselaer Co., NY. Owner code: JVC

Howland Mills (2) (est. 1873) at Sandy Hill, Washington Co., NY, made bogus manila on three Harper machines, a Gould machine, a cylinder-wire machine, and Brightman engines. The plant was powered by seven waterwheels. Owner code: JVA

Hubbard Mills (est. 1818) at Norwich, New London Co., CT, was a hand mill that installed two moving-wire machines in 1832. The plant was powered by a waterwheel and later specialized in colored mediums. Owner code: JVD.1

Hudson Mill (est. 1872) at Waterford, Saratoga Co., NY, made newsprint using the Jordan refiner. The plant had a moving-wire machine, two steam engines, and four waterwheels. Owner code: JQK

Huntington Mill (est. 1855) at Huntington, Fairfield Co., CT, made straw wrapping on a cylinder-wire machine. The plant was powered by two waterwheels. Owner code: JCY

Huntington Mills (circa 1880) at Shickshinny, Luzerne Co., PA, made straw wrapping on a cylinder-wire machine. Owner code: JYT

Hurlbut Mill (est. 1872) at South Lee, Berkshire Co., MA, made stationery, printing, and ledger paper on two moving-wire machines. The plant was powered by a steam engine and three water turbines. Owner codes: JNJ, JNI, JAY, JVO

Hydraulic Mill at Chillicothe, Ross Co., OH, made print and wrapping paper on two moving-wire machines. Owner code: JWA, JDD

Hydraulic Mill (est. 1875) at Piqua, Miami Co., OH, made straw wrapping on a cylinder-wire machine. The mill was powered by two steam engines and a waterwheel. Owner codes: KGO, KTS

Idler Mill (est. 1810) at Alsace Township, Berks Co., PA, operated continuously as a hand mill until 1850. Owner code: JVP

Illinois Valley Mill (est. 1884) at Marseilles, La Salle Co., IL, made straw wrapping paper on a cylinder-wire machine. The plant was powered by two waterwheels. Owner code: JVQ

Indian Kentuck Creek Mill (est. 1827) at Madison, Jefferson Co., IN, made rag paper in two hand vats. This mill closed in 1837. Owner code: KNC

Indiana Mill (circa 1883) at Indiana, Indiana Co., PA, made straw board on a cylinder-wire powered by a steam engine. Owner code: JTK

Indiana Straw Board Mill (est. 1853) at Indiana, Indiana Co., PA, made straw board on a cylinder-wire machine. The plant was powered by two steam engines. Owner code: KQE

Ingersoll Mill (est. 1833) at East Lee, Berkshire Co., MA, was a hand mill until 1839. Rebuilt in 1840 following a fire. Made straw wrapping on a cylinder-wire. Rebuilt again after a fire in 1884. Owner codes: JVV, KCM

Ingersoll Mill (circa 1880) at Pulaski, Oswego Co., NY, made manila paper on a cylinder-wire machine. Owner code: JVX

Ingham Mill (est. 1831) at Chillicothe, Ross Co., OH, made stationery, printing, and wrapping on a cylinder-wire machine. This plant later named the Hydraulic Mill. Owner code: JVY

Irving Mill (est. 1833) at South Lee, Berkshire Co., MA, made bond, stationery, printing, news, and ledger paper on a cylinder-wire machine. Facilities were powered by a single waterwheel. This was the former South Lee Mill. Plant burned down in 1876. Owner codes: KGT, KOT

Island Mill (est. 1863) at Ballston Spa, Saratoga Co., NY, made manila paper on a double-cylinder machine. The plant was powered by two steam engines and three waterwheels. Owner code: KVG

Island Mills (2) (est. 1788) at Newton Lower Falls, Middlesex Co., MA, was a hand mill until 1845. This plant later became Foster's Mill. Owner codes: JUI, JUJ, KBK

Ithaca Falls Mills (est. 1850) at Ithaca, Tompkins Co., NY, made straw wrapping on a cylinder-wire machine. This plant was powered by three waterwheels. Owner code: KSX

Ivanhoe Mill (est. 1850) at Paterson, Passaic Co., NJ, made book and stationery paper using a Jordan refiner on two moving-wire machines. The plant was powered by three waterwheels. Owner code: JWD

Ivy Mill (est. 1729) at Ivy Mills, Delaware Co., PA, made binder board, linen and linen composite stationery and bond paper by hand until 1866. Owner codes: KWX, KWW, KWV, KWL, KWU

Ivy Mill (est. 1860) at Reisterstown, Baltimore Co., MD, made news paper on a moving-wire machine. The plant was powered by two steam engines and two waterwheels. Owner code: JND

Jackson Mill (est. 1801) at Needham, Middlesex Co., MA, burned in 1815, was rebuilt and continued as a hand mill until 1845. It made rag paper on a moving wire machine until 1894, when the mill was rebuilt following a fire. Owner codes: JWJ, KBG, KTP, KKL, KKN, JEG

Jackson Mill (est. 1883) at Jackson, Jackson Co., MI, made manila-colored wood-pulp paper on a moving-wire machine. The plant was powered by two steam engines. Owner code: JWI

James River Mill (est. 1834) at Richmond, Henrico Co., VA, made book, blotter, and news paper using a Jordan. The mill had a single moving-wire machine. Owner code: KKT

Japanese Mill (est. 1847) at Westfield, Hampden Co., MA, made card, board, hardware, bogus manila, and stationery on a cylinder-wire and a moving-wire. The plant was powered by two steam engines and three water turbines. Owner codes: JKN, JKM

Japanese Paper Ware Mill (est. 1867) at Fairfield, Fairfield Co., CT, made hardware paper on a cylinder-wire machine. The plant was powered by two waterwheels. Owner code: JWL

Jarvis Mill (est. 1850) at Claremont, Sullivan Co., NH, made card, wall paper, newsprint, and bogus manila on a cylinder-wire machine. The plant was powered by two waterwheels. Owner code: JWK

Jay Bridge Mill (est. 1889) at Jay, ME, made paper on two moving-wire machines. Owner codes: KJZ, KGQ

Jersey City Mill (circa 1880) at Jersey City, Hudson Co., NJ, made manila tissue on two cylinder-wire machines. Owner code: JWM

Jessups Mill (est. 1835) at Westfield, Hampden Co., MA, made manila paper on a cylinder-wire and added a moving-wire machine in 1848. Owner code: KPB

Joncy Mill (est. 1854) at Angelica, Alleghany Co., NY, made manila rope on a double-cylinder machine powered by a waterwheel. Owner code: JFQ

Jones Mill (est. 1793) at West Manayank, Montgomery Co., PA, was a single vat mill that made binder's board and wrapping paper. Plant was closed after 1848. Owner codes: JWW, KTU, JTQ ('07)

Jones Mill (circa 1880) at Greenville, Greenville Co., SC, made manila news and wrapping paper on a cylinder-wire machine. Owner code: JWX

Jordan Mill (circa 1880) at Jordan, Onondaga Co., NY, made straw board on a cylinder-wire machine. Owner code: KJN

Kalamazoo Mill (est. 1866) at Kalamazoo, Kalamazoo Co., MI, made book and printing paper using a Jordan refiner on two moving-wire machines. The plant was powered by two steam engines and three waterwheels. Owner code: JXB

Kankakee Mill (est. 1873) at Kankakee, Kankakee Co., IL, made straw board on a cylinder-wire machine. The plant was powered by two steam engines and three waterwheels. Owner code: JXC

Katz Mill (est. 1750) at Whitemarsh Township, Montgomery Co., PA, was a hand mill that continued through 1845. Owner code: JXD

Kearney Mill (circa 1880) at Parkton, Baltimore Co., MD, made straw wrappers on a cylinder-wire machine. Owner code: JXE

Kearsage Mill (est. 1871) at Davisville, Merrimack Co., NH, made straw board on a moving-wire machine. The plant was powered by five waterwheels. Owner code: JLT

Keating Mill (est. 1768) at Burling Slip, New York, NY, was a hand mill that was removed to Peekskill in 1774. Owner code: JXF

Keck Mill (est. 1869) at Kecks Center, Onondaga Co., NY, made straw board on a cylinder-wire machine. The plant was powered by one steam engine and two waterwheels. Owner code: JXG

Keeney & Wood Mill (est. 1876) at N. Manchester, Hartford Co., CT, made straw card and board using a Jordan refiner on a wet machine. This plant was powered by two waterwheels. Owner code: JXH, JXI

Keith Mill (est. 1870) at Turner Falls, Franklin Co., MA, made stationery paper on two moving-wire machines. Owner code: JXM

Keller Mill (est. 1790) at Exeter Township, Berks Co., PA, was a hand mill that continued through 1852. Owner code: JXN

Kellogg Mill (est. 1843) at St. Louis, MO, made book and news paper on a moving-wire machine. Auxiliary equipment was powered by a steam engine. Owner codes: JXQ, JMZ

Kellogg Steam Mill (circa 1828) at Cleveland, Cuyahoga Co., OH, produced handmade paper until the plant burned in 1831. Auxiliary equipment was powered by a steam engine. Owner code: JXP

Kendall Mill (est. 1825) at N. Leominster, Worcester Co., MA, was a hand mill that made book paper; it later became the Wheelwright Mill. Owner codes: JXT, JKW

Keneetee Mill (est. 1864) at Brodlbin, Fulton Co., NY, made straw wrappers on a cylinder-wire machine. This plant was powered by two waterwheels. Owner code: KPN

Kentucky Mill (est. 1878) at Louisville, Jefferson Co., KY, made manila paper on a cylinder-wire machine. Owner code: JWC

Keuka Mill (circa 1880) at Penn Yan, Yates Co., NY, made straw wrappers on two cylinder-wire machines. The plant was powered by one steam engine and two waterwheels. Owner code: JOZ

Keystone Mill (est. 1851) at Upper Darby, Delaware Co., PA, made book and card paper using a Jordan refiner on a moving wire machine. The plant was powered by one steam engine and one waterwheel. Owner code: JPX

Killingsworth Mill (est. 1870) at Madison, New Haven Co., CT, made straw board on a cylinder-wire machine. The plant was powered by two steam engines and two waterwheels. Owner code: JJR

Kills Mill was established in 1845 at Sprintfield, Hamilton Co., OH. Owner code: JXZ

Kimberton Mill (est. 1820) at Kimberton, Chester Co., PA, made boards on two cylinder-wire machines. The plant was powered by one steam engine and two waterwheels. Owner code: KNB

Kinni Kinick Mill (est. 1812) at Chillicothe, Ross Co., OH, produced handmade rag paper through 1827.

Kirk Mill (est. 1814) at East Nottingham, Chester Co., PA, produced handmade rag paper. Auxiliary equipment was powered by a waterwheel. Reopened in 1838 as Stubbs Mill. Owner codes: JYK, JYL

Knowlton Bros. Mill (est. 1861) at Watertown, Jefferson Co., NY, made book and bogus manila on a moving-wire machine. This plant was powered by four waterwheels. Owner code: JYO

Knoxville Mill (est. 1835) at Knoxville, Knox Co., TN, made manila news and straw wrappers on a cylinder-wire machine. The plant was powered by two steam engines and a waterwheel. Owner code: KJX

Kownslar Mill (est. 1804) at Mill Creek, Berkeley Co., VA, was a hand mill that burned in 1852. Owner code: JYU

Kugler Mill was established in 1834 at Milford, Hamilton Co., OH. Owner code: JBV

Kutztown Mill (est. 1798) at Allen Township, Northampton Co., PA, was a hand mill that continued through 1832. Owner code: JZW

La Fayette Mill (est. 1882) at Lafayette, Tippecanoe Co., IN, made straw wrapping and building paper on a cylinder-wire machine. This plant was powered by two steam engines. Owner code: JZA

Lake George Pulp & Mill (est. 1882) at Ticonderoga, Essex Co., NY, made newsprint on a moving-wire machine. The plant was powered by a steam engine and seven waterwheels. Owner code: JZC

Lakeside Mill (est. 1868) at South Toledo, Lucas Co., OH, made news and wrapping paper on a cylinder-wire machine. The plant was powered by five waterwheels. Owner code: JIM

Lakeside Mill (est. 1883) at Skaneateles, Onondaga Co., NY, made straw wrapping and bogus manila on a double-cylinder machine. The plant was powered by two water turbines. Owner code: JZD

Lakeville Mill (est. 1868) at Shirley Village, Middlesex Co., MA, made leather board and carpet lining on a double-cylinder and a wet machine. This plant was powered by a steam engine and two waterwheels. Owner codes: JED, JEE, JMA, JMB

Lamb Mill (est. 1782) at Upper Darby Township, Delaware Co., PA, was a hand mill that continued through 1841, when it became the Clifton Mill. Owner codes: KAA, KAE, JNM

Lamden Mill (est. 1828) at Wheeling, VA, was a hand mill that was rebuilt in 1835 following a fire. Owner code: JZF

Lancaster Mill (est. 1864) at Lancaster, Coos. Co., NH, made manila tissue and straw wrapping on a cylinder-wire machine. This mill was powered by three waterwheels. Owner code: JZI

Lanvale Mill (circa 1880) at Ellicott City, Howard Co., MD, made manila-colored mediums on a moving-wire machine. Owner code: KDL

Laurel Mill (est. 1855) at Lee, Berkshire Co., MA, made rag paper on a moving-wire machine. Formerly the Castle Mill, this plant was powered by a waterwheel. Owner codes: KIX, KOB

Lawless Mill (est. 1881) at Penfield, Monroe Co., NY, made straw wrapping on a cylinder-wire machine. This mill was powered by two waterwheels. Owner code: JZL

Lawrence Mill (est. 1883) at Lawrence, Douglas

Co., KA made straw wrapping paper on a cylinder-wire machine. Owner code: JZN

Laws Mills (2) at Pembroke, Merrimack Co., NH, was established by 1823. Owner code: JZN.1

Lebanon Mill (est. 1889) at Lebanon, Linn Co., OR, made straw paper.

LeBourgois Mill (est. 1884) at Convent, St. James Parish, LA, made manila paper on a cylinder-wire machine. This plant was powered by two steam engines. Owner code: JZO

Lee Mill (est. 1873) at Manilus, Onondaga Co., NY, made carpet lining and building paper on a cylinder-wire machine, and straw board when in season. This plant was powered by a steam engine and a waterwheel. Owner code: JZP, JZQ

Leffingwell Mill (est. 1767) at Norwich, New London Co., CT, was a hand mill that added a moving-wire machine in 1828. Owner codes: JZT, JLF

Leland Mill (est. 1865) at Foxboro, Norfolk Co., MA, made leather card on a cylinder-wire machine. This plant was powered by a steam engine and a water turbine. Owner code: JUR

Leominster Mills (2) (est. 1796) at Leominster, Worcester Co., MA, was a hand mill that burned in 1810 and was rebuilt. The second mill sold in 1818 and was later renamed the Kendall Mill. Owner codes: KFX, JAR, KNQ, KRC

Lewis Mill (est. 1845) at Lyons Falls, Lewis Co., NY, made manila wrapping paper on a cylinder-wire machine. The plant was powered by two waterwheels. Owner codes: KNJ, KNK

Lewis Straw Board Mill (est. 1879) at Flint, Genesee Co., MI, made straw board and carpet lining on a double-cylinder machine. The plant was powered by a steam engine and two waterwheels. Owner code: KAG

Lexington Steam Mills (2) (est. 1815) at Lexington, Fayette Co., KY, produced handmade rag paper through 1839. Auxiliary equipment was powered by a steam engine.

Lick Mill (est. 1874) at Agnews, Santa Clara Co., CA made newsprint on a moving wire machine and manila paper on a cylinder-wire machine. This plant converted to wood-pulp in 1878. Auxiliary equipment was powered by two steam engines. The mill burned in 1882 and was later rebuilt as the Remington Mill. Owner codes: KAM, KKD

Lima Mills (est. 1870) at Lima, Allen Co., OH, made straw board on two cylinder-wire machines. The plant was powered by five steam engines. Owner code: KAN

Lincoln Mill (est. 1850) at N. Dighton, Bristol Co., MA, made manila wrappers on a double-cylinder machine. This was the former Park Mill. Owner code: KAO

Lincolnton Mill (est. 1861) at Lincolnton, Cleveland Co., NC, made rag paper on a cylinder-wire machine. See: KXN.1

Litchfield Mills (est. 1856) at E. Litchfield, Litchfield Co., CT, made manila tissue on a cylinder-wire machine. This plant was powered by three waterwheels. Owner code: KGW

Little Androscoggin Mill (est. 1876) at East Poland, Androscoggin Co., ME, made leather card and board on a cylinder-wire machine. This plant was powered by two waterwheels. Owner code: KEV

Little Badger Mill (est. 1874) at Neenah, Winnebago Co., WI, made newsprint using a Jordan refiner on a moving-wire machine. Owner code: JYC

Livingston Mill (circa 1880) at Dansville, Livingston Co., NY, made news paper on two cylinder-wires and one moving-wire machine. The plant was powered by three waterwheels and two steam engines. Owner code: KSW

Livingston Mills (2) (est. 1875) at Livingston, Columbia Co., NY, made manila tissue on a cylinder-wire machine. The plant was powered by three waterwheels. A second plant was erected in 1880. Owner codes: KAP, JNC

Lockland Mill (est. 1840) at Lockland, Hamilton Co., OH, made wrapping paper from straw or rag on a cylinder-wire and a moving-wire ('51) machine. The plant was rebuilt in 1841 following a fire. The facility was destroyed by fire in 1878. Owner codes: JBU, JPM, JYE

Lockport Mill (est. 1884) at Lockport, Niagara Co., NY, made straw wrapping, board, and building paper on a cylinder-wire machine. This mill was powered by a steam engine and two waterwheels. Owner code: KAQ

Lockport Mills (est. 1872) at Lockport, Will Co., IL, made straw board on two cylinder-wire machines. This plant was powered by four waterwheels. Owner code: KAR

Logansport Mill (est. 1864) at Logansport, Cass Co., IN, made straw wrapping paper on a cylinder-wire machine. This plant was powered by two waterwheels. Owner codes: JIQ, KAS

Long Mill (circa 1880) at Gargoa, Fulton Co., NY, made straw board on a cylinder-wire machine. Owner code: KAT

Longfellow Mill (est. 1831) at Needham, Middlesex Co., MA, made book, news, and wrapping paper on a moving-wire machine powered by a waterwheel. This was the former Lyon Mill and later Thomas Rice Mill. Owner code: KKI

Loomis & Norton Mill (est. 1871) at Suffield, Hartford Co., CT, made manila wrappers on a cylinder-wire machine. The plant closed in 1878. Owner code: KAV

Louisville Mill, Old (est. 1831) at Louisville, Jefferson Co., KY, made news paper on a cylinder-wire machine.

Louisville Mill (est. 1840) at Louisville, Jefferson Co., KY, made news paper with a Jordan refiner on three moving-wire machines. The plant was powered by four steam engines. Owner code: JMY

Lower Mill was established in 1855 at Lockland, Hamilton Co., OH. Owner codes: JAS, JRU

Loyalhana Mills (circa 1883) at Latrobe, Westmoreland Co., PA, made manila wrapping paper and building paper on a double-cylinder machine. Owner code: KIB

Lydall & Foulds Mill (est. 1879) at N. Manchester, Hartford Co., CT, made card and board stock using a Jordan refiner. This plant had a double-cylinder machine and was powered by a steam engine and two waterwheels. Owner code: KBB

Lydig Mill (est. 1808) at Bronx, Queens Co., NY, was a hand mill in Hemlock Grove that was destroyed by fire in 1822. Owner code: KBC

Lyndon Mill (circa 1880) at Lyndon, VT, made wood-pulp paper.

Lyndonette Mill, Old (est. 1790) at Atglen, Chester Co., PA, produced handmade paper until it installed a cylinder-wire machine. Afterwards, this mill made board stock. The plant was powered by a waterwheel. Owner code: KYM

Lyndonette Mill (est. 1892) at Atglen, Chester Co., PA, made wood-pulp board on a moving-wire machine. Owner code: JSR

Lyon Mill (est. 1810) at Needham, Middlesex Co., MA, produced handmade paper and installed a cylinder-wire machine in 1826. This plant was powered by a waterwheel. It burned in 1834 and was rebuilt as the Longfellow Mill. Owner codes: KBK, KBF

Lyon Mill (est. 1850) at Fitchburg, Worcester Co., MA, made news paper and card stock using a Jordan refiner on a moving-wire machine. The plant was powered by two steam engines and a waterwheel. Owner code: JKZ

Lyons Mill (est. 1873) at Lyons, Clinton Co., IA, made manila or straw board, hardware, and box board using a Jordan refiner on a double-cylinder and a triple-cylinder machine. This plant was powered by three steam engines. Owner code: KBM

Mad River Mill (est. 1834) at Dayton, Montgomery Co., OH, made straw wrapping, building, and carpet lining on two cylinder-wire machine. The plant was powered by three steam engines and a waterwheel. Owner code: JMS

Madison Mill (est. 1873) at Madison, Jefferson Co., IN, made wrapping paper on a cylinder-wire. This plant was powered by two steam engines. Owner codes: JCH, KUP

Mahaiwe Mill (est. 1853) at East Lee, Berkshire Co., MA, made stationery, printing, and bond paper on a moving-wire machine. The plant was powered by three steam engines and two waterwheels. The plant was rebuilt following a fire in 1878. Mill was closed in 1885. Owner codes: KCM, KLS, KCO

Malden Bridge Mill (est. 1845) at Malden Bridge, Columbia Co., NY, made straw wrapping and board on a cylinder-wire machine. The plant was powered by two water turbines and a waterwheel. Owner codes: KHT, KHU

Malone Mill (est. 1872) at Malone, Franklin Co., NY, made news on a moving-wire machine and used a Brightman engine. The plant also made manila paper on a cylinder-wire. The facilities were powered by seven water turbines and a steam engine. Owner codes: KUT

Mammoth Mill (est. 1831) at Chambersburg, Franklin Co., PA, made straw board and book paper on eight cylinder-wire machines powered by waterwheel(s). Plant was destroyed by war in 1864. Owner code: KNO

Manchester Mill (est. 1872) at N. Manchester, Hartford Co., CT, made straw board on a wet machine. The mill was powered by a steam engine. Owner code: JOY

Manchester Mill (circa 1880) at Richmond, Henrico Co., VA, made manila tissue on a cylinder-wire machine. Owner code: KBP

Manhan Mill (est. 1846) at Loudville, Hampshire Co., MA, made tissue paper on a cylinder-wire machine. The plant was powered by two steam engines and two waterwheels. Owner code: KAY

Manila Mill (circa 1880) at Brookville, Franklin Co., IN, made manila paper on a cylinder-wire machine.

Manila Mill (est. 1840) at Canal Village, Hampshire Co., MA, made manila wrappers on a cylinder-wire machine. The plant was powered by two waterwheels. This was the former South Hadley Mill that was dismantled in 1848. Owner codes: JWZ, KQS

Manilus Mill (circa 1880) at Manilus, Onondaga Co., NY, made bogus manila on a moving-wire machine. The plant was powered by two water turbines. Owner code: KBQ

Mansfield Mill at Mansfield, Richland Co., OH, made straw wrappers on a cylinder-wire machine. The plant was powered by two steam engines. Owner code: KBU

Marblevale Mill (est. 1828) at Ashland, Baltimore Co., MD, made newsprint on a moving-wire machine. The plant was powered by three steam engines and four waterwheels. Owner codes: JVK, JUH

Marcellus Mills (circa 1880) at Marcellus Falls, Onondaga Co., NY, made straw wrapping paper on a cylinder-wire machine. This plant powered by three waterwheels. Owner code: KNG

Marietta Mill (est. 1847) at Marietta, Cobb Co., GA, made book, news, and wrapping paper on a cylinder-wire. The plant was powered by two waterwheels. The facility was rebuilt in 1876. Owner codes: KBV, JGN

Marinette Mill (est. 1878) at Marinette, Marinette Co., WI, made news on a moving-wire machine and manila on a double-cylinder. Plant was equipped with a Jordan refiner, and was powered by one steam engine and nine waterwheels. Owner code: KBW

Marseilles Mill (circa 1880) at Marseilles, La Salle Co., IL, made wrapping paper on a cylinder-wire machine.

Marshall Mill (est. 1856) at Yorklyn, New Castle Co., DE, made wrapping paper on a cylinder-wire machine. The plant was powered by two steam engines and a waterwheel. Owner code: KCA

Marshall Mill (est. 1884) at Marshall, Calhoun Co., MI, made carpet lining from cedar bark pulp on a cylinder-wire machine. The plant was powered by two waterwheels. Owner codes: KDZ, KDY

Martin Mill (est. 1807) at Martinsburgh, Lewis Co., NY, was a hand mill that made stationery, wrapping, and hanging paper until its closure in 1826. Owner code: KCC

Marylandville Mill (circa 1880) at Philadelphia, Philadelphia Co., PA, made book paper with a Jordan refiner on a moving-wire machine. The plant was powered by two steam engines. Owner code: KCE

Massasoit Mill (est. 1872) at Holyoke, Hampden Co., MA, made ledger and stationery on two moving-wire machines. The plant was powered by three water turbines. Owner code: KCH

Massillon Mill (est. 1867) at Massillon, Stark Co., OH, made straw wrapping paper using a Brightman engine on two cylinder-wire machines. The plant was powered by two steam engines. Owner code: KCI

Mattapan Mill (circa 1880) at Milton, Norfolk Co., MA, made book, news, colored mediums, and printing paper on two moving-wire machines. Owner code: KRF

Mattson Mill (est. 1790) at Aston Township, Delaware Co., PA, was a hand mill that continued until 1848. Owner code: KCL

Mayville Mill (est. 1833) at Mayville, Jefferson Co., KY, made paper by hand and on a cylinder-wire machine.

McAllister Mill (est. 1883) at St. Johnsville, Montgomery Co., NY, made straw board on a cylinder-wire machine. The plant was powered by two waterwheels. Owner code: KCN

McCleaghan Mill (est. 1799) at Lower Merion Township, Montgomery Co., PA, was a hand mill that closed in 1848. Owner codes: KCQ, KTU, JWS

McKnight Mill (est. 1870) at Conyers, Rockdale Co., GA, made manila-colored wrapping paper on a moving-wire. The plant was powered by two waterwheels. Formerly named the Conyers Mill. Owner code: KDX

Meads Falls Mill (est. 1818) at Center Rutland, VT, was a hand mill that closed in 1846. Owner code: JRA

Medusa Mill (est. 1867) at Medusa, Albany Co., NY, made straw wrapping on a cylinder-wire machine. The plant was powered by a steam engine and a waterwheel. Owner code: KDW

Medway Mill at Medway, Clark Co., OH, made straw board on two board machines. Owner code: JUK

Meeter Mill (est. 1800) at Rockville, MD, was a hand mill that closed in 1842. Owner codes: KDE, JHI

Menasha Mill (est. 1882) at Menasha, Winnebago Co., WI, made book and news paper using a Jordan refiner on a moving-wire machine. The plant was powered by five waterwheels. Owner code: JQI

Menominee Mill (est. 1878) at Menominee, Marinette Co., WI, made manila news using a Jordan refiner on a double-cylinder machine. The plant converted to wood-pulp in 1891 and added two more moving-wires. Owner code: KBW

Meridian Mill (circa 1880) at Mishawaka, St. Joseph Co., IN, made straw wrapping on a cylinder-wire. The plant was powered by a steam engine and two waterwheels. Owner code: JVS

Merion Mill (circa 1880) at Abrams, Montgomery Co., PA, made box board on a cylinder-wire machine. Owner code: JAA

Merrill Mill (circa 1880) at Conway, Carroll Co., NH, made straw board on a cylinder-wire machine. Owner code: KDH

Merrimac Mill (circa 1880) at Lawrence, Essex Co., MA, made book and stationery on three moving-wire machines. Owner code: KDI

Merrimack Leather Board Mill (est. 1877) at South Lawrence, Essex Co., MA, made leather card on three wet machines powered by waterwheel(s). Owner code: JIZ

Messalonskee Mill was established in 1855 at Waterville, ME. Owner codes: KJZ, JRX

Miami Mill (est. 1849) at Hamilton, Butler Co., OH, made bogus manila, cigarette paper, and wrapping on two moving-wire machine. The plant was powered by five steam engines and three waterwheels. Rebuilt in 1888 and became the Beckett's Mill. Owner code: JDE

Miami Valley Mill at Miamisburgh, Montgomery Co., OH, made book and news paper on a moving-wire machine. Owner code: KDM

Michigan Paper Mills was established in 1870 at Gedds, MI. Owner code: KDN

Middle Brook Creek Mill was established circa 1820 at Knoxville, Knox Co., TN. Owner codes: KVZ, JTN

Middle Creek Mill (est. 1819) at Selinsgrove, Snyder Co., PA, was a hand mill that burned down in 1823. Owner code: KOR

Middle Mill (est. 1845) at East Lee, Berkshire Co., MA, made straw wrapping paper on a cylinder-wire machine. Owner code: KCM

Middleburgh Mill (est. 1844) at Middleburgh, Schoharie Co., NY, made straw wrapping on two cylinder-wire machines. The plant was destroyed by fire in the 1850s. Owner codes: JYX, JYY

Middlebury Mill was established in 1870 at Middlebury, VT, and it burned down in 1875.

Middlebury Pulp and Paper Mill were established around 1880 at Middlebury, VT.

Middlesex Mill (circa 1880) at Middlesex, Cumberland Co., PA, made straw board on a cylinder-wire machine. Owner codes: KXR, KXQ

Milan Mill (est. 1884) at Milan, Rock Island Co., IL, made straw wrapping paper on two cylinder-wire machines. The plant was powered by three steam engines and four waterwheels. Owner codes: JEJ, KDR

Milburn Mill (est. 1810) at Millburn, Essex Co., NJ, made paste board and wall paper. This plant later became the Short Hills Mill. Owner codes: JIT, JGU

Milford Mill (est. 1789) at Newark, New Castle Co., DE, was a hand mill that operated until 1847. It was renamed the Nonatum Mill. Owner code: KDE

Mill Grove Mill (est. 1812) at West Newton, Westmoreland Co., PA, was a hand mill that added a cylinder-wire machine in 1826. Afterward it made hardware and construction paper on a double-cylinder machine. Plant was powered by a steam engine and two waterwheels. Owner codes: KBX, KBY, KBZ

Mill Town Mill (est. 1866) at Springfield, Union Co., NJ, made paste board on a cylinder-wire machine. The plant was powered by three waterwheels. Owner code: KDC

Millburn Mill (est. 1872) at Millburn, Essex Co., NJ, made tissue paper on a cylinder-wire machine. The plant was powered by two steam engines and a waterwheel. Owner code: JML

Miller Mill (est. 1882) at St. Johnsville, Montgomery Co., NY, made straw board on a cylinder-wire machine. The plant was powered by two waterwheels. Owner code: KDU

Miller's River Mill (est. 1883) at Wendell Depot, Franklin Co., MA, made leather board on a board machine. The plant was powered by four waterwheels. Owner code: KDS

Milo Mill (est. 1871) at Penn Yan, Yates Co., NY, made straw wrapping paper on two cylinder-wire machines. The plant was powered by two steam engines and two waterwheels. Owner code: KLZ

Milton Mill (est. 1884) at Milton, Strafford Co., NH, made leather board on three wet machines. Owner code: JHB

Mineral Springs Mill (circa 1883) at Coatsville, Chester Co., PA, made paste board on a cylinder-wire machine. Owner code: JNF

Minneapolis Mill (est. 1868) at Minneapolis, Hennepin Co., MN, made news paper on three cylinder-wire machines. The plant was powered by three waterwheels. Owner code: KUD

Minneapolis Straw Board Mill (est. 1882) at Minneapolis, Hennepin Co., MN, made straw board and wrapping paper on a cylinder-wire machine. The plant was powered by two steam engines. Owner code: KDX

Mode Mill (est. 1810) at Colesville, Montgomery Co., PA, was a hand mill that closed in 1844.

Mohawk Mill (est. 1872) at Waterford, Saratoga Co., NY, made news paper using a Jordan refiner on a moving-wire machine. The plant was powered by two steam engines and four waterwheels. Owner code: JQK

Moline Mill (est. 1852) at Moline, Rock Island Co., IL, made press paper on two moving-wire machines. The plant was powered by two steam engines and four waterwheels. Owner code: KED

Monadnock Mill (est. 1872) at Bennington, Hillsborough Co., NH, made book paper on a moving-wire machine. Auxiliary equipment was powered by two steam engines. Owner code: JCJ

Monarch Mill (est. 1883) at Waterford, Saratoga Co., NY, made straw board on a cylinder-wire machine. The plant was powered by a steam engine and a waterwheel. This mill was destroyed in 1884. Owner code: KEI

Monattaquot Mill (circa 1880) at S. Braintree, Norfolk Co., MA, made bogus manila on three moving-wire machines. Owner code: JUP

Monroe Mfg. Mill (est. 1883) at Monroe, Monroe Co., MI, made wrapping paper on a cylinder-wire machine. The plant was powered by two steam engines. Owner codes: JAF, KEE

Monroe Mill (est. 1836) at Raisinville, Monroe Co., MI, made wrapping paper on a cylinder-wire machine. The plant was powered by two steam engines. It was rebuilt in 1863 and later became the Monroe Straw Mill. Owner codes: KEA, KEB

Monroe Straw Mill (est. 1873) at Monroe, Monroe Co., MI, made straw wrapping paper on a cylinder-wire machine. This was the former Monroe Mill. Owner code: KEF

Montague Mill (est. 1872) at Turner Falls, Franklin Co., MA, made wood-pulp paper on two moving-wire machines. The plant was powered by twelve waterwheels. Owner code: KEG

Montgomery Mill (circa 1880) at Montgomery, Orange Co., NY, made manila paper on two cylinder-wire machines. The plant had a Brightman engine and was powered by a steam engine and four waterwheels. Owner code: KTQ

Montpelier Mill (est. 1806) at Montpelier, VT, was a hand mill that closed in 1850. Owner code: JGH

Montville Mill (est. 1851) at Montville, New London Co., CT, made manila paper on a cylinder-wire machine. The plant had a Jordan refiner and was powered by two steam engines and two waterwheels. Owner code: KLG

Moore Mill (est. 1837) at Chatham, Columbia Co., NY, made straw wrapping on a cylinder-wire machine.

Moore Mill (circa 1880) at Bellows Falls, Windham Co., VT, made manila tissue on a cylinder-wire machine. Owner code: KEL

Moorehouse Brothers Mill (est. 1879) at Bridesburg, Philadelphia Co., PA, made book paper on two moving-wire machines. The plant had a Jordan refiner and was powered by two steam engines. Owner code: KEM

Moosalamoo Mill (est. 1879) at Salisbury, VT, made wood pulp and paper.

Morgan Mill (est. 1882) at Battle Creek, Calhoun Co., MI, made straw board and carpet lining on a double-cylinder machine. The plant was powered by four waterwheels. Owner codes: KEN, KEP

Moriches Mill (est. 1858) at Moriches, Suffolk Co., NY, made straw board on a cylinder-wire machine. The plant was powered by two waterwheels. Owner code: KOD

Morris Mill at Millburn, Essex Co., NJ, was built circa 1870. Owner code: JMF

Morris Mill (est. 1880) at Morris, Grundy Co., IL, made straw board on two cylinder-wire machines. The plant was powered by three steam engines. Facility was closed in 1880. Owner code: JAQ

Morris Plains Mill (est. 1861) at Morristown, Morris Co., NJ, made manila construction paper on a cylinder-wire machine powered by a waterwheel. Owner code: JFI

Morrison Mill (est. 1859) at Petersborough, Hillsborough Co., NH, made manila wrapping paper on a cylinder-wire machine. The plant was powered by a steam engine and a waterwheel. Owner code: KER

Morristown Mill (est. 1884) at Morristown, Morris Co., NJ, made leather board on a cylinder-wire machine. Owner code: KEY

Morrisville Mill (circa 1880) at Morrisville, VT, made wood-pulp paper.

Mosher Mill (circa 1880) at Stillwater, Saratoga Co., NY, made manila paper on a cylinder-wire machine. The plant had five waterwheels. Owner code: KET

Moshier Mill (est. 1831) in Connecticut was a rag mill until it converted to straw in 1837. The plant had a cylinder-wire machine. Owner code: KEU

Mount Holly Mill (est. 1796) at Mount Holly, Burlington Co., NJ, was a hand mill that installed a moving-wire machine in 1836. The mill then made newsprint until it was destroyed by fire in 1840. Owner codes: JBT, JKG, JDL

Mount Holly Mills (2) (est. 1858) at Mount Holly Springs, Cumberland Co., PA, made wrapping and safety paper on two moving-wire machines. The plant was powered by three steam engines and six waterwheels. Owner code: KEW

Mount Ida Mill (est. 1840) at Troy, Rensselaer Co., NY, made book, news, and wall paper on a cylinder-wire machine. The plant was powered by four waterwheels. Owner code: KBS

Mount Prospect Mill (circa 1880) at Ashland, Grafton Co., NH, made straw board then leather board on a cylinder-wire machine. Owner code: JCQ

Mount Vernon Mill, Old (est. 1810) at Mount Vernon, Chester Co., PA, was a hand mill that added a cylinder-wire in 1836. The plant made news and wrapping and was powered by a waterwheel. Facility closed in 1855. Owner codes: JMN, JMM, JMO

Mount Vernon Mill (est. 1835) at Troy, Rensselaer Co., NY, made book, news, and wall paper on a moving-wire machine. Auxiliary equipment was powered by a steam engine. Owner code: KGP

Mount Vernon Mill (est. 1872) at Mount Vernon, Chester Co., PA, made board stock on a wet machine powered by a waterwheel. The mill was rebuilt following a fire in 1878. Owner codes: KXI, KCY

Mountain Mill (est. 1854) at East Lee, Berkshire Co., MA, made book paper on a moving-wire machine. The plant had a Jordan refiner and was powered by two steam engines and two waterwheels. This facility later became the Greenwood Mill. Owner code: JDN

Mountain Spring Mill (est. 1863) at Lambertville, Hunterdon Co., NJ, made tissue paper on a cylinder-wire machine. The plant was powered by a steam engine and two waterwheels. Owner code: KUW

Mousam Mills (est. 1876) at Kennebunk, York Co., ME, made leather card and board on a cylinder-wire machine. This plant was powered by a steam engine and a waterwheel. Owner code: KEV

Muir's Mill (est. 1880) at Morristown, Morris Co., NJ, made leather board on a cylinder-wire machine. The plant was powered by two waterwheels. Owner code: KEX

Mullin Mill (circa 1880) at Mount Holly Springs, Cumberland Co., PA, made book paper on a moving-wire machine. Owner code: KFA

Munroe Mill (est. 1849) at Munroe Falls, Summit Co., OH, made bogus manila, hardware, and wrapping paper on two cylinder-wire machines. The plant was powered by four waterwheels.

Munroe Mill (est. 1881) at Lawrence, Essex Co., MA, made wall paper and construction paper on a cylinder-wire and a moving-wire machine. The plant had a Jordan refiner, and was powered by two steam engines and two water turbines. Owner code: KFB

Napanoch Mill (est. 1884) at Napanoch, Ulster Co., NY, made printing paper and curtain lining on a moving-wire machine. The plant had two waterwheels. Owner code: KFC

Nashville Mills at Nashville, Davidson Co., TN was established around 1850. Owner codes: KMS, KWC

Nassau Mill (est. 1883) at Brainard, Rensselaer Co., NY, made straw wrapping on a cylinder-wire machine. The plant had two waterwheels. Owner code: KNL

National Mill (est. 1855) at East Lee, Berkshire Co., MA, made paper collars and cover middles on a moving-wire machine. The plant was powered by a steam engine and a waterwheel. Owner codes: JJW, KBE, JBZ

Naubuc Mill (est. 1883) at S. Manchester, Hartford Co., CT, made board stock on two wet machines powered by a steam engine. Owner code: JHJ

Needham Mill (est. 1814) at Needham, Middlesex Co., MA, was a hand mill that became incorporated into the Curtis Mills around 1835. Owner code: JLJ

Neenah Mill (est. 1874) at Neenah, Winnebago Co., WI, made book, news, and manila paper (ca. '80) on a cylinder-wire. The plant used a Jordan refiner. Owner codes: JYC, KFF, KHK

Nehoiden Mill (est. 1794) at Newton Lower Falls, Middlesex Co., MA, was a hand mill that made stationery, book, and wrapping paper. In 1847 it added a moving-wire machine for the manufacturing of newsprint. This plant converted to textiles in 1870. Owner codes: JUI, KKM, JRP, JLI, KJI, JED, JMA

Neponset Mills (2) (est. 1728) at Milton, Norfolk Co., MA, was a hand-craft mill that added a second plant in 1765. Mill was rebuilt in 1817 and added a waterwheel. Facility later became the Eagle Mill. Owner codes: JTR, KOH, JER, KTC, KRF ('28)

Nestell Mill (circa 1880) at Fort Plain, Montgomery Co., NY, made straw board on a cylinder-wire machine. Plant was powered by two water turbines. Owner codes: KFH, KFG

New Baltimore Mill (est. 1826) at New Baltimore, Greene Co., NY, made wall paper on a cylinder-wire machine. Plant was destroyed by fire in the 1850s. Owner codes: JLB, JFW

New Castle Mill (est. 1882) at New Castle, Lawrence Co., PA, made manila wrapping on a cylinder-wire machine. Plant was powered by a steam engine and three waterwheels. Owner code: KFI

New Dominion Mill (est. 1883) at Shepardston, Jefferson Co., WV, made straw board on a cylinder-wire machine. The plant was powered by two steam engines. Owner code: KSO

New England Mills (est. 1856) at East Lee, Berkshire Co., MA, made stationery and ledger paper on a moving-wire machine. The plant was powered by two steam engines and a waterwheel. Owner codes: JHT, KSY

New Leeds Mill (est. 1854) at Elkton, Cecil Co., MD, made card board on a moving-wire machine. The plant was powered by a steam engine and a waterwheel. Owner code: JSS

New London Mills (est. 1883) at New London, New London Co., CT, made manila tissue on two cylinder-wire machines. The plant was powered by four waterwheels. Owner code: KLD

New Philadelphia Mill (est. 1873) at New Philadelphia, Tuscarawas Co., OH, made straw wrapping on a cylinder-wire machine. The plant was powered by two steam engines. Owner code: JXA

New Portage Mill (est. 1883) at New Portage, Summit Co., OH, made straw board on two cylinder-wire machines. The plant was powered by four steam engines. Owner code: KJB

New York Mill was established in 1793 at Troy, Rensselaer Co., NY. Owner code: KUV

New York Mill (circa 1880) at Chatham, Columbia Co., NY, made straw wrappers on a cylinder-wire machine. Owner code: JPW

Newark Board Mill (est. 1879) at Newark, Essex Co., NJ, made board stock on a wet machine powered by a steam engine. Owner code: JDY

Newark Paper Co. (est. 1882) at Newark, Licking Co., OH, made straw wrapping on a cylinder-wire machine. The plant was powered by two steam engines. Owner code: KFN

Newbury Mill (est. 1810) at Newbury, Orange Co., VT, was a hand mill that made book paper. Owner code: JEM, KVU

Newcomb Mill was established in 1883 at High Falls, Ulster Co., NY. Owner code: KFQ

Newton Falls Mill (circa 1832) at Newton Falls, Cuyahoga Co., OH, made rag paper on a cylinder-wire machine. Owner code: KRQ

Newton Mill (est. 1876) at Holyoke, Hampden Co., MA, made manila carpet lining and straw board on three cylinder-wire machines. The plant was powered by three steam engines and two water turbines. Owner code: KFT

Newton Mills (est. 1869) at Newton Upper Falls, Middlesex Co., MA, made manila carpet lining on two cylinder-wire machines. This plant was powered by four steam engines and five waterwheels. The facilities were rebuilt in 1881 for the manufacture of wood pulp. Owner codes: KUB, JIR

Newton Mills (est. 1871) at Sparta, Monroe Co., WI, made book and news on a cylinder-wire machine. Owner code: KFU

Niagara Falls Mill (circa 1880) at Niagara Falls, Niagara Co., NY, made news paper on a moving-wire. The plant had a Jordan refiner, and was powered by three waterwheels. Owner code: KFV

Niagara Mill at Middletown, Butler Co., OH, made manila paper on two double-cylinders. The mill also ran a Gould machine. Owner code: KUA

Niagara Mill (est. 1840) at Lee, Berkshire Co., MA, made colored mediums and book and news paper on a moving-wire machine. Owner code: KOB

Niagara Mill (est. 1873) at Lockport, Niagara Co., NY, made manila paper on a cylinder-wire. The plant had a Jordan refiner, and was powered by a steam engine and two water turbines. The mill was rebuilt in 1884 for the manufacture of wood-pulp. Owner code: KFE

Niagara Wood Mill (est. 1884) at Niagara Falls, Niagara Co., NY, made manila card and wrapping paper on a cylinder-wire machine. The plant was powered by three waterwheels. Owner code: KFW

Niles Mill (est. 1872) at Niles, Berrien Co., MI, made straw wrapping paper on two cylinder-wire machines. The plant was powered by five waterwheels. Owner code: KFZ

Nissitissitu Mills (est. 1834) at E. Pepperell, Middlesex Co., MA, made leather board and construction paper on a cylinder-wire machine. The plant was powered by two steam engines and seven waterwheels. Owner code: KHF

Nixon Mill (est. 1839) at Richmond, Wayne Co., IN, made manila paper on two cylinder-wire machines. The plant was powered by three steam engines and a water turbine. Owner code: KGC

Nonatum Mill (est. 1779) at Newton, Middlesex Co., MA, was a hand mill that closed in 1821. Owner codes: JLH, JQD

Nonatum Mill (est. 1848) at Newark, New Castle Co., DE, made envelopes and card middles and colored mediums on a moving-wire machine. The plant was powered by two steam engines and two waterwheels. Owner codes: JLH, KJL

Nonotuck Mill (est. 1880) at Holyoke, Hampden Co., MA, made book, envelope, and colored mediums on two moving-wire machines. The plant had a Jordan refiner, and was powered by three steam engines and two water turbines. Owner code: KGD

North Hoosick Mill (circa 1880) at North Hoosick, Rensselaer Co., NY, made news and wall paper on two moving-wire machines. The plant had a Jordan refiner. Owner code: KPR

North Mill (est. 1872) at Ypsilanti, Washtenaw Co., MI, made book, card, and colored mediums on a moving-wire machine. Owner code: KHX

Northampton Mill (circa 1880) at Northampton, Hampshire Co., MA, made bogus manila on a double-cylinder powered by a steam engine. Owner code: KSX

Northeast Mill was established in 1812 at Nottingham, Montgomery Co., PA, and later named the Stubbs Mill.

Northeast Mill (est. 1866) at Nottingham, Chester Co., PA, made straw board on a cylinder-wire machine powered by a waterwheel. This was the former Stubbs Mill that later burned in 1890 and was rebuilt as the Silvermere Mill. Owner code: JRJ

Northfield Mill at Northfield, Merrimack Co., NH, was established around 1823. Owner Code: JKJ.1

Northrup Mill was established in 1855 at East Lee, Berkshire Co., MA. This plant was rebuilt following fires in 1856, 1872, 1877, and 1881. Owner codes: KGE, JEN

Northumberland Mill (est. 1823) at Northumberland, Coos Co., NH, was a hand mill until it added a cylinder-wire in 1830. The mill made wrapping paper from rag, then straw, and then to manila in 1866 when it added a double-cylinder. The plant was powered by waterwheels, and finally succumbed to fire in 1883. Owner codes: KXP, JHX

Northville Mill (est. 1852) at Northville, Litchfield Co., CT, made book, card middles, and wall paper on a cylinder-wire machine powered by a waterwheel. Owner code: JTY

Norwich Mill (est. 1790) at Norwich, New London Co., CT, was a hand mill that operated until 1849. Owner code: JVL

Novelty Mill (est. 1864) at Zanesville, Muskingum Co., OH, made manila paper on a cylinder-wire machine. The plant was powered by two steam engines. Owner code: KCJ

Oak Grove Mill (est. 1848) at Trumbull, Fairfield Co., CT, made straw board on a cylinder-wire machine. The plant was powered by a steam engine and two waterwheels. Owner code: KQJ

Oak Grove Mills (est. 1873) at S. Manchester, Hartford Co., CT, made card and paste board on a cylinder-wire machine. The plant was powered by a steam engine and a waterwheel. The facility was rebuilt in 1880 for the manufacture of wood pulp. Owner code: KQL

Oak Mill (est. 1874) at N. Amherst, Hampshire Co., MA, made leather boards and straw wrapping on a cylinder-wire machine. The plant was powered by two waterwheels. Owner code: KLC

Oakland Mill (est. 1802) at N. Manchester, Hartford Co., CT, made stationery, bond, and printing paper by hand until installing a moving wire machine in 1828. The plant used a Jordan refiner, and was powered by two steam engines and three waterwheels. The facilities were rebuilt in 1830, 1866, and finally in 1880 upon conversion to wood pulp. Owner codes: JVG, KGH, JVH, JIB, JVO

Octoraro River Mill (est. 1830) at Bart Township, Lancaster Co., PA, was a hand mill that continued until 1838. Owner code: JNY

Oglesby-Moore Mills (2) (est. 1853) at Middletown, Butler Co., OH, made book and blotting paper on a moving-wire machine, as well as carpet lining and construction paper on a cylinder-wire machine. The plant was powered by three steam engines and four waterwheels. Owner code: KGI

Ohio Mills (2) (est. 1876) at Niles, Berrien Co., MI, made straw board and carpet lining on two cylinder-wire machines. The plant had a Jordan refiner, and was powered by a steam engine and nine waterwheels. A second facility was added in 1883 during conversion to wood-pulp. Owner code: KGJ

Ohio Mills (2) (est. 1880) at Middletown, Butler Co., OH, made paper on two moving-wire machines. The plant was powered by four steam engines and two waterwheels. Owner code: KGK

Oley Mill (est. 1812) at Manatawny, Berks Co., PA, was a hand mill that made book paper. The plant added a moving-wire machine in 1836. Owner codes: JJO, JFD

Ondawa Mill (est. 1882) at Middle Falls, Washington Co., NY, made manila paper on two double-cylinder machines. The plant used a Jordan refiner, and was powered by three waterwheels. Owner code: KGM

Onderdonk Mill (est. 1773) at Roslyn, Hempstead Township, Queens Co., NY, was a hand mill until 1831. This facility later became the Valentine Mill. Owner codes: KGN, JUT

Onondaga Mill (est. 1869) at Kecks Center, Onondaga Co., NY, made straw board on a cylinder-wire machine. The plant was powered by two waterwheels. Owner code: KEK

Ontario Mill (est. 1818) at Shortsville, Ontario Co., NY, made manila and straw wrapping paper and filter paper on a cylinder-wire machine, powered by two waterwheels. The facility was rebuilt in 1880 for the manufacture of wood pulp. Owner code: JWU

Oregon City Mill (est. 1866) at Oregon City, Clackamas Co., OR, made rag and straw paper on a cylinder-wire. Owner code: JFZ

Oronoque Mills (est. 1882) at Huntington, Fairfield Co., CT, made manila tissue on a cylinder-wire machine. The plant was powered by a steam engine and two waterwheels. Owner code: KLE

Orrs' Mill (est. 1866) at Troy, Rensselaer Co., NY, made manila paper on a double-cylinder machine. The plant was powered by four waterwheels. Owner code: KBR

Oswego Mill (circa 1880) at Cooperstown, Oswego Co., NY, made manila tissue on a cylinder-wire machine. Owner code: KGR

Oswego River Mill (est. 1870) at Oswego Falls, Oswego Co., NY, made manila tissue and wrapping paper on two cylinder-wire machines. The plant was powered by three waterwheels. Owner code: KUQ

Otagauma Pulp and Paper Mill (est. 1887) at Otagauma, WI, made wood-pulp paper.

Otis Board Mill (est. 1885) at Otis, La Port Co., IN, made paste board and card stock on a wet machine powered by a waterwheel. Owner code: KNT

Otis Falls Mill was established in 1888 at Otis Falls, ME. Owner code: KGQ

Otsego County Paper Works was established in 1872 at Cooperstown, Otsego Co., NY. Owner code: KXY

Owen Mill, Old (est. 1862) at Housatonic, Berkshire Co., MA, made stationery and folio paper on a moving-wire machine. This was the former Potter Mill. Owner code: KGU

Owl Mills (est. 1819) at Putney, Windham Co., VT, was a hand mill until installing two cylinder-wire machines in 1830. The plant later made manila tissue paper, and was powered by a steam engine and three waterwheels. The facilities were rebuilt in 1866. Owner codes: KPT, KLJ

Oxford Mill (est. 1779) at Fountain Green, Chester Co., PA, was a handmade mill with a waterwheel. This plant closed in 1849. Owner codes: JPR

Pacific Mill (est. 1850) at Windsor Locks, Hartford Co., CT, made book, colored mediums, and envelope paper on three moving-wire machines. The plant was powered by eight water turbines. Owner codes: JNU, JNV, KMY

Paddack Mill (est. 1865) at Elbridge, Onondaga Co., NY, made straw board on a cylinder-wire machine. The plant was powered by two waterwheels. Owner code: KGV

Paddack Mill (est. 1868) at Manilus, Onondaga Co., NY, made paste board on a cylinder-wire machine. The plant was powered by two waterwheels. Owner code: KGV

Palm Leaf Mill (est. 1850) at Sand Lake, Rensselaer Co., NY, made straw board on a cylinder-wire machine. The plant was powered by a steam engine and a waterwheel. Owner code: KNX

Palmer Falls Mill (est. 1873) at Corinth, Saratoga Co., NY, made newsprint using a Jordan refiner. The plant had three moving-wire machines, and was powered by nine waterwheels. Owner code: JVF

Palmer Mill (circa 1880) at Brooklyn, Kings Co., NY, made manila paper. Owner code: KGY

Palmyra Mill (est. 1874) at Palmyra, Lenawee Co., MI, made straw and manila wrapping paper on a cylinder-wire machine. The plant was powered by two waterwheels. Owner code: KEB

Pansler Mill (est. 1780) at Alsace Township, Berks Co., PA, was a hand mill that operated until 1839. Owner code: KHA

Papyrus Mill (est. 1865) at Shippensburgh, Cumberland Co., PA, made book paper and straw board on a cylinder-wire machine. Owner code: KNO

Paran Creek Mill (est. 1850) at North Bennington, Bennington Co., VT, made wall paper on a cylinder-wire machine. The plant was powered by three waterwheels. This was the former Hinsdill Mill. Owner code: KPI

Park Mill (est. 1809) at Taunton, Bristol Co., MA, was a hand mill that was removed to the Lincoln Mill. Owner codes: KHC, KHB, KVM, JUF

Park Mill (est. 1873) at Philadelphia, Philadelphia Co., PA, made wrapping paper and construction paper on a cylinder-wire machine. This plant was powered by five steam engines. Owner code: JZJ

Parsons Mill (est. 1853) at Holyoke, Hampden Co., MA, made bond, ledger, stationery, envelope, and card middles on four moving-wire machines. The plant was powered by five steam engines and engine seven waterwheels. Owner code: KHI

Passaic Mill (est. 1836) at Franklin, Essex Co., NJ, made stationery paper on a moving-wire machine. The plant had a Kingston refiner, and was powered by two steam engines and two waterwheels. Owner code: JYH

Passumpsic Mill was established at Passumpsic, VT, about 1880 for the manufacture of wood pulp.

Patchogue Mill (est. 1874) at Patchogue, Suffolk Co., NY, made straw board and wrapping paper on two cylinder-wire machines. The plant was powered by a steam engine and two waterwheels. Owner code: JTH

Patten Mill (est. 1882) at Appleton, Outagamie Co., WI, made book, news, and wrapping paper on two moving-wire machines. The plant had a Jordan refiner, and was powered by eight waterwheels. Owner code: KHO

Pearl Mill (est. 1883) at South Hadley, Hampshire Co., MA, made manila tissue on a double-cylinder machine. The plant was powered by a steam engine and two waterwheels. Owner code: KHR

Peekskill Mill (est. 1774) at Peekskill, Westchester Co., NY, was a hand mill that became the Putnam Mill in 1832. Owner code: JXF

Pembroke Mill at Pembroke, Merrimack Co.,

NH, was established by 1823. Owner code: KXB

Peninsular Mill (est. 1864) at Rochester, Oakland Co., MI, made book and news paper on a cylinder-wire machine. The plant was powered by three waterwheels. Owner code: JCK

Pennsylvania Pulp & Paper (2) (circa 1880) at Lock Haven, Clinton Co., PA, made book and news paper on two moving-wire machines and manila paper on two double-cylinders. The plant was powered by two steam engines and two waterwheels. Owner code: KHY

Pennypack Creek Mill (est. 1792) at Moreland Township, Montgomery Co., PA, was a hand mill that was rebuilt following a fire in 1809. By 1823 it produced high quality stationery, folio, and bond paper. The plant was destroyed by fire in 1858. Owner codes: KAU, KCU

Pepperell Mills (circa 1880) at Pepperell, Middlesex Co., MA, made book, card stock, and colored mediums on two moving wire machines. The plant had a Jordan refiner. Owner code: JOD

Pequea Creek Mill (est. 1853) at Compassville, Chester Co., PA, made straw board. The mill burned down around 1865. Owner code: JNL

Pequosette Mill (est. 1860) at Watertown, Middlesex Co., MA, made manila paper on two cylinder-wire machines, and rag paper on a moving-wire. Owner code: JUP

Perrine Mill (est. 1880) at Franklin, Warren Co., OH, made manila rope on a cylinder-wire machine powered by a steam engine. Owner code: KHZ

Perseverance Mill No. 1 (est. 1867) at Lambertville, Hunterdon Co., NJ, made manila wrapping on one cylinder-wire and one double-cylinder machine. Owner code: KCR

Perseverance Mill No.2 (est. 1876) at Lambertville, Hunterdon Co., NJ, made flour sack paper on a moving-wire machine. The plant had a Kingsland refiner, and the combined facilities were powered by seven steam engines and three water turbines. Owner code: KCR

Pettebone Mill (est. 1883) at Niagara Falls, Niagara Co., NY, made news paper on a moving-wire machine. The plant had a Brightman engine, and was powered by a waterwheel and a steam engine. Owner code: KIC

Philadelphia Inquirer Mill (est. 1865) at Manayank, Philadelphia Co., PA, made book and news on a moving-wire machine. The plant was powered by two steam engines. Owner code: JSP

Philadelphia Mill (circa 1880) at Philadelphia, Philadelphia Co., PA, made manila paper and construction paper on a double-cylinder machine. Owner code: JBX

Philadelphia Mill (est. 1884) at Philadelphia, Philadelphia Co., PA, made board stock on a cylinder-wire machine. The plant was powered by two steam engines. Owner code: KIG

Philmont Mill (circa 1880) at Philmont, Columbia Co., NY, made straw wrappers on two cylinder-wire machines. Owner code: KIL

Phoenix Mill (est. 1818) at South Orange, Essex Co., NJ, was a hand mill that made wrapping and hardware paper until adding a moving-wire machine in 1843. Owner codes: JMV, JQC, JCP, KFL

Phoenix Mill (est. 1822) at South Lee, Berkshire Co., MA, was a hand mill that made book and stationery paper until it closed in 1840. Owner code: KGT

Phoenix Mill (est. 1827) at Western Row, Cincinnati, Hamilton Co., OH, made card and stationery on a cylinder-wire machine powered by a steam engine. The plant closed in 1848 and equipment removed to the Hamilton Mill. Owner code: JRF

Phoenix Mill (est. 1850) at Cuyahoga Falls, Summit Co., OH, made stationery and wrapping paper on a cylinder-wire machine. Owner code: JLN

Phoenix Mill (est. 1872) at Battenville, Washington Co., NY, made straw wrapping on a cylinder-wire machine. The plant had two waterwheels. Owner code: KIE

Phoenix Mill (circa 1880) at Marcellus Falls, Onondaga Co., NY, made straw wrapping paper on two cylinder-wire machines. Owner code: JZM

Phoenix Mill (est. 1884) at Phoenix, Oswego Co., NY, made manila paper on a cylinder-wire machine. Owner code: KIF

Pickering Mill (est. 1827) at N. Windham, Tolend Co., CT, made rag paper on a moving-wire machine. Owner code: KIO

Pine Grove Mill (est. 1818) at Good Hope Township, Hocking Co., OH, was a hand mill that was rebuilt following a fire in 1830. The mill ultimately closed in 1846. Owner code: JRN

Pine Grove Mill (est. 1847) at Houcksville, Carroll Co., MD, made manila and straw wrapping paper on a cylinder-wire machine. The plant was powered by a steam engine and two waterwheels. This was the former Hoffman Mill. Owner code: JXO

Pioneer Mill (est. 1856) at Taylorville, Marin Co., CA, made book, news, and wrapping paper on a cylinder-wire machine. The plant was powered by two waterwheels converted to manila in 1883. Owner code: KQT

Pioneer Mill (est. 1863) at Ballston Spa, Saratoga Co., NY, made rag paper on a moving-wire machine. The plant was powered by two steam engines and three waterwheels. Owner code: KVG

Pioneer Mill (est. 1870) at Paperville, Clarke Co., GA, made rag paper on two moving-wire machines. The plant employed a Jordan refiner and was powered by two steam engines and two waterwheels. Owner code: KIR

Pioneer Mill (est. 1873) at Dalton, Berkshire Co., MA, made linen and linen composite stationery, ledger, and cartridge paper on a moving-wire machine. The plant was powered by two steam engines and two waterwheels. Owner code: JKL

Piqua Mill at Piqua, Miami Co., OH, made straw board on two wet machines. The plant was powered by four steam engines and one water turbine. Owner code: KIS

Pittsburgh Steam Mill (est. 1824) at Northern Liberties, Allegheny Co., PA, was a hand mill that closed in 1826. Owner code: KHN

Pittsford Mill (circa 1880) at Pittsford, VT, manufactured wood pulp paper.

Pleasant Garden Mill (est. 1870) at New London, Chester Co., PA, made manila rope on a double-cylinder machine. The mill was powered by two steam engines and a waterwheel. Owner code: KJV

Pleasant Mills (est. 1861) at Pleasant Mills, Atlantic Co., NJ, made manila paper on a double-cylinder machine. This mill was powered by a steam engine and three waterwheels. Owner code: KIZ

Pleasant Valley Mill (est. 1835) at Lenox, Berkshire Co., MA, made news paper on a cylinder-wire machine. Owner codes: KQY, KMV, KIX

Pleasant Valley Mill (est. 1883) at N. Lyme, New London Co., CT, made colored mediums on a moving-wire machine powered by a waterwheel. Owner codes: JEC, JLZ

Plover Mill (est. 1892) at Whiting, WI, made wood-pulp paper on two moving-wire machines. Owner code: KJA

Poland Mill (est. 1852) at Mechanic Falls, Androscoggin Co., ME, made book and news paper on a moving-wire machine. The plant was powered by a steam engine and three waterwheels. Owner code: JMG

Pond Lilly Mill (circa 1880) at Westville, New Haven Co., CT, made manila and news on a cylinder-wire machine powered by a waterwheel. Owner codes: JSU, JSV

Porter Mill (est. 1842) at E. Lampter Township, Lancaster Co., PA, was a hand mill that was rebuilt in 1850 following a fire. Owner code: JNY

Potomac Mills (est. 1876) at Washington, DC, made book and news on a moving-wire machine and bogus manila on a double-cylinder. The plant had a Jordan refiner and also repulped U.S. currency for papermaking in 1876. This was the former Hill Mill. Owner code: JUA

Potter Mill (est. 1854) at Housatonic, Berkshire Co., MA, made stationery and bond paper on a moving-wire machine. This plant later named the Owen Mill. Owner code: KGT

Powers Paper Mill (est. 1889) at Holyoke, Hampden Co., MA, made stationery, envelopes, and card stock from wood-pulp. Owner code: KJE

Providence Mill (est. 1881) at Elkton, Cecil Co., MD, made book and news paper on a moving-wire machine. The plant was powered by two steam engines and a waterwheel. Owner code: KNR

Puget Sound P&P Mill (est. 1892) at Everett, WA, made paper using a Jordan refiner on two moving-wire machines. The plant was powered by three steam engines. Owner code: KJJ

Pulaski Straw Board Mill (est. 1840) at Pulaski, Oswego Co., NY, made straw board on a cylinder-wire machine. The plant was powered by two waterwheels. Owner code: KGS

Putnam Mill (est. 1872) at Peekskill, Westchester Co., NY, made manila tissue on a cylinder-wire machine. The plant was powered by two steam engines and two waterwheels. This was the former Peekskill Mill. Owner code: JSI

Queechy Mill (est. 1858) at Canaan Four Corners, Columbia Co., NY, made straw wrapping and box board on a cylinder-wire machine. This plant was powered by two steam engines and two waterwheels. Owner code: KJP

Quinnesee Mill (est. 1892) at Quinnesee, Marinette Co., WI, made wood pulp paper. Owner code: KJR

Rahn Mill (est. 1809) at Maidencreek Township, Berks Co., PA, was a hand mill that continued through 1829. Owner code: KJS

Rahway River Mill (est. 1810) at Millburn, Essex Co., NJ, was a hand mill that began making board stock in 1860. Owner code: KTF, KQX, KVY

Rainbow Mills (est. 1864) at Poquonock, Hartford Co., CT, made book, printing, and colored mediums on two moving-wire machines. The plant had a Jordan refiner, and was powered by five waterwheels. Owner code: JTA

Rainbow Tissue Mill (est. 1860) at Rainbow, Hartford Co., CT, made tissue paper on a cylinder-wire machine. The plant was powered by two waterwheels. Owner code: JUE

Ravine Mill (circa 1880) at Coeymans, Albany Co., NY, made straw wrapping paper on a cylinder-wire machine.

Ravine Mill (est. 1883) at Appleton, Outagamie Co., WI, made book and stationery paper on a moving-wire machine. The plant hand a Jordan refiner and was powered by six waterwheels. Owner code: JPA

Reading Mill (circa 1880) at Reading, Berks Co., PA, made book and printing paper on three moving-wire machines. The plant had a Kingston refiner. Owner code: JCO

Readsboro Mill (circa 1880) at Readsboro, VT, made wood-pulp paper.

Rebecca Mill (est. 1881) at Bridgeport, Montgomery Co., PA, made manila-colored mediums on a moving-wire machine. The plant was powered by two steam engines. Owner code: KCZ

Record Leather Board Mills (2) (est. 1870) at Livermore Falls, Androscoggin Co., ME, made board and card stock on two cylinder-wire machines. The plant was powered by four waterwheels. Owner code: KJZ

Red Mill (est. 1801) at Dalton, Berkshire Co., MA, made hand-crafted stationery and printing paper until installing a waterwheel in 1826. The mill continued by hand until starting-up a cylinder-wire machine in 1831. This facility burned down in 1870. Owner codes: JKQ, JKL

Red Mill, Old (est. 1809) at Dalton, Berkshire Co., MA, hand-crafted stationery paper until installing a waterwheel in 1823. The mill continued by hand until installing a cylinder-wire machine in 1831. Owner code: JHF, JKQ

Red River Mill (est. 1883) at Fergus Falls, Otter Tail Co., MN, made carpet lining and straw wrapping on a board machine. The plant was powered by two waterwheels. Owner code: KKA

Redstone Mill (est. 1796) at Brownsville, Fayette Co., PA, was a hand mill until it installed a cylinder-wire machine in 1825. The mill made news paper until destroyed by fire in 1844. Owner codes: JWH, KMZ

Redstone Mill (est. 1808) at St. Louis, MO, was a hand mill that made news paper until 1829. Owner code: JKJ

Remington Mill (est. 1865) at Watertown, Jefferson Co., NY, made news paper on two moving-wire machines. The plant was powered by ten waterwheels. Owner code: KKC

Remington Mill (est. 1881) at Brownville, Jefferson Co., NY, made news paper on a moving-wire machine. The plant operated a Brightman engine, and was powered by seven water turbines. Owner code: KKE

Remington Mill (est. 1883) at Agnews, Santa Clara Co., CA, made wood pulp paper on a moving-wire. The plant had Jordan refiner, and was powered by two steam engines. This mill closed in 1898. Owner code: KKD

Rensselaer Mill (est. 1846) at Nassau, Rensselaer Co., NY, made straw wrappers on two cylinder-wire machines. Owner code: KQZ

Rialto Mill (est. 1848) at Rialto, Hamilton Co., OH, made book and wrapping paper on a moving-wire machine. The plant was powered by a steam engine and a waterwheel. By 1868 the plant was listed at Lockland, Hamilton Co., OH, and was making straw paper on a cylinder-wire machine. Owner code: JPM

Rice & Garfield Mill (est. 1831) at Needham, Middlesex Co., MA, made paper collars, book, and news paper on a cylinder-wire machine. Mill was torn down in 1886. Owner codes: KBK, JVM, KKJ, KKN, JOX

Richardson Mill (est. 1883) at Monroe, Monroe Co., MI, made straw wrapping on a double-cylinder machine. The plant was powered by two steam engines. Owner code: KKR

Richmond Mill (est. 1831) at Richmond, Wayne Co., IN, was a hand mill that continued through 1839. Owner code: JZR

Richmond Mill (est. 1854) at Appleton, Outagamie Co., WI, made manila paper on a cylinder-wire machine. The plant was powered by six waterwheels. Owner code: KKU

Richmond Mill (est. 1882) at Philadelphia, Philadelphia Co., PA, made book paper on two moving-wire machines. The plant was powered by three steam engines. Owner code: JCE

Richmond Mill (est. 1883) at Providence, Providence Co., RI, made newsprint on two moving-wire machines. Owner code: KKS

Ridley Creek Mill (est. 1765) at Upper Providence Township, Delaware Co., PA, was a hand mill that converted to textile manufacturing in 1818. Owner codes: KAZ, KWV, KAE, KCD

Riege Mill (est. 1866) at Riegelsville, Warren Co., NJ, made manila paper on a cylinder-wire machine. The plant was powered by two steam engines. Owner code: KKW

Rimmon Falls Mill (est. 1803) at Seymour, New Haven Co., CT, was a hand mill that installed a cylinder-wire machine powered by a waterwheel in 1840. Owner code: JVI, KKX

Rising Sun Mill (est. 1871) at Rising Sun, Cecil Co., MD, made colored mediums on a moving-wire machine. The plant was powered by two waterwheels. This was the former Cecil Mill. Owner code: KVA

Rittenhouse Mill (est. 1690) at Germantown, Philadelphia Co., PA, was a hand-craft mill that made printing paper. Installed a waterwheel around 1855 and continued until about 1855. Owner code: KKZ

Riverdale Mill (circa 1880) at S. Manchester, Hartford Co., CT, made card stock and board stock on a wet machine. The mill was powered by a steam engine and a waterwheel. Owner code: JHB.1

Riverside Mill (est. 1867) at Holyoke, Hampden Co., MA, made card stock and stationery on two moving-wire machines. The plant was powered by three steam engines and two water turbines. Owner code: KLA

Riverside Mill (est. 1857) at Manayank (aka Lafayette Station), Philadelphia Co., PA, made book, envelope, and stationery on two moving-wire machines. The plant had a Jordan refiner, Holland refiner and a Gould machine, and was powered by two steam engines. Owner code: JSD

Riverside Mill (circa 1880) at Castleton, Rensselaer Co., NY, made straw wrappers on two cylinder-wire machines. Owner code: JHL

Riverside Paper Mill was established in 1891 at Berlin, NH. Owner code: JDP

Riverton Mill (est. 1870) at Springfield, Sangamon Co., IL, made straw wrapping on a cylinder-wire machine. The plant was powered by a steam engine. Owner code: JEI

Roaring Brook Mill (est. 1872) at East Glastonbury, Hartford Co., CT, made card stock on a cylinder-wire machine powered by a waterwheel. Mill was rebuilt in 1884 and converted to wood pulp. Owner code: KLB

Roaring Springs Mill (est. 1866) at Roaring Springs, Blair Co., PA, made book and news paper on two moving-wire machines. The plant had a Jordan refiner, and was powered by three steam engines and a waterwheel. Owner code: KES

Roberts Mill (circa 1880) at Waltham, Middlesex Co., MA, made wrapping paper on two cylinder-wire machines. Owner code: KLD

Robertson Brothers Mill (est. 1882) at Hinsdale, Cheshire Co., NH, made manila tissue on a cylinder-wire machine. The mill was powered by six waterwheels. Owner code: KLF

Robertson Mill (est. 1880) at Bellows Falls, Windham Co., VT, made manila tissue on a cylinder-wire. The mill was powered by two water turbines. Owner code: KLI

Rochester Mill (est. 1818) at Rochester, Monroe Co., NY, was a hand mill that was destroyed by fire in 1827. Owner codes: JQL, KHW

Rock Creek Mill (est. 1805) at P Street, Washington, DC, was a hand mill that continued until 1848. Owner code: KHM

Rock Falls Mill (est. 1882) at Rock Falls, Whiteside Co., IL, made straw wrapping on a cylinder-wire machine. The plant was powered by a steam engine and two waterwheels. Owner code: JIK

Rock Falls Mills (est. 1853) at Rock City Falls, Saratoga Co., NY, made newsprint on a Harper machine. The mill was powered by two steam engines and three water turbines. Owner codes: JYA, JFY

Rock Mill (est. 1870) at Kecks Center, Onondaga Co., NY, made straw board on a cylinder-wire machine. The plant was powered by two waterwheels. Owner code: JJX

Rock Mill (est. 1872) at Little Falls, Herkimer Co., NY, made wrapping paper, construction paper, and carpet lining on a cylinder-wire machine. The plant was powered by three waterwheels. Plant was rebuilt in 1884 and converted to wood pulp. Owner code: JYJ

Rock River Mills (3) (circa 1880) at Beloit, Rock Co., WI, made construction paper and both manila and straw board and wrapping on five cylinder-wire machines. Owner code: KLN

Rockdale Mill (est. 1871) at Newington, Hartford Co., CT, made collar paper and card stock on a cylinder-wire machine. The plant was powered by a steam engine and waterwheel. Owner code: KVN

Rockdale Mills (2) (est. 1859) at Hoffmanville, Baltimore Co., MD, made book, news, and writing on a moving-wire machine. The plant was powered by a steam engine and two waterwheels. A second facility was erected in 1865. Owner code: JUH

Rockdale Mills (circa 1880) at Hamilton, Butler Co., OH, made straw board on two double-cylinder machines. Owner code: KQO

Rockford Mill (est. 1880) at Rockford, Winnebago Co., IL, made straw wrappers on a cylinder-wire machine. Owner code: JXI, KKH

Rockford Mill (est. 1882) at Rockford, Floyd Co., IA, made wrapping paper on a cylinder-wire machine. The plant was powered by two waterwheels. Owner code: JNN

Rockingham Mill (est. 1802) at Bellows Falls, Windham Co., VT, was a hand mill that was rebuilt after a fire in 1812. This mill continued until 1845. Owner codes: JYG, JLM, JEM

Rockland Mill (est. 1851) at Montville, New London Co., CT, made manila paper on a cylinder-wire machine. This plant had a Jordan refiner, and was powered by a steam engine and two waterwheels. Owner code: KLG

Rockland Mill (circa 1880) at Wilmington, New Castle Co., DE, made news paper on three moving-wire machines. Owner code: JWN

Rockwood Mill (est. 1877) at Rockwood, Fulton Co., NY, made straw board on a cylinder-wire machine. This plant was powered by a steam engine and a waterwheel. Owner codes: KPE, KPF

Rogers Mill (est. 1850) at Philmont, Columbia Co., NY, made straw wrapping paper on a cylinder-wire machine. This plant was powered by a steam engine and two waterwheels. Owner code: KLR

Rokerby Mill (est. 1866) at Ercildoun, Chester Co., PA, made board stock on a wet machine. The plant was powered by a steam engine and a water turbine. Owner code: JFP

Rollstone Mill, Old (est. 1864) at Fitchburg, Worcester Co., MA, made news paper on a moving-wire machine. This plant had a Jordan refiner, and was powered by two steam engines and a waterwheel. Owner code: KVS

Rollstone Mill (est. 1884) at Fitchburg, Worcester Co., MA, made news paper on a moving-wire machine. The plant had a Jordan refiner, and was powered by two steam engines and a waterwheel. Owner code: KVS

Royer Ford Mill (circa 1880) at Spring City, Chester Co., PA, made card stock and filter paper on a cylinder-wire and two moving-wire machines. The plant was powered by five steam engines. Owner code: JAX

Rozet Mill (est. 1868) at Three Rivers, St. Joseph Co., MI, made book and news paper on a cylinder-wire machine. The plant had a Brightman engine. Owner code: JPK

Ruckawa Mill (est. 1885) at Ruckawa, Montgomery Co., OH, made board stock on two wet machines powered by a steam engine. Owner code: JYM

Rumford Falls Mill (est. 1897) at Rumford Falls, ME, made wood-pulp paper on a moving-wire machine. Owner code: KLY

Russell Mill (est. 1840) at Russell, Hampden Co., MA, made stationery paper on two moving-wire machines. The plant was powered by two steam engines. Owner codes: KOA, KSX

Russell Mill, Old (est. 1864) at Lawrence, Essex Co., MA, made book and stationery paper on a moving-wire machine. Owner code: KMB

Russell Mill (est. 1880) at Lawrence, Essex Co., MA, made wood-pulp news and book paper on four moving-wire machines. The mill was powered by eight steam engines and eight water turbines. Owner code: KMB

Russell Willard & Co. Mills (2) (circa 1880) at Bellows Falls, Windham Co., VT, made manila paper on a double-cylinder machine. The plant had a Jordan refiner and was powered by four waterwheels. Owner code: KMC

Rutland Mill (est. 1818) at Rutland, Rutland Co.,

VT, made paper by hand until about 1845. Owner code: JRA

Saccarapa Mills (est. 1862) at Saccarapa, Cumberland Co., ME, made leather card and board stock on a cylinder-wire machine. The plant was powered by a steam engine and five waterwheels. Owner code: JLV

Sadsbury Mill (est. 1860) at Atglen, Chester Co., PA, made straw board on a cylinder-wire machine powered by a waterwheel. This facility later became the Fern Rock Mill. Owner code: KDG, JDT

St. Charles Mill (est. 1840) at St. Charles, Kane Co., IL, made straw wrapping on a cylinder-wire machine. This plant was powered by three waterwheels. Owner code: JFV

St. Johnsville Straw Wrapping Mill (est. 1873) at St. Johnsville, Montgomery Co., NY, made straw wrapping on a cylinder-wire machine. This plant was powered by a steam engine and two waterwheels. Owner code: JKD

St. Joseph Mills (2) (est. 1881) at St. Joseph, Berrien Co., MI, made straw wrapping, board, and bogus manila on a cylinder-wire machine and a wet machine powered by a steam engine. Owner code: KPD

St. Joseph Valley Mill (est. 1873) at Elkhart, Elkhart Co., IN, made stationery paper on a moving-wire machine. The plant was powered by seven waterwheels. Owner code: JNX

St. Papyrus Mill (est. 1790) at Shippensburgh, Cumberland Co., PA, made straw board and wrapping paper on a cylinder-wire machine. The plant was powered by two steam engines and two waterwheels. Owner code: KNN

Salem Mill (est. 1791) at Salem, Stokes Co., NC, made paper by hand until about 1855.

Salmon Falls Mill was established in 1850 at Rollinsford, NH. Owner code: JRW

Salmon River Board Mill (est. 1880) at Pulaski, Oswego Co., NY, made board stock on a wet machine. This mill was powered by four waterwheels. Owner code: KQH

Sammonsville Mill (est. 1881) at Sammonsville, Fulton Co., NY, made straw board on a cylinder-wire machine. This plant was powered by a steam engine and two waterwheels. Owner code: KVB

Sampson Mill (circa 1880) at Charles River Village, Norfolk Co., MA, made leather board on two wet machines. Owner code: KME

San Geronimo Mill (est. 1884) at Taylorville, Marin Co., CA, made wood-pulp book, news, and colored mediums on a moving-wire machine. This plant was powered by a steam engine and two waterwheels. Owner code: KQT

Sandusky Mill (est. 1880) at Sandusky, Erie Co., OH, made straw wrappers on a cylinder-wire machine powered by a steam engine. Owner code: KMF

Sasco Creek Mill (est. 1870) at Southport, Fairfield Co., CT, made straw board on a cylinder-wire machine. The plant was powered by a steam engine and two waterwheels. Owner codes: JOS, KTL

Satterly Mill (est. 1850) at Little Falls, Herkimer Co., NY, was partially burned in 1853. Owner code: KMH

Saugerties Mill (est. 1827) at Saugerties, Ulster Co., NY, made news paper on two moving-wire machines. This plant burned down in 1872. Owner codes: JCI, JCT, KND

Saunderson Mill (est. 1801) at Milton, Norfolk Co., MA, was a hand mill until it installed a cylinder-wire machine in 1829. Plant was powered by a single waterwheel and closed in 1834. Owner code: KMI

Savannah Mill (circa 1880) at Savannah, Chatham Co., GA, made wood pulp paper on two moving-wire machines. The plant had a Jordan refiner, and was powered by two steam engines. Owner codes: JSC, KMJ

Scandaga Mill (est. 1836) at Broadalbin, Fulton Co., NY, made straw wrapping in a cylinder-wire machine. The plant was powered by two waterwheels. Owner code: JEL

Schauck Mill (est. 1824) at Bentley Springs, Baltimore Co., MD, was a hand mill that continued until around 1854, when it became the Eagle Mills. Owner code: KMM

Schenectady Mill (est. 1808) at Schenectady, Schenectady Co., NY, was a hand mill that continued until about 1831. Owner code: KSU

Schmidt & Ault Mill (est. 1897) at York, York Co., PA, made wood-pulp board stock. Owner code: KMN

Schutz Mills (2) (est. 1778) at Lower Merion Township, Montgomery Co., PA, was a hand mill until about 1835, when it became the Dove Mill. Owner codes: KMO, JTQ

Schuylerville Mill (est. 1863) at Schuylerville, Saratoga Co., NY, made book, card, news, and stationery on two moving-wire machines. This plant was powered by two steam engines and three waterwheels. Owner code: JGC

Schuylkill Mill (est. 1825) at E. Manayank, Philadelphia Co., PA, made book and news paper on a moving-wire machine. The plant had a Jordan refiner, and was powered by both a steam engine and a waterwheel. Owner codes: JNR, KCS

Scioto Mill (est. 1882) at Kenton, Harding Co., OH, made straw board on a cylinder-wire machine. Owner code: KMQ

Seaman Mill (circa 1880) at Ridgewood, Queens Co., NY, made straw board on a cylinder-wire machine. Owner code: KMU

Seneca Mill (est. 1884) at Penn Yan, Yates Co., NY, made news paper on two moving-wire machines. The plant had a Jordan refiner, and was powered by three steam engines and four waterwheels. Owner code: JBH

Severance Mill (est. 1853) at Manayank, Philadelphia Co., PA, made carpet lining, construction paper, and board stock on two double-cylinder machines. The plant was powered by two steam engines. Owner code: KPO

Seymour Mill (est. 1801) at Seymour, New Haven Co., CT, was a hand mill until the 1820s or 1830s, after which it made news and colored mediums on a cylinder-wire machine. The plant was rebuilt in 1850. Owner codes: KJL, JQJ, JCV

Sharon Mill (est. 1801) at Sharon, VT, was a handmade plant that continued until around 1844. Auxiliary equipment was powered by a waterwheel. Owner code: JGF

Shawshin River Mill (est. 1789) at Andover, Essex Co., MA, was a hand mill that was sold for other uses in 1820. Owner code: KII

Sheffield Mills (2) (est. 1828) at Saugerties, Ulster Co., NY, made stationery paper on a cylinder-wire and two moving-wire machines. The plant had two Jordan refiners, and was powered by two steam engines and two water turbines. Owner code: KND

Shelby Mill (est. 1851) at Shelby, Orleans Co., NY, made manila or straw wrapping paper on a cylinder-wire machine. This plant was powered by a steam engine and a water turbine. Owner code: KOS

Sherman Mill (est. 1870) at Belfast, Waldo Co., ME, made leather card on a cylinder-wire machine. The plant was powered by two waterwheels. Owner code: KNF

Shiawassee Mill (est. 1866) at Vernon, Shiawassee Co., MI, made straw wrapping paper on a cylinder-wire machine. This plant was powered by a steam engine and two waterwheels. Owner code: KFO

Shirley Mill (est. 1883) at Bridgeton, Cumberland Co., NJ, made colored mediums on a moving-wire machine. This plant was powered by two steam engines. Owner code: KEZ

Shober Mill (est. 1806) at North Salem, Forsyth Co., NC, was a hand mill that closed in 1838. Owner code: KNH

Shober Mill was established in 1860 at Eden, Lancaster Co., PA.

Short Creek Mill (est. 1802) at Mt. Pleasant Township, Jefferson Co., OH, was a hand mill that continued until around 1847. Owner code: KSE

Short Hills Mill (est. 1840) at Short Hills, Essex Co., NJ, made printing and wall paper on a Harper machine. This plant was powered by two steam engines and two waterwheels. Owner code: JGV

Silver Lake Mill (est. 1834) at Gibsonville, Columbia Co., NY, made manila wrapping on a cylinder-wire and a double-cylinder machine. The plant was powered by two water turbines. Owner code: KVG

Silver Spring Mill was established at Millburn, Essex Co., NJ, around 1870.

Silver Stream Mill at Norwalk, Fairfield Co., CT, made manila construction paper on a cylinder-wire machine. Owner code: JVE

Silvermere Mill (est. 1890) at Nottingham, Chester Co., PA, made straw board on a cylinder-wire machine powered by a waterwheel. This was the former Northeast Mill. Owner code: JRK

Skaneateles Mill (est. 1865) at Skaneateles, Onondaga Co., NY, made book and news paper on a moving-wire machine. The plant had a Kingston refiner and was powered by three waterwheels. The facilities were rebuilt in 1875. Owner code: KNS

Skowhegan Mill (est. 1848) at Skowhegan, Somerset Co., ME, made manila paper on a cylinder-wire machine. The plant was powered by a steam engine and three waterwheels. Owner code: JRX, KDK

Slackwater Mill (est. 1866) at Lancaster, Lancaster Co., PA, made book and news paper on a moving-wire machine. This plant had a Jordan refiner and was powered by five waterwheels. Owner code: KNI

Slee's Mill (est. 1815) at Poughkeepsie, Duchess Co., NY, was a hand mill that continued until around 1838. Owner code: KQF

Smith & Bassett (est. 1834) at Seymour, New Haven Co., CT, made straw wrapping on a cylinder-wire machine. Owner code: KNZ

Smith Brothers Mill (est. 1822) at Millburn, Essex Co., NJ, was a hand mill that turned to making board stock. Plant was abandoned in 1872. Owner code: KON

Smith Mill (circa 1880) at Hempstead, Queens Co., NY, made straw board on a cylinder-wire machine. Owner code: KOM

Smiths Mill (est. 1831) at Seymour, New Haven Co., CT, made manila paper on a double-cylinder machine. The plant was powered by three waterwheels. Owner code: KOO

Smithton Mill (est. 1868) at Smithton, Westmoreland Co., PA, made straw wrapping on a cylinder-wire machine. The plant was powered by two steam engines. Owner code: KOG

Snow Falls Mill (est. 1881) at Snow Falls, ME, made wood pulp paper.

Snow Mill (est. 1850) at Fitchburg, Worcester Co., MA, made news and card stock on a moving-wire machine. The plant had a Jordan refiner, and was powered by two steam engines and two waterwheels. Owner code: JKZ

Snow Mill (est. 1884) at S. Fitchburg, Worcester Co., MA, made manila wall paper on three double-cylinder-machines. The plant was powered by a steam engine and three waterwheels. Owner code: KOQ

Solitude Mill (est. 1881) at Downington, Chester Co., PA, made board stock on a cylinder-wire machine. The plant was powered by a steam engine and two waterwheels. Owner code: KDT

South Bend Mill (est. 1869) at South Bend, St. Joseph Co., IN, made book and news paper on a moving-wire machine. Mill was powered by two waterwheels. Owner code: KKF

South Coast Mill (est. 1880) at Soquel, Santa Cruz Co., CA, made straw wrapping on a cylinder-wire machine. This mill was powered by two steam engines and two waterwheels. Owner code: KGG

South Hadley Mill (est. 1824) at Canal Village, Hampshire Co., MA, made book and stationery on three cylinder-wire machines. The plant was powered by waterwheel(s). Sold and later became the Manila Mill. Owner code: JBB

South Huntington Mill was established in 1846 at South Huntington Township, Westmoreland Co., PA. Owner code: KBY

South Lee Mill (est. 1804) at South Lee, Berkshire Co., MA, began as a hand-craft mill until expanding and adding a waterwheel in 1808. The plant made book, ledger, printing, envelope, and stationery paper. It was later renamed the Irving Mill. Owner codes: JIJ, KGT, KOT

South Mill (est. 1872) at Ypsilanti, Washtenaw Co., MI, made book, card, and colored mediums on a moving-wire machine. Owner code: KHX

Southford Mill (est. 1881) at Southford, New Haven Co., CT, made board stock on a cylinder-wire machine. This mill was powered by two steam engines and four water turbines. Owner code: KOV

Southwestern Mill (circa 1880) at Lawrence, Douglas Co., KS, made straw wrapping paper. Owner code: KOW

Southworth Mill (est. 1839) at Mittineague, Hampden Co., MA, made linen and linen composite bond, ledger, and stationery paper on two moving-wire machines. The plant was powered by a steam engine and three water turbines. Owner code: KOX

Spaulding Brothers Mill (est. 1877) at Towsend Harbor, Middlesex Co., MA, made leather board on a cylinder-wire machine. Owner code: KOY

Spring Grove Mill (est. 1864) at Spring Forge, York Co., ME, made newsprint on three moving-wire machines. The plant was powered by four steam engines and two waterwheels. Owner code: JQN

Spring Hill Mill (est. 1790) at Middletown Township, Delaware Co., PA, was a hand mill that continued until around 1854. Owner codes: JZZ, JZY

Spring Lawn Mill (est. 1829) at Lewisville, Chester Co., PA, made book and colored mediums on a moving-wire machine. The mill had a Brightman engine, and converted to wood pulp in 1870. Owner codes: KNE, KDF, JAV, JUL

Spring Valley Mill (est. 1874) at Otis, La Port Co., IN, was powered by two steam engines. Owner code: JHO

Springdale Mill (est. 1882) at Westfield, Hampden Co., MA, made stationery paper on a moving-wire machine. The plant was powered by a steam engine and a waterwheel. Owner code: KLL

Springfield Mill (est. 1790) at Springfield Township, Delaware Co., PA, was a hand mill that closed in 1838. Owner code: KAD

Springfield Mill (est. 1800) at Springfield, Hampshire Co., MA, handmade printing and stationery paper. The plant was powered by a waterwheel. Owner codes: JBA, JBB

Springfield Mill (est. 1838) at Rainbow, Hartford Co., CT, made blotting paper and linen composite ledger paper on a moving-wire machine. The plant had a Jordan refiner, and was powered by a steam engine and three waterwheels. Owner code: KPC

Springfield Mill was established in 1866 at Springfield, Union Co., NJ. Owner code: JRO

Springfield Mill (est. 1875) at Springfield, Sangamon Co., IL, made straw wrapping on a cylinder-wire machine. The plant was powered by two steam engines. Owner code: JEU

Squampaumick Valley Mill (est. 1867) at Mellenville, Columbia Co., NY, made stationery and wrapping paper on a cylinder-wire machine. The plant was powered by two waterwheels. Owner code: KOI

Squannacook Mill (est. 1868) at West Groton, Middlesex Co., MA, made leather board on two wet machines. This plant was powered by a steam engine and three waterwheels. Owner code: JUA

Star Mill (est. 1851) at Lockland, Hamilton Co., OH, made manila construction, hardware, wrapping, and carpet lining on a cylinder-wire machine. The plant was powered by two steam engines and a waterwheel. Owner code: KQP

Star Mill (est. 1852) at Mechanic Falls, Androscoggin Co., ME, made book and news paper on a moving-wire machine. The plant was powered by a steam engine and three waterwheels. Owner code: JMG

Star Mill (circa 1880) at Shortsville, Ontario Co., NY, made news and wrapping paper on a cylinder-wire machine. Owner code: KPG

Starin Mill (est. 1810) at Esperance, Schoharie Co., NY, was a hand mill that continued until around 1846. Owner code: KPH

State Line Mill (est. 1874) at North Bennington, Bennington Co., VT, made wall paper on a cylinder-wire machine. This plant was powered by a steam engine and three waterwheels. Owner code: KPI

State Mill (circa 1880) at South Bend, St. Joseph Co., IN, made book and news paper on two moving-wire machines. This mill had a Jordan refiner, and was powered by four steam engines. Owner code: JVS

Steelville Mill (est. 1788) at West Fallow Township, Chester Co., PA, was a hand mill that added a cylinder-wire machine and waterwheel in the 1830s. Plant was demolished in 1855. Owner codes: KPL, KPM, JDS

Sterling Mill (circa 1880) at Sterling, Whiteside Co., IL, made straw wrapping on a double-cylinder machine. Owner code: KPP

Stewart Mill (circa 1880) at Brookville, Franklin Co., IN, made news paper on a moving-wire machine.

Stockport Mills (2) (est. 1809) at Stockport, Columbia Co., NY, made straw wrapping on three cylinder-wire machines. The plant was powered by ten waterwheels. Owner code: KLU, JIG

Stone Fort Mill, Old (est. 1861) at Manchester, Coffee Co., TN, made book and printing paper on a cylinder-wire machine. Owner code: KWC

Stone Fort Mill (est. 1868) at Manchester, Coffee Co., TN made book, news, and wrapping paper on a cylinder-wire and a moving-wire machine. The plant was powered by two steam engines and three waterwheels. Mill was rebuilt in 1879 and again in 1884, when it converted to wood pulp. Owner code: KPT.1

Stone Mill (est. 1810) at Mount Vernon, Chester Co., PA, produced handmade news paper. Auxiliary equipment was powered by a waterwheel. Mill failed in 1864 and reopened in 1870. Owner codes: JMO, JMM, JMN, KXX, KCW

Stone Mill (est. 1844) at Dalton, Berkshire Co., MA, made linen composite ledger and manuscript paper on a moving-wire machine. This plant was powered by two steam engines and two waterwheels. Facilities were destroyed by fire in 1869. Owner code: JKL

Stone Mill (est. 1850) at Fitchburg, Worcester

Co., MA, made news and card middles on a moving-wire. This mill had a Jordan refiner, and was powered by two waterwheels and four steam engines. Owner code: JKZ

Stone Mill (est. 1857) at Adams, Berkshire Co., MA, produced handmade ledger, stationery, and folio paper. Owner code: JFU

Stone Mill (est. 1866) at Stanfordville, Duchess Co., NY, made straw wrapping on a cylinder-wire machine. The plant was powered by two steam engines and four waterwheels. Owner code: JXY

Stone Mill (est. 1875) at Chatham, Columbia Co., NY, made bogus manila and wrapping paper on a cylinder-wire machine. The plant was powered by two waterwheels. Owner code: KOZ

Stone Ridge Mill (est. 1875) at Stone Bridge, Ulster Co., NY, made straw wrapping paper on a cylinder-wire machine. The plant was powered by two steam engines and a waterwheel. Owner code: KFP

Stony Brook Mill (circa 1880) at Chatham, Columbia Co., NY, made straw wrapping on a cylinder-wire machine. Owner code: KHQ

Straw Board Mill (est. 1881) at Beloit, Rock Co., WI, made construction paper on a board machine. This plant was powered by four waterwheels. Owner code: JDJ

Straw Pulp Mill was established at Dansville, Livingston Co., NY, circa 1870. Owner code: KXU

Streator Mill (circa 1880) at Streator, La Salle Co., IL, made straw wrapping paper. Owner code: KXB

Strong's Mill (est. 1870) at N. Westchester, New London Co., CT, made construction paper on a wet machine. Owner code: KPV, KGF

Stubbs Mill (est. 1838) at East Nottingham, Chester Co., PA, made rag paper on a cylinder-wire machine, converting to straw board in 1851. This mill was powered by a waterwheel, and was the former Northeast Mill. Owner code: KPW

Stuyvesant Falls Mill (est. 1813) at Stuyvesant Falls, Columbia Co., NY, made rag paper on a cylinder wire machine. The mill converted to straw production in 1840 and was destroyed by fire in 1857. Owner code: JUG, KRK

Sudbury Mill was established in 1876 at W. Cummington, Hampshire Co., MA. Owner code: KOU

Suffield Mill (est. 1848) at Scitico, Hartford Co., CT, made manila paper on a cylinder-wire machine. This plant was powered by a steam engine and two waterwheels. Owner code: KPU

Suffield Mill (est. 1866) at Suffield, Hartford Co., CT, made colored mediums and envelope paper on a moving-wire machine. This mill used a Jordan refiner, and was powered by a steam engine and two water turbines. Owner code: KVX

Suffolk Mill (est. 1807) at Dorchester, Suffolk Co., MA, was a hand mill that continued until 1827. Owner code: JCA

Sugar Creek Mill (circa 1880) at Atlanta, Fulton Co., GA, made manila paper on a cylinder-wire machine. Owner code: KPY

Sugar House Mill (est. 1861) at Salt Lake City, Salt Lake Co., UT, made book, news, printing, and wrapping paper on a cylinder-wire machine. This was later the Deseret Mill. Owner code: JUY

Sugar River Mill (est. 1866) at Claremont, Sullivan Co., NH, made book and news paper on two moving-wire machines. This plant was powered by two steam engines and three waterwheels. Owner code: KPZ

Suncook Falls Mills at Pembroke, Merrimack Co., NH, was established by 1823. Owner code: KJG

Sunny Side Mill (est. 1882) at Unionville, Hartford Co., CT, made leather cards and boards on a cylinder-wire machine and a wet machine. Owner code: KKY

Sunnydale Mill (est. 1815) at Beaver Valley, Delaware Co., PA, made manila tissue on a cylinder-wire machine. This plant was powered by a steam engine and a waterwheel. The mill was rebuilt in 1824 following a fire. Owner code: KQT.1

Superior Mill (circa 1880) at Ypsilanti, Washtenaw Co., MI, made book and news paper on two cylinder-wire machines. This plant had a Jordan refiner. Owner code: KYP

Susquehanna Water Power & Mill (est. 1883) at Conowingo, Cecil Co., MD, made news and wall paper on two moving-wire machines. This plant had a Jordan refiner and was powered by six waterwheels. Owner code: KQC

Sutphin-Wrenn Mill at Middletown, Butler Co., OH, made book, news, and blotting paper on a moving-wire machine. Owner code: KQD

Syms & Dudley Mill (est. 1881) at Holyoke,

Hampden Co., MA, made stationery paper on three moving-wire machines. This plant had a Jordan refiner, and was powered by three steam engines and four water turbines. Owner code: KQG

Taggart Brothers Mill (est. 1865) at Watertown, Jefferson Co., NY, made manila and twine on a cylinder-wire machine and news paper on a moving-wire machine. The facilities were powered by thirteen waterwheels. Owner code: KQI

Talbot County Mills (est. 1880) at Easton, Talbot Co., MD, made straw board on a cylinder-wire machine. This plant was powered by two steam engines. Owner code: KQK

Tama Mill (circa 1880) at Tama City, Tama Co., IA, made manila and straw wrappers on a cylinder-wire machine. Owner code: KQM

Tarrentown Mill (est. 1884) at Tarrentown, Allegheny Co., PA, made manila wrapping paper on a cylinder-wire machine powered by a steam engine. Owner code: JQX

Taunton Mill (est. 1809) at Taunton, Bristol Co., MA, was a hand mill that continued until 1834. Owner code: KVH

Tecumseh Mill (est. 1868) at Tecumseh, Lewanee Co., MI, made straw wrapping paper on a cylinder-wire machine. This plant was powered by a steam engine and a waterwheel. Owner code: KOF

Tennessee Mill (circa 1880) at Nashville, Davidson Co., TN, made straw wrapping on a cylinder-wire machine. Owner code: KQT.2

Tenny Mill at Bloomfield, Morris Co., NJ, made manila book paper. Plant was destroyed by fire and rag boiler explosion in 1870.

Thatcher Mill (est. 1824) at East Lee, Berkshire Co., MA, was a hand mill that installed a moving-wire machine in 1852. Mill burned down in 1874. Owner codes: KQV, KQU, KOE, JEZ, JON

Thilmany Mill (est. 1883) at Kaukauna, Outagamie Co., WI, made wood pulp paper on four moving-wire machines. Owner code: KQW

Third River Mill (est. 1809) at Franklin, Essex Co., NJ, was a hand mill that later became the Passaic Mill. Owner code: JEF, JLK

Thistle Mill (est. 1797) at Millburn, Essex Co., NJ, was a hand mill that made bond, news, and stationery paper. Rebuilt in 1805 following a fire. Added a cylinder-wire machine in 1830. Destroyed by fire in 1857. Owner code: JGT

Thomas Rice Mill (est. 1836) at Needham, Middlesex Co., MA, made news paper on a moving-wire machine. Plant closed in 1865. Owner codes: KKO, KKN

Thompson Mill (est. 1829) at Westville, New Haven Co., CT, made hardware paper and wrapping on a cylinder-wire machine. Plant was powered by a steam engine and four waterwheels. Owner code: KRB

Three Rivers Mill (est. 1868) at Three Rivers, St. Joseph Co., MI, made book and news on a moving-wire machine and manila on a cylinder-wire. This plant was equipped with a Brightman engine. Owner code: JPK

Thurber Mill (est. 1780) at Providence, Providence Co., RI, made was a hand mill that closed in 1812. Owner code: KRD

Thylers Paper Mill (est. 1810) at Millburn, Essex Co., NJ, was a hand mill that made paste board. Owner code: KTE

Ticonderoga Pulp & Mill (est. 1884) at Ticonderoga, Essex Co., NY, made book paper on two moving-wire machines. This plant was powered by two steam engines and three waterwheels. Owner code: KRE

Tileston & Hollingsworth Mill (est. 1810) at Dorchester, Suffolk Co., MA, was a hand mill that closed around 1852. Owner code: KRF

Tioga Mill (est. 1883) at Appleton, Outagamie Co., WI, made book and news on two moving-wire machines. The plant was equipped with a Jordan refiner. Owner code: JYC

Tipp Mill (est. 1882) at Tippecanoe City, Miami Co., OH, made straw board on a double-cylinder machine. This plant was powered by three steam engines. Owner code: KRH

Tippecanoe Mills (est. 1880) at Monticello, White Co., IN, made news paper on a double-cylinder machine. This plant was powered by seven waterwheels. Owner code: KRI

Tompkins Mill (est. 1850) at Hulls Mills, Duchess Co., NY, made straw wrapping on a cylinder-wire machine powered by a waterwheel. Owner code: KRL

Tondes Mill (est. 1842) at Lafayette, Tippecanoe Co., IN, made rag paper on a cylinder-wire machine. Owner code: KRM

Tracy's Mill (est. 1882) at Lineboro, Carroll Co., MD, made straw wrapping paper on a cylinder-wire machine. This plant was powered by a steam engine and two waterwheels. Owner code: KRP

Trimble Mill (est. 1799) at Chester Creek, Delaware Co., PA, was a hand mill that converted to a textile factory in 1813. Owner code: KRS

Trout Creek Mill (est. 1746) at Lower Merion, Montgomery Co., PA, was a hand mill that made linen and linen composite bond and book paper. This facility became one of the Hagey Mills, and later the Cope Mill. Owner code: KFR, JJS

Troy City Mill (est. 1800) at Troy, Rensselaer Co., NY, made rag paper on a cylinder-wire machine. The plant flooded and was rebuilt on adjoining land in 1814. Owner codes: KNY, KXC, JOP, JGA, KEO

Troy Mill (est. 1835) at Troy, Rensselaer Co., NY, made book, news, and wall paper on a moving-wire machine. This plant was powered by a steam engine and a waterwheel. Owner code: KGP

Trumbull Mill (est. 1850) at Gargoa, Fulton Co., NY, made straw board on a cylinder-wire machine. This plant was powered by two waterwheels. Owner code: KRU

Tunxis Mill (est. 1847) at Unionville, Hartford Co., CT, made linen composite stationery on a moving-wire machine. The plant was powered by two steam engines and three water turbines. Owner code: KIW

Turkey Mill (est. 1828) at Tyringham, Berkshire Co., MA, made wrapping paper on a cylinder-wire machine. This facility closed in 1877. Owner codes: KOC, KIV

Turner's Falls Mill (est. 1877) at Turner Falls, Franklin Co., MA, made news paper on a moving-wire. This plant was powered by three waterwheels. Owner code: KRW

Tuscarora Mills (est. 1778) at Upper Darby Township, Delaware Co., PA, was a hand mill that continued until around 1836. Owner code: KAA, KAE

Tyrone Mill (est. 1880) at Tyrone, Blair Co., PA, made book and news paper on a moving-wire machine. This plant was powered by four steam engines. The facility was rebuilt following a fire in 1884. Owner code: KES

Tytus Mill (est. 1873) at Middletown, Butler Co., OH, made colored mediums on three moving-wire machines. This plant was powered by three steam engines and a waterwheel. Owner code: KRX

Uhler Mill (circa 1880) at Uhlersville, Northampton Co., PA, made manila-colored mediums on a moving-wire machine and manila paper on a double-cylinder. Owner code: KRY

Ulster Mill (circa 1880) at Napanoch, Ulster Co., NY, made straw wrapping paper on a cylinder-wire machine. Owner code: KYI

Underhill Mill (est. 1820) at Urbana, NY, was a hand mill that burned down in 1838. Owner code: KRZ

Union Mill (est. 1799) at Newton Square, Delaware Co., PA, was a hand mill (on Darby Creek) that made stationery and folio paper in two vats. By 1867 it made card middles and wall paper on a moving-wire machine. This plant had a Kingston refiner, and was powered by two steam engines and a waterwheel. Owner code: KCK, JPX

Union Mill (est. 1808) at Lee, Berkshire Co., MA, was a hand mill until installing a cylinder-wire machine in 1838. Mill was rebuilt following a boiler explosion in 1859. Owner codes: JIJ, KIX

Union Mill (est. 1810) at Mill Grove, Warren Co., OH, was a hand mill that continued until around 1847. Owner code: JLA

Union Mill (est. 1819) at New Hope, Bucks Co., PA, was a hand mill that installed a cylinder-wire machine in 1840. The plant then made news paper, later installing a Kingston refiner. The facilities were rebuilt following a fire in 1846 and added a moving wire machine. The plant was later closed, then reopened in 1880 for the manufacture of colored tissue and manila paper. The facilities were powered by this time by two steam engines and three waterwheels. Plant was destroyed by fire in 1895. Owner codes: KFA, KSA, JYR

Union Mill (est. 1840) at Hohokus, Bergen Co., NJ, made manila tissue paper on a cylinder-wire machine. This plant was powered by two waterwheels. Owner code: KVT

Union Mill (est. 1852) at Mechanic Falls, Androscoggin Co., ME, made book and news paper on a moving-wire machine. This plant was powered by a steam engine and three waterwheels. Owner code: JMG

Union Mill (est. 1863) at Freeport, Queens Co., NY, made straw board on a cylinder-wire machine powered by a water turbine. Owner code: KOJ

Union Mill (est. 1870) at Holyoke, Hampden

Co., MA, made stationery paper on two moving-wire machines. This plant was powered by two steam engines and two water turbines. Owner code: KSB

Union Pacific Mills (est. 1857) at Monongahela City, Washington Co., PA, made straw board on a cylinder-wire machine. This plant was powered by two steam engines. Owner code: KIG

Union Village Mill (est. 1799) at Unionville, Hartford Co., CT, made book paper on two moving-wire machines. This plant was powered by a steam engine and two water turbines. The plant was rebuilt in 1847. Owner codes: KIW, JGJ, JIN

Unionville Mill (est. 1868) at Unionville, Hartford Co., CT, made collar paper, and card middles on a moving-wire machine. This plant was powered by a steam engine and two water turbines. Owner code: JMB

Unkley Mill (est. 1802) at Lower Saucon Township, Northampton Co., PA, was a hand mill that continued until about 1846. Owner code: KSD

Upper Beard Mill (est. 1854) at Fayetteville, Onondaga Co., NY, made manila paper on a cylinder-wire machine. This plant was powered by a steam engine and two waterwheels. Owner code: JHN

Upper Sandusky Mill (est. 1885) at Upper Sandusky, Wyandot Co., OH, made straw board on a board machine. Owner codes: KGL, KMG

Upper Valley Mill (circa 1880) at Lee, Berkshire Co., MA, made collar paper and board stock on a cylinder-wire machine. This plant was powered by a steam engine and waterwheel. Owner code: JBZ

Valentine Mill at Roslyn, Queens Co., NY, made straw board on a cylinder-wire machine. This plant was powered by steam engine and waterwheel. This facility closed in 1890. Owner codes: KSJ, KSH

Valley City Mills (circa 1880) at Dayton, Montgomery Co., OH, made book and news paper on a moving-wire. Owner code: KAK

Valley Falls Mill (est. 1865) at St. Johnsbury, Caledonia Co., VT, made straw board on a cylinder-wire machine. This plant was powered by two waterwheels. Owner code: KIQ

Valley Falls Mill (est. 1873) at Valley Falls, Rensselaer Co., NY, made straw wrapping paper on a cylinder-wire machine. This plant was powered by three waterwheels. Owner code: KSK

Valley Mill at Middletown, Butler Co., OH, made manila paper on two cylinder-wire machines. This mill also had a Gould machine. Owner code: KDQ

Valley Mill (est. 1810) at Oxford, Chester Co., PA, was a hand mill that added a cylinder-wire machine in 1834. Mill eventually made manila wrapping on a double-cylinder machine. This plant was powered by a turbine wheel and two steam engines. Owner codes: KVA, JMR, KUZ

Valley Mill (est. 1833) at Bentley Springs, Baltimore Co., MD, made straw wrapping on a cylinder-wire machine installed in 1834. After 1847 the mill made manila paper. This plant was powered by a steam engine and two water turbines. Plant was rebuilt in 1862 following a fire. Owner codes: JRC, JMR, KUZ, KVA

Valley Mill (est. 1840) at Lee, Berkshire Co., MA, made book and news on two moving-wire machines. Owner code: KOB

Valley Mill (est. 1847) at Cleveland, Cuyahoga Co., OH, made news paper on two moving-wire machines. Auxiliary equipment was powered by a steam engine. Owner code: JJA

Valley Mill (est. 1850) at Coeymans, Albany Co., NY, made straw wrapping on two cylinder-wire machines. This plant was powered by a steam engine and two waterwheels. Owner code: JFN

Valley Mill (circa 1880) at Ashland, Grafton Co., NH, made leather board on a triple-cylinder machine. Owner code: JCD

Valley Mill (circa 1880) at South Windsor, Hartford Co., CT, made board stock. Owner code: JQB

Valley Mills (2) (est. 1866) at Holyoke, Hampden Co., MA, made stationery and envelope paper on two moving-wire machines. This plant was powered by three steam engines and two water turbines. A second facility was erected in 1877. Owner code: KSL

Valley Paper Mill was established in 1876 at Hulton, PA. Owner code: KSM

Valley Pulp & Paper Mill (est. 1880) at Appleton, Outagamie Co., WI, made wood-pulp newsprint on a cylinder-wire and a moving-wire machine. This plant was powered by five waterwheels. Owner code: KSN, KHL

Valparaiso Mill (circa 1880) at Valparaiso, Porto

Co., IN, made straw wrapping on a cylinder-wire machine. Owner code: KNP

Van Nortwick Mills (est. 1869) at Batavia, Kane Co., IL, made news paper on three moving-wire machines. This plant had a Jordan refiner, and was powered by three steam engines and seven water turbines. Owner code: KSR

Van Reed Mill (est. 1812) at Reading, Berks Co., PA, was a hand mill that made book and news paper. The mill added a cylinder-wire machine in 1826. Owner codes: KST, KSS

VanCourtland Mill (est. 1819) at Annsville, Westchester Co., NY, was a hand mill that continued until around 1849. Owner code: KSP

Vandalia Mill (est. 1884) at Vandalia, Fayette Co., IL, made straw board, card, and wrapping paper on a cylinder-wire machine. This plant was powered by three steam engines. Owner code: KSV

Vermont Gazette Mill was a mill at Bennington, Bennington Co., VT, established circa 1885. Owner code: KMA

Victoria Mill (est. 1880) at Fulton, Oswego Co., NY, made manila paper on a cylinder-wire machine. This plant was powered by three waterwheels. Owner code: KSZ

Vinton Mill (circa 1880) at Brattleboro, Windham Co., VT, made news paper on a moving-wire machine. This plant had a Jordan refiner, and was powered by two waterwheels. Owner code: KTA

Virginia Cane Fiber Co. at Fredericksburg, VA, made construction paper and wrappers.

Virginia Mill (est. 1835) at Wheeling, Ohio Co., WV, made wrapping paper on a cylinder-wire machine. Owner code: JZG

Vulcan Mill (est. 1881) at Appleton, Outagamie Co., WI, made book and news paper on a moving-wire machine. The plant had a Jordan refiner, and was rebuilt in 1859 following a fire. Owner code: JYC

Wade Mill (est. 1835) at Millburn, Essex Co., NJ, made board stock on a board machine. This plant was powered by a steam engine and waterwheel. Owner code: KTG

Wade Mill was established in 1835 at Millburn, Essex Co., NJ. Owner code: KTD

Wading River Mill (est. 1880) at W. Mansfield, Bristol Co., MA, made leather board on a cylinder-wire machine powered by a waterwheel. Owner code: KLL

Wait Mills (2) (est. 1850) at Sandy Hill, Washington Co., NY, made wall paper on a moving-wire machine. This plant had a Brightman engine, and was powered by nine water turbines. Owner code: KTJ

Waite Mill (est. 1858) at Little Falls, Herkimer Co., NY, made manila paper on a cylinder-wire machine. This plant was powered by four waterwheels. Owner code: KTK

Walbridge Mill (est. 1786) at Brattleboro, Windham Co., VT, was a hand mill that continued until around 1843. Owner code: KTM, KTU

Waldron Mill (est. 1872) at Waldron, Kankakee Co., IL, made straw wrapping paper on a cylinder-wire machine. This plant was powered by two steam engines and three waterwheels. Owner code: KTN

Waldschmidt Mill (est. 1810) at Sycamore Township, Hamilton Co., OH, was a hand mill that made stationery and book paper. This mill continued until around 1848. Owner codes: KTO, JYZ

Walesville Mill (est. 1850) at Walesville, Oneida Co., NY, made straw wrapping on a double-cylinder machine. This plant was powered by two water turbines. Owner code: JFG

Walkill Mill (est. 1875) at Walkill, Ulster Co., NY, made book and news paper on a moving-wire machine. This plant was powered by a steam engine and three waterwheels. Owner code: JCU

Wallingford Mill (circa 1880) at South Wallingford, VT, made wood pulp paper.

Walloomsac Mill (est. 1874) at North Hoosick, Rensselaer Co., NY, made wall paper on two moving-wire machines. This plant had a Jordan refiner, and was powered by six waterwheels and three steam engines. Owner code: KTT

Walpole Mill (est. 1881) at Walpole, Norfolk Co., MA, made board stock on two wet machines. This plant was powered by a steam engine. Owner code: KAL

Walsh Mill (est. 1793) at Newburgh, Orange Co., NY, made stationery paper by hand. The mill later added a cylinder-wire machine powered by a waterwheel. Owner codes: KTV, KTW, KQF, KTX

Wampler Mill (est. 1818) at Baltimore, Baltimore Co., MD, was a hand mill that continued until the 1840s, when it was converted to making straw wrapping on a cylinder-wire machine. Owner code: KTY

Ward Mill (rebuilt 1883) (est. 1878) at Riverton, Litchfield Co., CT, made manila paper on a double-cylinder machine. This plant was powered by three waterwheels. Owner code: KTZ

Ware Mill (est. 1790) at Needham, Middlesex Co., MA, was a hand mill that added a cylinder-wire machine in 1825. The plant was rebuilt following a fire in 1834, and added a moving-wire machine in 1854. This facility was later renamed the Crehore Mill. Owner codes: KUC, JVN, JVM, KKJ

Warren Mill was established in 1850 at Maylandville, PA.

Warren Mill (est. 1869) at Gardiner, Kennebec Co., ME, made book and stationery paper on a moving-wire machine. This was the former Copesecook Mill that had a Jordan refiner and converted to wood-pulp shortly after opening. This mill was powered by seven waterwheels. Owner code: KUG

Warren Mill (est. 1873) at Riegelsville, Warren Co., NJ, made paper on two moving-wire machines. This plant was powered by three steam engines and two waterwheels. Owner code: KUE

Washington Mill (est. 1836) at East Lee, Berkshire Co., MA, made book and wrapping paper on a cylinder-wire machine. This plant had a Jordan refiner, and was powered by a steam engine and a water turbine. The plant was rebuilt in 1883 following a fire. Owner codes: JAO, KIW, JED, KBA, JMA

Watehung Mill (circa 1880) at Montclair, Essex Co., NJ, made straw board on a cylinder-wire machine. Owner code: KVQ

Water Shops Mill (est. 1827) at Springfield, Hampshire Co., MA, made construction paper on a cylinder-wire machine powered by a waterwheel. Owner code: JBB

Waterbury Mill (circa 1880) at Brooklyn, Kings Co., NY, made colored mediums on a moving-wire machine and a double-cylinder. Owner code: KUI

Waterbury Mill (est. 1883) at Hotchkissville, Litchfield Co., CT, made straw board on a board machine. Owner code: KXM

Waterford Mill, Old (est. 1832) at New London, New London Co., CT, made paper on two cylinder-wire machines. This plant was powered by two waterwheels. Owner code: KXW

Watertown Mill (circa 1880) at Watertown, Jefferson Co., NY, made book and news paper on a moving-wire machine. Owner code: KUJ

Watson Mill (circa 1880) at Canaan Four Corners, Columbia Co., NY, made straw wrapping on a cylinder-wire machine. Owner codes: KUM, KUL

Watson Mill (est. 1884) at Loudville, Hampshire Co., MA, made tissue paper on a cylinder-wire machine. This plant was powered by a steam engine and two waterwheels. Owner codes: KUK, KUO

Wauregan Mill (est. 1879) at Holyoke, Hampden Co., MA, made stationery, ledger, and envelope paper on a moving-wire. The plant had a Jordan refiner, and was powered by a steam engine, eight waterwheels, and four water turbines. Owner code: KUR

Waverley Mill (est. 1829) at East Lee, Berkshire Co., MA, made book and stationery paper by hand before installing a cylinder-wire in 1840. This plant later was named the New England Mill. Owner codes: JII, KQU, JCF

Waverly Mill (circa 1880) at Buckland, Hartford Co., CT, made book paper on two moving-wire machines. This plant had a Jordan refiner, and was powered by three steam engines and a water turbine. Owner code: JAH

Waverly Mill (circa 1880) at Waverly, Tioga Co., NY, made book and news paper on a moving-wire machine. Owner code: KUS

Webb Mill (est. 1799) at Kennett Square, Chester Co., PA, was a hand mill that continued until around 1833. Owner codes: KUU, JFJ ('21)

Wells River Mill (est. 1857) at Wells River, Orange Co., VT, made manila tissue paper on a cylinder-wire machine. This plant was powered by a steam engine and three waterwheels. Owner code: JME

West Dudley Mill (est. 1881) at W. Dudley, Worcester Co., MA, made manila wrapping paper on a cylinder-wire machine. This plant was powered by a steam engine and three waterwheels. Owner code: KUH

West Groton Mill (circa 1850) at West Groton, Middlesex Co., MA, made colored mediums on a moving-wire machine. Owner code: JUO

West Jersey Mill (circa 1880) at Camden, Camden Co., NJ, made manila cartridge paper on a triple-cylinder machine. Owner code: KVC

West Medway Mill (est. 1880) at W. Medway, Norfolk Co., MA, made manila paper on a cylinder-wire machine. Owner code: JRQ, JGR

West Mill (circa 1880) at Hadley, Saratoga Co., NY, had a moving-wire machine and a double-cylinder. The plant used a Jordan refiner. Owner code: KVG

West Newton Mill, Old, was established in 1859 at West Newton, Westmoreland Co., PA. Owner code: KBY

West Newton Mills (2) at West Newton, Westmoreland Co., PA, made wood-pulp news paper on four wet machines. Owner code: KVD

West Point Mill (est. 1876) at West Point, Cuming Co., NE made wrapping paper on a cylinder-wire machine. The plant was rebuilt in 1882 and commenced using wood pulp. Owner code: KVE

West Rock Mill (est. 1778) at Westville, New Haven Co., CT, made blotting paper on a moving-wire machine. This plant was powered by three steam engines and a waterwheel. Owner code: KHG, JGD

West Side Mill (est. 1848) at St. Charles, Kane Co., IL, made wrapping and news paper on a cylinder-wire machine powered by a waterwheel. Following a fire in 1866 this mill was rebuilt and renamed the St. Charles Mill. Owner code: JGK, JWF

West Townsend Leather Board Mill (circa 1880) at West Townsend, Middlesex Co., MA, made leather board on a cylinder-wire machine. This plant was powered by two waterwheels. Owner code: KOY, JOJ

West Ware Mill (est. 1884) at W. Ware, Hampshire Co., MA, made book and news paper on a moving-wire machine. Owner code: KVF

Western Reserve Mill (circa 1880) at Cahagrin Falls, Cuyahoga Co., OH, made manila paper on a double-cylinder machine. Owner code: JGP

Westfield Mill (circa 1880) at Westfield, Chantanqua Co., NY, made manila construction and wrapping paper on a double-cylinder machine. This plant was powered by a steam engine and two waterwheels. Owner codes: JXW, KQR

Weston Mill (est. 1840) at Saugerties, Ulster Co., NY, made rag paper on a cylinder-wire machine. This mill later became part of the Sheffield Mills. Owner code: KVK

Weymouth Mill (est. 1860) at Weymouth, Atlantic Co., NJ, made manila twine on double-cylinder machine. Owner code: KVO

Wheeler Mill (est. 1835) at Chatham, Columbia Co., NY, made straw wrapping and bogus manila on a cylinder-wire machine. This plant was powered by a steam engine and two waterwheels. Owner code: KVR

Wheeler Mill was established in 1846 in Connecticut. Owner code: KVP

Wheelwright Mill (est. 1863) at N. Leominster, Worcester Co., MA, made book and card middles on a cylinder-wire and moving-wire machines. This plant had a Jordan refiner, and was powered by two steam engines and two waterwheels. This was the former Kendall Mill. Owner code: KVS

Whetstone Creek Mill (est. 1811) at Brattleboro, Windham Co., VT, was a hand mill that made stationery, paste board, and printing paper. This mill burned down in 1815. Owner codes: JOO, JUN

White Mill (est. 1870) at Newton Upper Falls, Middlesex Co., MA, made manila wrapping and wall paper on a cylinder-wire machine. Owner code: KVX

White Mills (circa 1880) at Chatham, Columbia Co., NY, made straw wrapping on a double-cylinder machine. Owner code: KVW

White River Mill (est. 1821) at Berlin, VT, was a hand mill that was destroyed by a flood in 1840. Owner code: JRD

Whitehall Mill (circa 1880) at White Hall, Baltimore Co., MD, made straw wrapping paper on a cylinder-wire machine. Owner code: KXN

Whites Creek Mill (est. 1815) at Nashville, Davidson Co., TN was a hand mill that made news, book, and wrapping paper until its closure about 1836. Owner code: JSY, KWB

Whitewater Mill (circa 1880) at Whitewater, Walworth Co., WI, made news and wrapping paper. Owner code: KWE

Whiting Mill No. 1 (est. 1865) at Holyoke, Hampden Co., MA, made stationery on two moving-wire machines. This plant was powered by four waterwheels. Owner code: KWF

Whiting Mill No. 2 (est. 1873) at Holyoke, Hampden Co., MA, made stationery, ledger, and envelope paper on three moving-wire machines. This plant was powered by a steam engine and four waterwheels. Owner code: KWF

Whitmore Mill (est. 1881) at Holyoke, Hampden Co., MA, made wood-pulp board and bogus manila. Owner code: KWH

Whittlesey's Tissue Mill (est. 1810) at Windsor Locks, Hartford Co., CT, was a hand mill that converted to manila tissue on a double-cylinder. This plant was powered by two water turbines. The facilities were rebuilt in 1872. Owner codes: KWK, KWJ

Wilcox Mill (est. 1874) at Rochester, Oakland Co., MI, made manila wrappers on a cylinder-wire machine. This plant was powered by a steam engine and two waterwheels. Owner code: JEK

Wilder Mills (2) (est. 1881) at Ashland, Grafton Co., NH, made newsprint on a moving-wire machine, a Harper machine, and a double-cylinder. Second plant built in 1884. Owner code: KWM

Wiley Mill at Portsmouth, Scioto Co., OH, made straw wrapping on a cylinder-wire machine. Owner code: KWP

Wilkes Barre Mill (est. 1883) at Wilkes Barre, Luzerne Co., PA, made straw or manila wrapping and printing paper on a cylinder-wire machine. This plant was powered by two steam engines. Owner code: KWN

Wilkinson Mill (est. 1884) at Banning, Carroll Co., GA, made manila printing on a cylinder-wire machine. This plant was powered by two waterwheels. Owner codes: KWQ, KWS

Willamette Mill was established in 1889 at Oregon City, Clackamas Co., OR. Owner code: KWT

Williams Mill (est. 1860) at Schuylkill, Chester Co., PA, made board stock on a wet machine powered by a waterwheel. Owner code: KXA, KWZ

Williamsville Mill (est. 1848) at W. Stockbridge, Berkshire Co., MA, made news and straw wrapping on a cylinder-wire machine. This plant was powered by a steam engine and two waterwheels. Owner code: JJM

Williston Mill (est. 1850) at Cheney, Delaware Co., PA, made manila tissue on a cylinder-wire machine. This plant was powered by two steam engines and a waterwheel. Owner code: JPZ

Willow Brook Mill (est. 1823) at Kennett Square, Chester Co., PA, made paste board on a cylinder-wire machine. This plant was powered by a steam engine and waterwheel. The facilities were rebuilt following a fire in 1824. The mill was abandoned in 1834. Owner codes: KYA, JFJ

Willow Dale Mill (est. 1854) at Leona, Chautauqua Co., NY, made newsprint on a cylinder-wire machine. This plant was powered by a steam engine and two waterwheels. Owner codes: KNU, JAN

Willow Grove Straw Board Mill (est. 1856) at Chambersburg, Franklin Co., PA, made straw board on two cylinder-wire machines. This plant was powered by a steam engine and two waterwheels. Owner code: JZU

Wilson Mill (circa 1883) at Ercildoun, Chester Co., PA, made straw board on a cylinder-wire machine. Owner code: KXD

Winchester Mill (est. 1870) at Winchester, Frederick Co., VT, made straw board on a cylinder-wire machine. This plant was powered by three steam engines. Owner code: KXG

Windham Mill (circa 1880) at Bellows Falls, Windham Co., VT, made manila paper on a cylinder-wire and wood pulp paper on a moving-wire machine. Owner code: KEJ

Windham Paper Mill was established in 1876 at Chaplin, Windham Co., CT. Owner code: KXH

Winnebago Mills (est. 1854) at Rockton, Winnebago Co., IN, made manila wrapping on two cylinder-wire machines. This plant was powered by seven waterwheels. Owner code: JFH

Winnebago Mills (est. 1874) at Neenah, Winnebago Co., WI, made book and news paper on a cylinder-wire machine. This plant was powered by five waterwheels. Owner code: KXJ

Winnegar Mill (est. 1837) at Queechy Lake, Columbia Co., NY, made straw wrapping on a cylinder-wire machine.

Winnesheik Mill (est. 1871) at Freeport, Winnesheik Co., IA, made straw wrapping on a cylinder-wire machine. This plant was powered by four waterwheels. Owner code: JEX

Winnipiseogee Mill (est. 1870) at Franklin, Merrimack Co., NH, made newsprint on four moving-wire machines. This plant was powered by fifteen waterwheels. Owner code: KXK

Winona Mill (est. 1880) at Holyoke, Hampden

Co., MA, made book, envelope, and stationery on two moving-wire machines. This plant was powered by five steam engines and three water turbines. Owner codes: KXL, JHZ

Winters Run Mill (est. 1810) at Bel Air, Harford Co., MD, was a hand mill that continued until around 1849. Owner code: JZE

Wissahiccon Mill (est. 1823) at Lower Merion, Montgomery Co., PA, made printing and wrapping paper on a cylinder-wire machine. Owner code: JDL, JJS

Wiswall's Mill (est. 1869) at Newton Lower Falls, Middlesex Co., MA, made wall paper, bogus manila, and manila wrapping on a cylinder-wire machine. This plant was powered by two steam engines and four waterwheels. This was formerly Foster's Mills, and it burned down sometime after 1892. Owner code: KXO

Wolverine Mill (est. 1881) at Detroit, Wayne Co., MI, made construction paper on a double-cylinder machine. Plant was rebuilt in 1883. Owner code: KXS

Woodbine Mill (est. 1800) at Morgan, Carroll Co., MD, was a hand mill until 1840. It then made straw wrapping paper on a cylinder-wire machine. This plant was powered by two steam engines and a waterwheel. Owner code: JND

Woodland Mills (est. 1882) at E. Hartford, Hartford Co., CT, made envelope and wrapping paper on two moving-wire machines. This plant was powered by a steam engine and four waterwheels. Owner code: JSZ

Woodstock Mill (circa 1880) at Cairo, Greene Co., NY, made straw wrapping paper on two cylinder-wire machines. Owner code: JHP

Woodville Mill (est. 1848) at Woodville, Jefferson Co., NY, made straw wrapping on a cylinder-wire machine. This plant was powered by two waterwheels. Plant was destroyed by fire in the 1850s. Owner code: JIS

Worthy Mill (est. 1872) at Mittineague, Hampden Co., MA, made stationery paper on a moving-wire machine. This plant was powered by a steam engine and three waterwheels. Owner code: KXZ

Wyman, Franklin Mills (circa 1880) at Westminster, Worcester Co., MA, made book and news on two moving-wire machines. Owner code: KYB

Wyoming Valley Mill (circa 1880) at Pittston, Luzerne Co., PA, made news and bogus manila on a cylinder-wire machine. The plant had a Kingston refiner, and was powered by two steam engines. Owner code: KLT

Xenia Mill (est. 1882) at Xenia, Green Co., OH, made straw and manila wrapping and hardware on a cylinder-wire machine. This plant was powered by two steam engines. Owner code: KYC

XX Mill (est. 1870) at Brainard, Rensselaer Co., NY, made straw wrapping on a cylinder-wire machine. This plant had a Kingston refiner, and was powered by two waterwheels. Owner code: KRJ

Yantic Mill (est. 1873) at Greenville, New London Co., CT, made collar paper on a cylinder-wire machine. This plant was powered by a steam engine and two waterwheels. Owner code: KYD

Yarmouth Mill at Bowdoin, ME, was established around 1880, and made paper on a cylinder-wire machine and four moving-wires. Owner code: JOU

Yarnall Mills (2) (est. 1800) at Great Crossing, Franklin Co., KY, were hand mills that continued until around 1836. Owner code: KYE

Yoran Mill (est. 1884) at Fort Plain, Montgomery Co., NY, made straw board on a cylinder-wire machine. This plant was powered by two water turbines. Owner code: KYF

York Haven Mill (est. 1885) at Harrisburgh, Dauphin Co., PA, made book and news paper on two moving-wires. This plant also had a Gould machine. Owner code: KYH

York Mill (est. 1829) at York, York Co., PA, was a hand mill that made book paper. A cylinder-wire was added later. The plant flooded in 1884, and later became the York Wall Paper Mill. Owner code: JYF, JWN

York Wall Paper Mill (est. 1888) at York, York Co., PA, made wall paper. This is the former York Mill and later Schmidt & Ault Mill. Owner codes: KYG, KSC, KFD, KLV

Yorkville Mill (est. 1859) at Yorkville, Kendall Co., IL, made straw wrapping on a cylinder-wire machine. This plant had a Jordan refiner, and was powered by a steam engine and six waterwheels. Owner code: JPD

Youghiogheny River Mill (est. 1812) at Connellsville Township, Franklin Co., PA, was a hand mill that continued until about 1828. Owner codes: KLP, JYS

Young Mill (est. 1881) at Bentley Springs, Baltimore Co., MD, made construction paper on a cylinder-wire machine. This plant was powered by two waterwheels. Owner code: KYJ

Ypsilanti Mill (circa 1880) at Ypsilanti, Washtenaw Co., MI, made book and news on two moving-wire machines. The plant had a Jordan refiner. Owner code: KYP

Zellerbach Mill (est. 1899) at Lebanon, Linn Co., OR made wood-pulp paper. Owner code: KYQ

Appendix II: Directory of Paper Mill Owners

A paper mill could have several owners over the course of a century, so a separate listing seemed a natural extension of the census. The historical record sometimes records the name of a company over the location of the mill, or vice versa. This directory also includes some closely associated classes of merchants such as early printers, booksellers, and stationers (code: STA), who may have had a stake in a local paper mill.

The least common denominator between hand and machine production essentially comes down to daily capacity and total employees. Hand mills were capable of just a few hundred of pounds of paper a day, while mills with the cylinder-wire measured daily output in the thousands. Mills employing the moving-wire machine enjoyed production capacities in excess of two tons a day. Another easily recognizable trend is the number of employees at "refined" mills versus those operations with just a few workers making wrapping or board.

Where the historical name of a mill is not known, a name is assigned based on a nearby feature such as a waterway or village. In cases with numerous plants in a given area, names are assigned after the owner or company. In an effort to reduce duplication, facilities of a combined or consolidated nature are grouped together under a single listing.

Partnerships were often quite fluid, so to reduce overly-redundant listings for essentially the same firm, minor shareholders are restricted to the comment line. As well, firms changed addresses repeatedly over the decades, so to avoid excessive listings the addresses selected here are for identification purposes only.

JAA: Abraham, Thomas D., had offices in Abrams, PA, and made 5,000 pounds a day. See: Merion Mill

JAB: Acushnet Paper Co. (est. 1883) at Acushnet, MA, employed 8 and made 5,000 pounds a day. See: Acushnet Mill

JAC: Adam, John, was established in 1783 at Salisbury, CT. See: Adam Mill

JAD: Adams & Bishop at Newburgh, NY, made 5,000 pounds a day. See: Grove Mill

JAE: Adams & Co. (est. 1858) at Chagrin Falls, OH, employed 51 and made 4,500 pounds a day. See: Adams & Co. Mill

JAF: Adams & Terry (est. 1883) at Monroe, MI, employed 25 and made 8,000 pounds a day. See: Monroe Mfg. Mill

JAG: Adams, Ebenezer, at Chatham, NY, made 6,000 pounds a day. See: Eagle Mill

JAH: Adams, Peter, Co. at Buckland, CT, made 12,000 pounds a day. See: Waverly Mill

JAI: Agawam Paper Co. (est. 1865) at Mittineague, MA, employed 161 and made 5,000 pounds a day. See: Agawam Mill

JAJ: Ager, J.W. (est. 1848), at Lyonsdale, NY, employed 6 and made 2,000 pounds a day. See: Ager Mill

JAK: Akron Straw Board Co. (est. 1872) at Akron, OH, employed 75 and made 20 tons a day. See: Akron Straw Mill

JAL: Albion Paper Co. (est. 1869) at Holyoke, MA, employed 280 and made 10 tons a day. Principals were David H. and John C. Newton. See: Albion Mill

JAM: Aldrich & Hayden at Bristol, NH, employed 29 and made 4,500 pounds a day. See: Grafton Mill

Directory of Paper Mill Owners

JAN: Alexander, P.B. (Bros.) (est. 1853), at Leona, NY, employed 6 and made 1,200 pounds a day. See: Willow Dale Mill

JAO: Allen & Co. was established in 1836 at East Lee, MA. Principals were Joseph B. Allen, Leander Backus, and William H. Allen. Firm failed in 1840 and again in 1847. See: Washington Mill

JAP: Allen Brothers (est. 1857) at Sandy Hill, NY, employed 80 and made 28,000 pounds a day. See: Allen Brothers Mill

JAQ: Allen Paper Car Wheel Co. was established in 1880 at Morris, IL. See: Morris Mill

JAR: Allen, Andrew J., was established in 1818 at Boston, MA. See: Leominster Mills, Allen Mill

JAS: Allen, Barker & Servis was established in 1855 at Lockland, OH. See: Lower Mill

JAT: Allen, Joseph, was established in 1848 at Lockland, OH. See: Athern Upper Mill

JAU: Alling, Frank P. (est. 1883), at Bangall, NY, employed 3 and made 1,500 pounds a day. See: Alling Mill

JAV: Allison, Dr. Robert, was established in 1846 at Jennersville, PA. See: Spring Lawn Mill

JAW: Alston, M.P., was established in 1857 at Hamilton, OH. See: Graham Mill

JAX: American Wood Paper Co. at 140 Nassau St., New York, NY, made 18,000 pounds a day. See: Royer Ford Mill

JAY: American Writing Paper Co. was established in 1888 at Holyoke, MA. See: Hurlbut Mill

JAZ: Ames & Gaskill was established in 1838 at Upper Darby Township, PA. Principals were Israel Ames and Benjamin Gaskill. See: Glenwood Mills

JBA: Ames, Col. David, was established in 1802 at Springfield, MA. See: Springfield Mill

JBB: Ames, D. & J. (est. 1822), at Springfield, MA, employed 200 and made 4,000 pounds a day. See: Water Shops Mill, Cox Mill, Eagle Mill, Old, South Hadley Mill, Springfield Mill

JBC: Ames, William, & Co. was established in 1828 at Upper Darby Township, PA. See: Clifton Mill

JBD: Amies, Thomas, was established in 1794 at Lower Merion Township, PA. See: Dove Mills

JBF: Amies, Thomas, Jr., & Son (est. 1823) at Lower Merion Township, PA, produced 400 pounds a day. Later leased mill of Samuel Levis in Upper Darby Township, Delaware Co., PA, from 1828 to 1838. See: Glenwood Mills, Dove Mills

JBG: Anderson & Yeatman (est. 1867) at Columbus, OH, employed 35 and made 4,000 pounds a day. See: Columbus Mill

JBH: Andrews & Co. (est. 1883) at Penn Yan, NY, employed 40 and made 12,000 pounds a day. See: Seneca Mill

JBI: Angell & Langdon (est. 1884) at Greenwich, NY, made 6,000 pounds a day. See: Central Fall Mill

JBJ: Angell & Safford (est. 1863) at Greenwich, NY, employed 16 and made 3,200 pounds a day. See: Greenwich Mill

JBK: Angell, Edwin, at Chatham, NY, employed 10 and made 2.5 tons a day. See: High Bridge Mill

JBL: Antietam Paper Co. (est. 1856) at Hagerstown, MD, employed 42 and made 5,000 pounds a day. See: Antietam Mill

JBM: Appleton Paper & Pulp Co. (est. 1873) at Appleton, WI, employed 40 and made 16,000 pounds a day. See: Appleton Paper & Pulp Mills

JBN: Appleton Straw Board Co. (est. 1885) at Appleton, WI, employed 25 and made 20,000 pounds a day. See: Fox River Straw Board Mill

JBO: Armour Bros. & Co. at Stanley, NJ, made 18 tons a day. See: Armour Bros. Mill

JBP: Athern & Bachelor, est. 1846 at Lockland, OH. See: Athern Upper Mill

JBR: Atlas Paper Co. (est. 1878) at Appleton, WI, employed 150 and made 4 tons a day. See: Atlas Mill

JBS: Austin & Austin (est. 1873) at Catskill, NY, employed 5 and made 500 pounds a day. See: Hope Mill

JBT: Austin & Carr, est. 1796 at Mount Holly, NJ. Principals were Cyrus Austin and Issac Carr. See: Mt. Holly Mill

JBU: Bachelor, Charles & Brother was established in 1851 at Lockland, OH. See: Lockland Mill

JBV: Bachelor, Chas. P., was established in 1834 at Millford, OH. See: Kugler Mill

JBW: Bacon Paper Co. was established in 1852 at 352 Washington St. in Boston, MA. The company employed 105 and made 12 tons a day. See: Bacon Mill

JBX: Baeder, Adamson & Co. had offices in Philadelphia, PA. See: Philadelphia Mill

JBY: Bailey, E.F., was established in 1867 at Ashland, NH. The company employed 10 and made 1 ton a day. See: Bailey's Leather Board Mill

JBZ: Baird, Prentiss C., was established in 1859 at East Lee, MA. The company employed 35 and made 1,000 pounds a day. See: Upper Valley Mill, Congress Mill, Forest Mill, Upper, Greenwater Mill, National Mill

JCA: Baker & Cox was established in 1807 at Dorchester, MA. See: Suffolk Mill

JCB: Baker, Charles, & Co. was established in 1819 at Dorchester, MA. See: Baker Mill

JCC: Baker, John, & Co. was established in 1838 at East Lee, MA. The "and company" was George Wilson and Ira Van Bergan. See: Baker Mill

JCD: Baker, S.C., Agent was established at Ashland, NH. The company employed 12 and made 3,000 pounds a day. See: Valley Mill

JCE: Balfour, Alexander, was established in 1882 at 18 Decatur St., Philadelphia, PA. The company employed 100 and made 20,000 pounds a day. See: Richmond Mill

JCF: Ballard, Charles, & Co. was established in 1851 at East Lee, MA. The "and company" was Werden ('53), William P. Hamblin ('55), and Prentiss Chaffee ('56). See: Waverley Mill

JCG: Bannister & Co. was established in 1807 at Petersburg, VA. See: Bannister Mill

JCH: Barber & Cravens was established in 1873 at Madison, IN. The company employed 11 and made 3,000 pounds a day. See: Madison Mill

JCI: Barclay, Henry, was established in 1820 at Saugerties, NY. See: Saugerties Mill

JCJ: Barker, W.T., & Co. was established in 1881 at 221 Devonshire St., Boston, MA. The company employed 50 and made 7,000 a day. See: Monadnock Mill

JCK: Barnes Brothers was established in 1864 at Rochester, MI. The company employed 14 and made 3,000 pounds a day. See: Peninsular Mill

JCL: Barnitz Paper Co. had offices in Middletown, OH, and made 5.5,000 pounds a day. See: Globe Mill

JCM: Barret Mfg. Co. had offices in Bath, SC. The company employed 50 and made 6,000 pounds a day. See: Bath Mills

JCN: Barrett & Swan was established in 1828 at Camden, ME. See: Camden Mill

JCO: Bashong Paper Co. had offices in Reading, PA, and made 6,000 pounds a day. See: Reading Mill

JCP: Bastet & Bain was established at South Orange, NJ, and made 8,000 pounds a day. See: Phoenix Mill

JCQ: Batchelder, Nathaniel P., had offices in Ashland, NH, and made 1,500 pounds a day. See: Mt. Prospect Mill

JCR: Bates, F.A., was established in 1876 at Hampshire Co., MA. See: Bates Mill

JCS: Baxter & Charmer was established in 1873 at Pittsfield, MA. See: Colt Mill

JCT: Beach, Hommerkin & Kearney was established in 1827 at Saugerties, NY. See: Saugerties Mill

JCU: Beach, J.C., & Bro. was established in 1875 at 239 Broadway New York. The company employed 40 and made 3 tons a day. See: Walkill Mill

JCV: Beach, S.Y., Paper Co. was established in 1880 at Seymour, CT. The company employed 7 and made 1 ton a day. See: Seymour Mill

JCW: Beard, Crouse & Co. had offices in Fayetteville, NY, and made 8,000 pounds a day. See: Fayetteville Mill

JCX: Beard, Henry L., & Sons had offices in Chittenango, NY, and made 10,500 pounds a day. See: Fayetteville Mill, Central New York Mill

JCY: Beard, T.E., was established in 1855 at Huntington, CT, and made 2,000 pounds a day. See: Huntington Mill

JCZ: Beardsley, J.R., was established in 1880 at Elkhart, IN. The company employed 25 and made 6,000 pounds a day. See: Beardsley Mill

JDA: Beaver Falls Paper Co. was established in 1883 at Beaver Falls, PA. The company employed 20 and made 5,000 pounds a day. See: Beaver Falls Mill

JDB: Bechtel, Peter, Sr., and Peter, Jr. was established in 1800 at Germantown, PA. See: Betchel Mill

JDC: Bechtel, Samuel, was established in 1820 at Oxford, PA. See: Coates Mill

JDD: Beckett & Gridley was established in 1859 at Delphi, IN. See: Hydraulic Mill

JDE: Beckett, Laurie & Co. was established in 1849 at Hamilton, OH. The company employed 60 and made 3 tons a day. Principals were William Beckett, Calvin Reily, and Adam Laurie. See: Miami Mill

JDF: Beckley, Daniel, had offices in Beckleysville, MD, and made 4,000 pounds a day. See: Beckley Mill

JDG: Beckley, George, was established in 1800 at Beckleysville, MD.

JDH: Bedford Paper Co. was established in 1898 at Big Island, VA. See: Bedford Mill

JDI: Beebe & Holbrook was established in 1871 at Holyoke, MA. The company employed 115 and made 6,000 pounds a day. Principals were Jared Beebe and G.B. Holbrook. See: Beebe & Holbrook Mill

JDJ: Beloit Straw Board Co. was established in 1881 at Beloit, WI. The company employed 18 and made 24,000 pounds a day. See: Straw Board Mill

JDK: Belvedere Manufacturing Co. was established in 1851 at Richmond, VA. The company employed 50 people. See: Franklin Mill

STA: Benjamin, Nathan, was established in 1813 at Catskill, NY.

JDL: Bennett & Walton was established in 1823 at Philadelphia and moved to Mount Holly, NJ, around 1843. Principals were Titus Bennett and Joseph Walton. See: Mount Holly Mill, Wissahiccon Mill

JDM: Bennett, D.C., was established in 1867 at Southington, CT. The company employed 2 and made 1,000 pounds a day. See: Excelsior Mill

JDN: Benton & Garfield was established in 1833 at East Lee, MA. The company employed 25 and made 2,000 pounds a day. See: Mountain Mill, Forest Mill, Forest Mill, Upper

JDO: Benton Bros. was established in 1867 at East Lee, MA. See: Greenwood Mill

JDO.1: Berkshire Paper Co. was established in Berkshire Co., MA, around 1880. The company employed 40 and made 10,000 pounds a day. See: Carroll Mills

JDP: Berlin Mills Company was established in 1888 at Berlin, NH. See: Riverside Paper Mill

JDQ: Berstler, John and Jacob, was established in 1810 at Uwchlan Township, MA. See: Buck Run Mill, Old

JDR: Bicking, David, was established in 1845 at Upper Oxford Township, PA. See: Falls Mill

JDS: Bicking, Edmond, was established in 1845 at Steelville, PA. See: Steelville Mill

JDT: Bicking, Edmund, Jr., was established in 1867 at Atglen, PA. See: Sadsbury Mill

JDU: Bicking, Frederick, was established in 1769 at Lower Merion Township, PA. See: Bicking Mill

JDV: Bicking, John, Sr., and John, Jr., was established in 1791 at Downington, PA. See: Bicking Mill

JDW: Bicking, Joseph, was established in 1806 at Coatsville, PA. See: Coatsville Mill

JDX: Bicking, S. Austin, was established in 1881 at Downington, PA. The company employed 6 and made 3,000 pounds a day. See: Bicking Mill

JDY: Bidwell, H., was established in 1879 at Newark, NJ. The company employed 25 and made 5 tons a day. See: Newark Board Mill

JDZ: Billings & Morrison was established in 1882 at Appleton, WI. The company employed 25 and made 20,000 pounds a day. See: Fox River Straw Board Mill

JEA: Billings Brothers was established in 1810 at Montpelier, VT. Principals were James and Stephen Billings. See: Billings Mill

JEB: Bingham, C.E., was established in 1850 at Germantown, NY. The company employed 40 and made 8 tons a day. See: Eureka Mill, Crescent Mill, DeMeza Mill

JEC: Bingham, James, was established in 1883 at North Lyme, CT, and made 3,000 pounds a day. See: Pleasant Valley Mill, Five Miles Mill

JED: Binney, B.S., was established in 1871 at Shirley Village, MA. The company employed 8 and made 16,000 pounds a day. See: Nehoiden Mill, Lakeville Mill, Washington Mill

JEE: Bird, F.W., & Son was established in 1795 at East Walpole, MA, and made 7 tons a day. The principal, William Bird, was five time congressman and one time senator from Massachusetts. See: Gillespie Mill, Bird's Upper Mill, Lakeville Mill

JEF: Bird, Hopkins & Whiting was established in 1809 at Franklin, NJ. See: Third River Mill

JEG: Bishop, Robert, was established in 1888 at Needham, MA. See: Jackson Mill

JEH: Bissell & Pease was established in 1801 at Suffield, CT. See: Franklin Mill

JEI: Black, John, was established in 1870 at Springfield, IL. The company employed 18 and made 40,000 pounds a day. See: Riverton Mill

JEJ: Blackhawk Paper Co. was established in 1884 at Milan, IL. The company employed 40 and made 14,000 pounds a day. See: Milan Mill

JEK: Blackmar, Frank L. was established in 1874

at Rochester, MI. The company employed 14 and made 3,000 pounds a day. See: Wilcox Mill

JEL: Blair Bros. was established in 1836 at Brodlbin, NY. The company employed 4 and made 2,000 pounds a day. See: Scandaga Mill

JEM: Blake, Cutler & Fleming was established in 1802 at Bellows Falls, VT. Principals were Bill Blake, John Atkinson (1804), Alexander Fleming (1821), and James I. Cutler. See: Rockingham Mill, Newbury Mill

JEN: Blauvelt & Co. was established in 1864 at Elizabeth, NJ. Principals were James and Fredrick Gilmore, John Trimble, Albert C. Sparks (1878), George Tanner, and John T. Faxon (1880). See: Northrup Mill

JEO: Blauvet & Gilmor was established in 1861 at Lee, MA. See: Blauvet & Gilmor Mill

JEP: Bleything, William H., was established in 1832 at Hanover, NJ. See: Bleything Mill

JEQ: Bloomsburg Paper Co. was established in 1882 on Light Street in Philadelphia, PA, and made 2,000 pounds a day. See: Bloomsburgh Mill

JER: Boies, James, & Co. was established in 1765 at Milton, MA. See: Neponset Mills, Boies Family Mill, Boies Mill

JES: Boies, Jeremiah Smith, was established in 1792 at Milton, MA.

JET: Bollinger, Valentine, was established in 1780 at Shamburg, MD. The company employed 6 and made 1 ton a day. See: Gunpowder Mill, Old

JEU: Bolton & Bacon was established in 1875 at Springfield, IL. The company employed 25 and made 5 tons a day. See: Springfield Mill

JEV: Boonton Paper Co. was established in 1880 at Boonton, NJ, and made 20,000 pounds a day. See: Boonton Mill

JEW: Boorum & Pease was established in 1881 at 28-34 Reade St., New York, NY, and made 3 tons a day. See: Antelope Mill

JEX: Booth, J.R., Winnesheik Paper Mill was established in 1871 at Decorah, IA. The company employed 20 and made 7,000 pounds a day. See: Winnesheik Mill

JEY: Boswell Keene & Co. was established in 1851 at Hartford, CT. See: Hartford Mill

JEZ: Bottomly, John, & Co. was established in 1863 at East Lee, MA. The "and company" was James Toole. See: Thatcher Mill

JFA: Bowdoin Paper Mfg. Co. was established in 1868 at Brunswick, ME. The company employed 110 and made 10 tons a day. See: Bowdoin Mills

JFB: Bowers, L had offices in West Chester, PA, and made 3,000 pounds a day. See: Hilldale Mill

JFC: Boyce, John, had offices in Franklin, NY, and made 600 pounds a day. See: Boyce Mill

JFD: Boyer, Solomon & Samuel was established in 1812 at Manatawny, PA. See: Oley Mill

JFE: Bradford Paper Mfg. Co. had offices in Bradford, VT, and made 1 ton a day. See: Bradford Mill

JFF: Bradley, A., & Sons was established in 1838 at Dansville, NY. See: Bradley Mill

JFG: Bradley, Horace, was established in 1850 at Walesville, NY. The company employed 18 and made 9,000 pounds a day. See: Walesville Mill

JFH: Bradner Smith & Co. was established in 1854 at Rockton, IL. The company employed 36 and made 8 tons a day. See: Winnebago Mills

JFI: Brandt, E.R., & Co. was established in 1861 at Morristown, NJ. The company employed 6 and made 4,000 pounds a day. See: Morris Plains Mill

JFJ: Bratton, Bishop & Co. was established in 1822 at Kennett Square, PA. See: Willow Brook Mill, Webb Mill

JFK: Bremaker-Moore Co. was established in 1863 at Louisville, KY. The company employed 150 and made 14,000 pounds a day. See: Falls City Mill

JFL: Bridge Mill Paper Co. had offices in Pawtucket, RI, and made 6,000 pounds a day. See: Bridge Mill

JFM: Bridgeport Paper Co. was established in 1883 at Bridgeport, CT. The company employed 50 and made 12,000 pounds a day. See: Bridgeport Mill

JFN: Briggs A.D., was established in 1850 at Alcove, NY. The company employed 25 and made 7,000 pounds a day. See: Valley Mill

JFO: Brookside Paper Mfg. Co. was established in New York. See: Brookside Mill

JFP: Broomell, Joshua B., was established in 1866 at Ercildoun, PA. The company employed 14 and made 6,000 pounds a day. See: Rokerby Mill

JFQ: Brown & Lockhart was established in 1854

at Angelica, NY. The company employed 16 and made 3,000 pounds a day. See: Joncy Mill

JFR: Brown Bros. was established in 1870 at Comstock Bridge, CT. The company employed 20 and made 4,000 pounds a day. See: Comstock Mills

JFS: Brown Bros. & Watson was established in 1879 at Corralitos, CA. See: Corralitos Mill

JFT: Brown, James S., was established in 1857 at Monongahela City, PA, and made 13,000 pounds a day.

JFU: Brown, L.L., Paper Co. was established in 1850 at Adams, MA. The company employed 200 and made 10,000 pounds a day by machine and 2,000 pounds a day of handmade. Principals were Levi L. Brown and uncles, Daniel and William Jenks. See: Stone Mill, Cummington Mill, Greylock Mill

JFV: Brownell & Miller was established in 1840 at St. Charles, IL, and made 2 tons a day. See: St. Charles Mill

JFW: Bruce, Nathaniel, and Hiram Seely was established in 1829 at New Baltimore, NY. See: New Baltimore Mill

JFX: Bryand Paper Co. was established in 1898 at Kalamazoo, MI. See: Bryand Mill

JFY: Buchanan & Kilmer was established in 1853 at Rock City Falls, NY. The company employed 30 and made 8,500 pounds a day. Principals were C.S. Buchanan and Chauncey Kilmer. See: Rock Falls Mills

JFZ: Buck, William, was established in 1866 at Oregon City, OR. See: Oregon City Mill, Bancroft Mills

JGA: Buel, David, was established in 1800 at Troy, NY. See: Troy City Mill

JGB: Bulkley, Dunton & Co. was established in 1865 at Bancroft, MA. The company employed 60 and made 10,000 pounds a day. See: Bancroft Mills

JGC: Bullard, D.A., & Sons was established in 1863 at Schuylerville, NY. The company employed 52 and made 4.5 tons a day. See: Schuylerville Mill

JGD: Bunce, Charles, was established in 1778 at Westville, CT. See: West Rock Mill

JGE: Burbank, Abijah & Sons was established in 1776 at Sutton, MA. The "and sons" were Caleb, Elijah, and Abijah, Jr.

JGF: Burbank, Abijah, Jr., was established in 1801 at Sharon, VT. See: Sharon Mill

JGG: Burbank, General Caleb, and Co. was established in 1796 at Boston, MA. The "and company" was Elijah Burbank. See: Burbank Mill

JGH: Burbank, Silas, was established in 1806 at Montpelier, VT. See: Montpelier Mill

JGI: Burke, A.J., was established in 1852 at White Hall, MD. The company employed 8 and made 2,000 pounds a day. See: Excelsior Mill

JGJ: Butler & Hudson was established in 1799 at Unionville, CT. See: Union Village Mill

JGK: Butler & Hunt was established in 1842 at St. Charles. The company employed 80 and made 4,000 pounds a day. Principals were Oliver Butler and Joseph Hunt. See: West Side Mill, Devitt Mill

JGL: Butler & Ward was established in 1818 at Suffield, CT. See: Eagle Mill, Old

JGM: Butler, Simeon and Asa, was established in 1789 at Suffield, CT. See: Eagle Mill, Old

JGN: Byrd, James, was established in 1847 at Marietta, GA, and made 1,000 pounds a day. See: Marietta Mill

JGO: Cadwallader & McCahan was established in 1800 at Birmingham, PA. See: Cadwallader Mill

JGP: Chagrin Falls Paper Co. was established at Chagrin Falls, OH, and made 3,500 pounds a day. See: Western Reserve Mill

JGQ: California Paper Co. was established in 1883 at 4 California St., San Francisco, CA, and made 16,000 pounds a day. See: California Mills

JGR: Campbell, Eden, was established in 1883 at W. Medway, MA, and made 3,000 pounds a day. See: West Medway Mill

JGS: Campbell, George, was established in 1850 at Norfolk, MA. The company employed 11 and made 4,000 pounds a day. See: Campbell Mill

JGT: Campbell, Samuel, was established in 1796 at New York. See: Thistle Mill

JGU: Campbell, Thomas, was established in 1817 at Millburn, NJ. See: Milburn Mill

JGV: Campbell, Wellington & Son (Charles) was established in 1840 at Short Hills, NJ. The company employed 35 and made 5 tons a day. See: Short Hills Mill

JGW: Cannelton Paper Co. was established in 1873 at Cannelton, IN. The company employed 2 and made 2 tons a day. See: Cannelton Mill

JGX: Canney, J.M., was established in 1857 at Center Ossipee, NH, and made 1,500 pounds a day. See: Canney Mill

JGY: Canton Paper Co. was established in 1863 at Canton, OH. The company employed 30 and made 6,000 pounds a day. See: Canton Mill

JGZ: Carew Mfg. Co. was established in 1848 at S. Hadley Falls, MA. The company employed 125 and made 6,000 pounds a day. See: Carew Mills

JHA: Carlyle Paper Co. was established in 1898 at Carlyle, IL, made 15 tons a day. See: Carlyle Mill

JHB: Carrecabe, J.M., was established in 1850 at Lynn, MA. The company employed 15 and made 4,000 pounds a day. See: Milton Mill, Flax Leather Board Mill

JHB.1: Carrier, E.T., was established at Manchester, CT. The company employed 20 and made 3 tons a day. See: Riverdale Mill

JHC: Carroll, John, was established in New Marlborough, Berkshire Co., MA, about 1870. See: Carroll Mills

JHC.1: Carroll, Theron G. was established in Berkshire Co., MA, in 1883. The company employed 40 and made 10,000 pounds a day. See: Carroll Mills

JHD: Carson & Brown was established in 1868 at Dalton, MA. See: Berkshire Mill

JHE: Carson & Chamberlin was established in 1822 at Dalton, MA. Principals were David Carson and Joseph Chamberlin. See: Defiance Mill

JHF: Carson, David, & Sons was established in 1812 at Dalton, MA. The company employed 100 and made 4,000 pounds a day. The "and sons" were Thomas G. and William W. Carson. See: Red Mill, Old

JHG: Carson, David, Jr., had offices in Newburgh, NY. See: Highland Mill

JHH: Carsons Paper Co. was established in 1866 at Dalton, MA. See: Berkshire Mill

JHI: Carter, Robert, was established in 1816 at Rockville, MD. See: Meeter Mill

JHJ: Case Bros. was established in 1883 at Chaplin, CT. The company employed 17 and made 8,000 pounds a day. See: Arlington Mill, Chaplin Mill, Forest Mill, Highland Mill, Naubuc Mill

JHK: Case, F.L., was established in 1884 at S. Manchester, CT. The company employed 10 and made 4 tons a day. See: Brookside Mill

JHL: Castleton Paper Co. had offices in Castleton, NY, made 5 tons a day. See: Riverside Mill

JHM: Caswell, Gurdon, was established in 1808 at Watertown, NY. See: Black River Mill

JHN: Cataract Paper Co. was established in 1884 at Fayetteville, NY. The company employed 15 and made 2 tons a day. See: Upper Beard Mill

JHO: Cattron, C.F., was established in 1874 at Otis, IN, and continued to about 1883. See: Spring Valley Mill

JHP: Cave, Charles J., had offices in Cairo, NY, and made 6 tons a day. See: Woodstock Mill, Cascade Mill

JHQ: Cecil Paper Co. was established in 1836 at Elkton, MD, and made 7,000 pounds a day. See: Cecil County Mills, Cecil Mill

JHR: Cedar Falls Paper Mfg. Co. was established in 1882 at Cedar Falls, IA. The company employed 27 and made 14,000 pounds a day. See: Cedar Falls Mills

JHS: Chaffee & Callender was established in 1835 at Glendale, MA. The company employed 53 and made 5 tons a day. See: Glendale Mill

JHT: Chaffee, Prentiss & Co. was established in 1863 at East Lee, MA. The "and company" was William H. Hamblin. See: New England Mills

JHU: Chamberlin, H., & Co. was established in 1853 at Dalton, MA. The "and company" was Henry, Albert S., and Burr Chamberlin. See: Defiance Mill

JHV: Chapin and Gould was established in 1858 at Springfield, MA. The company employed 125 and made 6,000 pounds a day. See: Chapin and Gould Crescent Mill

JHW: Chase & Linton was established in 1847 at Athens, GA, and made 1,000 pounds a day. See: Athens Mill

JHX: Chase, Robert, & Co. was established in 1865 at Northumberland, NH. The company employed 11 and made 1,000 pounds a day. See: Northumberland Mill

JHY: Chelsea Paper Mfg. Co. was established in 1834 at Greenville, CT, and made 22,000 pounds a day. See: Chelsea Mill

JHZ: Chemical Paper Co. was established in 1880 at Holyoke, MA. The company employed 225 and made 25 tons a day. See: Winona Mill, Chemical Paper Co. Mill

JIA: Chemical Pulp Co. was established at Straudsburgh, PA, and made 6,000 pounds a day. See: Delaware Water Gap Pulp & Mill

JIB: Cheney Paper Co. was established in 1873 at N. Manchester, CT. See: Oakland Mill

JIC: Cheney, P.C., Co. had offices in Manchester, NH. The company employed 125 and made 14,000 pounds a day. See: Cheney Mill

JID: Chester Paper Co. was established in 1853 at Huntington, MA. The company employed 75 and made 4,000 pounds a day. See: Chester Mill

JIE: Chestertown Straw Board and Mfg. Co. was established in 1882 at Chestertown, MD. The company employed 30 and made 12,000 pounds a day. See: Chestertown Straw Board Mill

JIF: Childs & Carper was established in 1884 at Rockford, MI. The company employed 24 and made 6,000 pounds a day. See: Childs Mill

JIG: Chittenden, George & Sons was established in 1809 at Stockport, NY. See: Stockport Mills

JIH: Church, Henry & Co. was established in 1838 at Rochester, NY. See: Church Mill

JII: Church, Luman, was established in 1819 at Lee, MA. See: Waverley Mill, Enterprise Mill, Forest Mill, Old

JIJ: Church, Samuel, was established in 1806 at Great Barrington, MA. See: Union Mill, Castle Mill, South Lee Mill

JIK: Church, Utley & Co. was established in 1882 at Rock Falls, IL. The company employed 35 and made 12,000 pounds a day. See: Rock Falls Mill

JIL: Ciscle & Mchannon was established in 1852 at Lockland, OH. See: Athern Lower Mill

JIM: Claflin Paper Co. was established in 1868 at South Toledo, OH. The company employed 45 and made 3 tons a day. See: Lakeside Mill

JIN: Clapp, Keeney & Co. was established in 1838 at Unionville, CT. The "and company" was Timothy Keeney, James B. Wood, and Sandford Buckland. See: Union Village Mill

JIO: Clark & Bailey was established in 1820 at Millburn, NJ.

JIP: Clark & Gray was established in 1807 at Chaplin, CT. See: Chaplin Mill

STA: Clark & Sharpless was established in 1820 at Delaware Co., PA.

JIQ: Clark, Charles A., was established in 1884 at Logansport, IN. The company employed 10 and made 8,000 pounds a day. See: Logansport Mill

JIR: Clark, Chas. P., Jr., was established in 1881 at Newton Upper Falls, MA. The company employed 25 and made 7,000 pounds a day. See: Newton Mills

JIS: Clark, James, was established in 1848 at Woodville, NY. The company employed 9 and made 2,500 pounds a day. See: Woodville Mill

JIT: Clark, John, was established in 1810 at Millburn, NJ. See: Milburn Mill

JIU: Clark, John F., was established in 1883 at Marseilles, IL, and made 14,000 pounds a day. See: Clark Mill

JIV: Clark, W.I., was established at York, PA, and made 4,000 pounds a day. See: Codorus Mill

JIW: Clark, William, was established in 1859 at Dayton, OH. See: Clark Mill

JIX: Clarksville Paper Co. was established in 1880 at Clarksville, MS. The company employed 30 and made 7,500 pounds a day. See: Calumet Mill

JIY: Clear Springs Paper Co. at Lambertville, NJ (circa 1880), made 3,500 pounds a day. See: Great Spring Mill

JIZ: Clegg & Fisher was established in 1877 at South Lawrence, MA, and continued until about 1883. See: Merrimack Leather Board Mill

JJA: Cleveland Paper Co. was established in 1847 at Cleveland, OH. The company employed 300 and made 12 tons a day. See: Valley Mill, Broadway Mill, Forest Street Mill

JJB: Clifton Paper Co. was established in 1866 at Clifton, OH, made 4,000 pounds a day. See: Clifton Mill

JJC: Clinton Paper Co. had offices in Clinton, IA, made 12,000 pounds a day. See: Clinton Mill

JJD: Clyde River Paper Co. was established in 1872 at West Derby, VT. The company employed 20 and made 5,000 pounds a day. See: Clyde River Mill

JJE: Coates, Warrick, was established in 1847 at Russellville, PA. The company employed 2 and made 1,800 pounds a day. See: Coates Mill

JJF: Coghlan, Robert J., was established at Whippany, NJ, made 1 ton a day. See: Caledonia Mill

JJG: Cole & Gough was established in 1880 at Putney, VT. The company employed 10 and made 1,400 pounds a day. See: Eagle Mill

STA: Coleman, Ezra, was established in 1845 at Philadelphia, PA, as a stationer.

JJH: Collins Mfg. Co. was established in 1872 at N. Wilbraham, MA. The company employed 150 and made 12,000 pounds a day. See: Collins Mill

JJI: Colt, Thomas, was established in 1851 at Pittsfield, MA. See: Colt Mill

JJJ: Columbia Paper Co. was established in 1834 at Rock Bridge, MO. See: Columbia Mill

JJK: Columbia River Paper Co. was established in 1884 at Portland, OR. The company employed 30 and made 8,000 pounds a day. Principals were J.K. Gill, H.L. Pittock, and William Lewthwaite. See: Columbia Mill, Clackamas Mill

JJL: Combined Locks Paper and Pulp Mills was established in 1891 at Outagamie Co., WI. See: Combined Locks Paper and Pulp Mills

JJM: Comstock, P.G., was established in 1848 at W. Stockbridge, MA. The company employed 7 and made 3,000 pounds a day. See: Williamsville Mill

JJN: Condit, Israel & Co. was established in 1856 at Millburn, NJ. The "and company" was Amzi Condit and R.D. Traphagen.

JJO: Connard, G.B., was established in 1836 at Redding, PA, and made 2,000 pounds a day. See: Oley Mill

JJP: Contoocook Valley Paper Co. was established in 1871 at West Henniker, NH. The company employed 40 and made 5,000 pounds a day. See: Contoocook Valley Mill

JJQ: Cook, A.S.C., was established in 1883 at Bridgeport, CT, and made 2,000 pounds a day.

JJR: Cooper, E.W., & Son was established in 1870 at Madison, CT. The company employed 37 and made 5 tons a day. See: Hammanossette Mill, Killingsworth Mill

JJS: Cope, Edwin, was established in 1856 at Philadelphia, PA. The company employed 30 and made 3,000 pounds a day. See: Trout Creek Mill (for Cope Mill), Wissahiccon Mill

JJT: Corralitos Paper Mill Co. was established in 1880 at Corralitos, CA, and made 4 tons a day.

JJU: Corwell & Bro. was established in 1853 at Ann Arbor, MI. The company employed 12 and made 4,000 pounds a day. See: Hoosic Mill

JJV: Coshocton Paper Co. was established in 1865 at Coshocton, OH. The company employed 20 and made 6,000 pounds a day. See: Coshocton Mill

JJW: Couch & Worden was established in 1852 at East Lee, MA. The firm later became Bradford Couch and E.T. Clark, which failed in 1857. See: National Mill, Congress Mill

JJX: Coughnet, James H., was established in 1870 at Kecks Center, NY. The company employed 6 and made 1 ton a day. See: Rock Mill

JJY: Coulter, Bever & Bowman was established in 1806 at Beaver Creek, OH. See: Beaver Creek Mill

JJZ: Cowles Paper Co. was established in 1866 at Unionville, CT. The company employed 22 and made 6,000 pounds a day. See: Cowles Mill

JKA: Cox & Baker was established in 1807 at Dorchester, MA. Principals were Edmund Baker and Henry Cox. See: Baker Mill

JKB: Cox & Thorp was established in 1813 at Dorchester, MA. Principals were Henry Cox and Eliab Thorp. See: Dorchester Mill, Baker Mill

JKC: Cox, Benjamin & Co. was established in 1806 at Chicopee, MA. See: Cox Mill

JKD: Cox, Dewitt C., was established in 1873 at St. Johnsville, NY. The company employed 8 and made 2,000 pounds a day. See: St. Johnsville Straw Wrapping Mill

JKE: Cox, E.T., & Co. was established in 1819 at Zanesville, OH. This was the firm of Ezekiel Taylor Cox. See: Cox Mill

JKF: Cox, H. & J, & Co. was established in 1846 at Zanesville, OH. The company employed 20 people. Principals were Horatio J. and Jones L. Cox. See: Cox Mill

JKG: Cox, J., & Son was established in 1811 at Mount Holly, NJ. See: Mount Holly Mill

JKH: Coy Paper Co. was established in 1884 at West Claremont, NH. The company employed 15 and made 6,000 pounds a day. See: Coy Mill

JKI: Craig, Elijah, the Rev. was established in 1793 at Louisville, KY. See: Craig Mill

JKJ: Craig, Parkers & Co. was established in 1792 at Georgetown, KY. See: Redstone Mill, Craig Mill

JKJ.1: Crain, N.T., was established around 1823 in Northfield, NH. See: Northfield Mill

JKK: Crane & Carson was established in 1825 at Dalton, MA. Principals were Zenas Crane and David Carson. See: Castle Mill

Directory of Paper Mill Owners

JKL: Crane & Co. was established in 1842 at Dalton, MA. The company employed 65 and made 3,000 pounds a day. Principals were Zenas Marshall Crane and James Brewer Crane. See: Stone Mill, Bay State Mill, Old, Pioneer Mill, Red Mill

JKM: Crane Brothers was established in 1867 at Westfield, MA. The company employed 100 and made 2 tons a day. Principals were James A. and Robert B. Crane. See: Crane Bros. Mill, Japanese Mill

JKN: Crane, James Brewer, was established in 1847 at Westfield, MA. The company employed 100 and made 2 tons a day. See: Japanese Mill, Crane Bros. Mill

JKO: Crane, S. & L., was established at Ballston Spa, NY. Principals were Seymour and Lindley Crane. See: Crane Mill

JKP: Crane, Z., Jr., & Bro. was established in 1865 at Dalton, MA. The company employed 125 and made 5,000 pounds a day. See: Bay State Mill

JKQ: Crane, Zenas, & Co. was established in 1801 at Dalton, MA. The "and company" was Daniel Gilbert and Henry Wiswell. See: Red Mill, Old, Berkshire Mill, Old, Red Mill

JKR: Crane, Zenas, Jr., was established in 1885 at Dalton, MA. See: Berkshire Mill

JKS: Crehore & Neal was established in 1834 at Needham, MA. Crehore Mill

JKT: Crehore Paper Co. was established in 1845 at Needham, MA. Principals included Charles Crehore and family. See: Crehore Mill

JKU: Crehore, C.F., & Son was established at 87 Milk St., MA, and made 1 ton a day.

JKV: Crehore, Charles F., was established at 87 Milk St., Boston, and made 1 ton a day. See: Crehore Mill

JKW: Creshore, Edward, was established in 1845 at Leominster, MA. See: Kendall Mill

JKX: Crichton, A., had offices in Chittenango, NY, and made 3,500 pounds a day. See: Crichton Mills

JKY: Crocker Mfg. Co. was established in 1871 at Holyoke, MA. The company employed 85 and made 4 tons a day. See: Crocker Mill No. 2, Crocker Mill No. 1

JKZ: Crocker, Burbank & Co. was established in 1850 at Fitchburg, MA. The company employed 185 and made 35,000 pounds a day. See: Stone Mill, Brick Mill, Hanna Mill, Lyon Mill, Snow Mill

JLA: Cross & Earenfight was established in 1810 at Mill Grove, OH. Principals were John Cross and C. Earenfight. See: Union Mill

JLB: Croswell & Parsons had offices in New Baltimore, NY, and made 2,000 pounds a day. Principals were William Croswell and Stephen Parsons. See: New Baltimore Mill

JLC: Crouse, David, & Son was established in 1812 at Chillicothe, OH. See: Crouse Mill

JLD: Crown Paper Company was established in 1890 at West Linn, OR. See: Crown Mill

JLE: Cullen, L.C., was established in 1873 at Cheney, PA. The company employed 6 and made 2 tons a day. See: Forest Mill

JLF: Culver & Mitchell was established in 1851 at Norwich, CT. See: Leffingwell Mill

JLG: Cummings, Benjamin, was established in 1819 at Schoharie River, NY. See: Cummings Mill

JLH: Curtis Brothers had offices in Newark, DE. The company employed 18 and made 3,000 pounds a day. See: Nonatum Mill

JLI: Curtis, A.C. & W., & Co. was established in 1818 at Needham, MA. Principals were Allen, Crocker and William (son, George B. Curtis, 1851). The firm went out of business in 1860. See: Nehoiden Mill, Curtis Mills

JLJ: Curtis, Solomon, was established in 1789 at Needham, MA. Sons of the principal, Allen and William, were admitted in 1818. See: Needham Mill, Curtis Mills

JLK: Curtis, Warren & Melville was established in 1830 at Franklin, NJ. See: Third River Mill

JLL: Cushman, A.R., was established in 1835 at N. Amherst, MA. The company employed 20 and made 4,000 pounds a day. See: Cushman Mills

JLM: Cutler, James I., & Co. was established in 1821 at Bellows Falls, VT. See: Rockingham Mill

JLN: Cuyahoga Paper Co. was established in 1850 at Cuyahoga Falls, OH. The company employed 75 and made 10,000 pounds a day. See: Phoenix Mill, Cuyahoga Mill, Empire Mill

JLO: Dager & Cox was established in 1882 at Bridgeport, PA. The company employed 21 and made 2.5 tons a day. See: Eureka Mill

JLP: Davenport, Jas. B., was established in 1876 at Hamburg, NJ. See: Davenport Mill

JLQ: Davey, E.H., was established in 1841 at Bloomfield, NJ. The company employed 15 and made 3,000 pounds a day. See: Davey's Board Mill

JLR: Davey, W.O., & Sons had offices in 115 Wall St. New York, NY. The company employed 60 and made 12,000 pounds a day. See: Davey Mill

JLS: Davis & Young was established in 1878 at Accord, NY. The company employed 10 and made 4,000 pounds a day. See: Accord Mills

JLT: Davis Brothers was established in 1871 at Contoocook, NH. The company employed 14 and made 10,000 pounds a day. See: Kearsage Mill

JLU: Davis, C.F., was established in 1872 at Valatie, NY. The company employed 23 and made 10,000 pounds a day. See: Centennial Mill

JLV: Davis, George E., was established in 1862 at Saccarapa, ME. The company employed 41 and made 12,000 pounds a day. See: Saccarapa Mills

JLW: Dawson, Seth F., was established in 1880 at 37 High St., Boston, MA. The company employed 27 and made 6,000 pounds a day. See: Essex Mills

JLX: Day, Alfred and Dan, was established in 1813 at Burlington, VT. See: Day Mill

JLY: DeMeza, H., had offices in Chittenango, NY. The company employed 8 and made 2,000 pounds a day. See: Garlock Mill, Crescent Mill, DeMeza Mill

JLZ: Debit, William, was established in 1829 at Hartford, CT. See: Five Miles Mill, Pleasant Valley Mill

JMA: Decker & Sabin was established in 1882 at East Lee, MA. The company employed 30 and made 3 tons a day. Principals were John A. Decker and T.G. Sabin. See: Washington Mill, Lakeville Mill, Nehoiden Mill

STA: DeForest, Smith & Bassett was established in 1834 at Seymour, CT. Firm later became Smith & Bassett.

JMB: Delaney & Munson Mfg. Co. was established in 1868 at Unionville, CT. The company employed 30 and made 3 tons a day. See: Unionville Mill, Lakeville Mill

JMC: Delaware Water Gap Pulp & Paper Co. was established in 1884 at Stroudsburg, PA. The company made 6,000 pounds a day. See: Delaware Water Gap Pulp & Mill

JMD: Delphos Paper Co. was established in 1878 at Delphos, OH. The company employed 15 and made 5,000 pounds a day. See: Delphos Mill

JME: Deming, Learned & Co. was established in 1883 at Wells River, VT. The company employed 12 and made 1,200 pounds a day. See: Wells River Mill

JMF: Denman & Ayers had offices in Millburn, NJ. See: Morris Mill

JMG: Dennison Paper Mfg. Co. was established in 1852 at Mechanic Falls, ME. The company employed 225 and made 12 tons a day. See: Union Mill, Androscoggin Mill, Diamond Mill, Eagle Mill, Poland Mill, Star Mill

JMH: Deseret News Co. was established in 1880 at Butlerville, UT. The company made 5 tons a day. See: Granite Mill, Deseret Mill

JMI: Devitt, M., was established in 1839 at Chicago, IL. See: Devitt Mill

JMJ: Devries & Son was established in 1818 at Elkridge, MD. Principal was Christian Devries. See: Elkridge Mill

JMK: Dexter, C.H., & Sons was established in 1836 at Windsor Locks, CT. Firm went out of business by 1880. See: Dexter Mills

JML: Diamond Mills Paper Co. was established in 1872 at 44 Murray St., New York. The company employed 40 and made 3 tons a day. See: Millburn Mill, Diamond Mills, Eden Mill

JMM: Dickey & LeFever was established in 1820 at Mount Vernon, PA. See: Stone Mill, Mt. Vernon Mill, Old

JMN: Dickey & Lysle was established in 1836 at Mount Vernon, PA. See: Stone Mill, Mt. Vernon Mill, Old

JMO: Dickey, Joseph, was established in 1810 at Mount Vernon, PA. See: Stone Mill, Mt. Vernon Mill, Old

JMP: Dickinson & Clark Paper Co. was established in 1869 at Holyoke, MA. The company employed 50 and made 4 tons a day. See: Dickinson & Clark Mill

JMQ: Dickinson, George R., Paper Co. was established in 1883 at Holyoke, MA. The company employed 175 and made 24 tons a day. See: Dickinson Mill

JMR: Dickinson, Maurice, was established in 1847 at Bentley Springs, MD. See: Valley Mill (1810), Valley Mill (1833)

JMS: Diem F.J., & Co. was established in 1834 at Dayton, OH. The company employed 27 and made 9,000 pounds a day. See: Mad River Mill

JMT: Dodge, Levi, had offices in Constantine, MI. See: Dodge Mill

JMU: Donaldson & Geer was established in 1855 at Ballston Spa, NY. The company employed 35 and made 6,000 pounds a day. See: Hollister Mill

JMV: Donaldson, Robert, was established in 1818 at South Orange, NJ. The company employed 19 people. See: Phoenix Mill

JMW: Dorlan, Samuel B., was established in 1834 at Dorlans Mill, PA. The company employed 8 and made 1200 pounds a day. See: Dorlans Mill

JMX: Doud, William H., had offices in Chesterville, PA. The company employed 5 and made 1,000 pounds a day. See: Auburn Mill (1883)

JMY: Du Pont & Co. was established in 1840 at Louisville, KY. The company employed 185 and made 11 tons a day. See: Louisville Mill

JMZ: Du Pont Co. had offices in Wilmington, DE. See: Kellogg Mill

JNA: Dubuque Paper Co. was established in 1883 at Dubuque, IA. The company employed 30 and made 6 tons a day. See: Dubuque Mill

JNB: Duckett, John B., was established in 1825 at Upper Providence Township, PA. See: Duckett Mill, Chester Creek Mill

JNC: Dumond, George N., was established in 1875 at Linlithgo, NY. The company employed 10 and made 2,000 pounds a day. See: Livingston Mills

STA: Dunham & Davis was established in 1801 at 26 Moore St., New York as stationers. Principals were David Dunham and Matthew L. Davis. See: Woodbine Mill, Caledonia Mill, Eagle Mills, Ivy Mill

JND: Dushane, J.A., & Co. was established in 1800 at Baltimore, MD. The company employed 15 and made 5,000 pounds a day. See: Grove Mill, Caledonia Mill, Eagle Mills (2), Ivy Mill, Woodbine Mill

JNE: Duval, Joseph, was established in 1813 at Lebanon, OH. See: Grove Mill

STA: Dwight, William W., was established in 1846 at Suffield, CT.

JNF: Eachus, James, had offices in Coatsville, PA. The company made 1.5 tons a day. See: Mineral Springs Mill

JNG: Eagle Paper Co. was established in 1883 at Franklin, OH. The company made 6 tons a day. See: Eagle Mill

JNH: East Hartford Mfg. Co. was established in 1865 at Burnside, CT. The company employed 60 and made 2,000 pounds a day. See: Charter Oak Mill

JNI: Eaton, Crane & Pike was established in 1893 at Pittsfield, MA. See: Hurlbut Mill

JNJ: Eaton, Hurlbut Paper Co. was established in 1889 at Pittsfield, MA. Principal was Arthur W. Eaton. See: Hurlbut Mill

JNK: Eau Claire Pulp & Paper Co. was established in 1882 at Eau Claire, WI. The company employed 60 and made 7,000 pounds a day. See: Eau Claire Pulp & Mill

JNL: Eby, George S., was established in 1853 at Compassville, PA. See: Pequea Creek Mill

JNM: Eckstein, Samuel, was established in 1825 at Philadelphia, PA, as a paper merchant. See: Lamb Mill, Buck Run Mill, Old

JNN: Eggert, Robert, was established in 1882 at Rockford, IA. The company employed 17 and made 8,000 pounds a day. See: Rockford Mill

JNO: Elkhart Combination Board Co. was established in 1870 at Elkhart, IN. The company employed 22 and made 32,000 pounds a day. See: Elkhart Combination Board Mill

JNP: Elkhart Paper Co. was established in 1876 at Elkhart, IN. See: Elkhart Mill

JNQ: Elkhart Tissue Paper Co. was established in 1876 at Elkhart, IN. The company employed 25 and made 2,000 pounds a day. See: Elkhart Tissue Mill

JNR: Elliot, W.J., had offices in Philadelphia, PA. The company employed 80 and made 6,000 pounds a day. See: Schuylkill Mill

JNS: Elliott & Curtis was established in 1794 at Needham, MA. Principals were Simon Elliot and Solomon Curtis. See: Curtis Mills

JNT: Ellis, Isaac, was established in 1832 at Norwood, MA. The company employed 18 and made 4,000 pounds a day. See: Ellis Mill

JNU: English, James W., was established in 1850 at Windsor Locks, CT. The company employed 42 and made 10,000 pounds a day. See: Pacific Mill

JNV: English, William, was established in 1850 at Windsor Locks, CT. The company employed 42 and made 10,000 pounds a day. See: Pacific Mill

JNW: Erwin, John, was established in 1848 at Hamilton, OH. See: Hamilton Mill

JNX: Erwin-Lane Paper Co. was established in

1873 at Elkhart, IN. The company employed 140 and made 5,000 pounds a day. See: St. Joseph Valley Mill

JNY: Eshelman, B.B., was established in 1830 at Bart Township, PA. See: Porter Mill, Octoraro River Mill

JNZ: Essex Paper Co. had offices in Bloomfield, NJ. The company made 20,000 pounds a day. See: Essex Mill

JOA: Excelsior Paper Co. was established in 1873 at Holyoke, MA. The company employed 50 and made 4 tons a day. Principals were Daniel H. and John C. Newton. See: Excelsior Mill

JOB: Eyster & Son was established in 1869 at Halltown, WV. The company employed 50 and made 15,000 pounds a day. See: Eyster Mill

JOC: Ezera Baldwin had offices in Millburn, NJ. See: Canoe Creek Leather Mill

JOD: Fairchild Paper Co. had offices in 24 Equitable Building, Boston, MA. The company employed 175 and made 13,000 lbs. a day. See: Pepperell Mills

JOE: Fairchild, D. & P.N., & Co. was established in 1827 at Bridgeport, CT. The company employed 18 and made 2,000 pounds a day. See: Fairchild Mill

JOF: Fall Mountain Paper Co. was established in 1875 in Vermont. The company employed 425 and made 15 tons a day. See: Fall Mountain Mills (7)

JOG: Falls, James, was established in 1829 at Upper Oxford Township, PA. See: Falls Mill

JOH: Fargo Paper Mill Co. was established in 1883 at Fargo, DK. The company made 24,000 pounds a day. See: Fargo Mill

JOI: Farley Paper Co. was established in 1881 at Wendell Depot, MA. The company employed 19 and made 2 tons a day. See: Farley Mill

JOJ: Farnsworth, E.M., had offices in West Townsend, MA. The company employed 7 and made 2,000 pounds a day. See: West Townsend Leather Board Mill

JOK: Farra, John, & Son was established in 1824 at Beaver Valley, PA. The "and son" was Daniel Farra from 1832 to 1860, and Francis Tempist from 1860 to 1901.

JOL: Featherman, Michael, was established in 1816 at West Nottingham, PA. See: Featherman Mill

JOM: Felt Mills was established around 1870 at 28-34 Reade St., New York. The company made 3 tons a day. See: Antelope Mill

JON: Ferry & Wrinkle was established in 1873 at East Lee, MA. See: Thatcher Mill

JOO: Fessenden & Fessenden was established in 1811 at Brattleboro, VT. Principals were Joseph and William Fessenden. See: Whetstone Creek Mill

JOP: Field & Stone was established in 1819 at Troy, NY. See: Troy City Mill

JOQ: Fisk, George C., was established in 1873 at Hinsdale, NH. The company employed 17 and made 2,200 pounds a day. See: Brightwood Mill

JOR: Fitchburg Paper Co. was established in 1864 at W. Fitchburg, MA. The company employed 70 and made 8 tons a day. See: Fitchburg Mills

JOS: Fletcher & McEwan was established in 1870 at Southport, CT, and made 3,000 pounds a day. See: Sasco Creek Mill

JOT: Flint Wyman & Sons was established in 1872 at Bellows Falls, VT. The company employed 15 and made 2,400 pounds a day. See: Bellows Falls Mill

JOU: Forest Paper Co. was established in 1880 at Bowdoin, ME. See: Yarmouth Mill

JOV: Fort Madison Paper Co. was established in 1882 at Fort Madison, IA. The company employed 35 and made 16,000 pounds a day. See: Fort Madison Mill

JOW: Fort Orange Paper Co. was established in 1881 at Castleton, NY. The company employed 75 and made 9 tons a day. See: Fort Orange Mill

JOX: Foster, Bart & Son was established in 1818 at Newton, MA. See: Foster Mill

JOY: Foulds, William, was established in 1872 at N. Manchester, CT. The company employed 6 and made 3,000 pounds a day. See: Manchester Mill

JOZ: Fox & Curtis had offices in Penn Yan, NY. The company employed 25 and made 5 tons a day. See: Keuka Mill

JPA: Fox River Flour & Paper Co. was established in 1883 at Appleton, WI. The company employed 70 and made 3 tons a day. See: Ravine Mill

JPB: Fox River Paper Co. was established in 1882 at Yorkville, IL. The company employed 16 and made 11,000 pounds a day. See: Fox River Mill

JPC: Fox, E.S., had offices in Sauquiot, NY. The company made 3 tons a day. See: Fox Mill

JPD: Fram, R.S., was established in 1859 at Yorkville, IL. The company employed 12 and made 8,000 pounds a day. See: Yorkville Mill

JPE: Frambach Paper Co. was established in 1883 at Kaukauna, WI. The company employed 60 and made 16,000 pounds a day. See: Badger Mill

JPF: Franklin Manufacturing Co. was established in 1845 at Richmond, VA. The company employed 50 people. Principals were Conway Robinson and Thomas Ritchie. See: Franklin Mill

JPG: Franklin Paper Co. was established in 1866 at Holyoke, MA. The company employed 60 and made 6,000 pounds a day. Principal was James A. Newton. See: Franklin Mill

JPH: Franklin Paper Co. was established in 1848 at Newton Upper Falls, MA. Principal was Joseph Daniel Stowe. See: Franklin Mill

JPI: Franklin Paper Co. had offices in Franklin, OH. The company employed 85 and made 6 tons a day. See: Franklin Mills (2)

JPJ: Franklin Paper Mill was established in 1872 at Northeast, PA. The company made 4,000 pounds a day. See: Franklin Mill

JPK: French, J.W., Mfg. Co. was established in 1868 at Three Rivers, MI. The company made 10,000 pounds a day. See: Three Rivers Mill, Rozet Mill

JPL: Friend & Forgy Paper Co. had offices in Franklin, OH. The company made 6,000 pounds a day. See: Friend & Forgy Mill

JPM: Friend & Fox Paper Co. was established in 1848 at Lockland, OH. The company employed 225 and made 25,000 pounds a day. See: Rialto Mill, Crescent Mill, Friend & Fox Mill, Lockland Mill

JPN: Friend & French was established in 1859 at Lockland, OH. See: Athern Lower Mill

JPO: Friend & Tangeman was established in 1851 at Lockland, OH. See: Friend & Tangeman Mill, Athern Upper Mill

JPP: Friend, Col. C.W., was established in 1864 at Lockland, OH. See: Crescent Mill

JPQ: Friend, George H., was established in 1864 at Lockland, OH. The company made 9 tons a day. See: Excelsior Mill, Athern Lower Mill, Buckeye Mill, Carrollton Mill, Eagle Mill

JPR: Fulton, James and Miller, was established in 1779 at Fountain Green, PA. See: Oxford Mill

JPQ: Fulton, John (See John, Fulton), was established in 1809 at Oxford, PA. See: Fulton Mill

JPS: Funke, Ferdinand, was established in 1858 at Evansville, IN. The company employed 7 and made 2,000 pounds a day. See: Evansville Mill

JPT: Gardner, H., & Co. was established in 1873 at Callicoon Depot, NY. The company employed 17 and made 5,000 pounds a day. See: Callicoon Mill

JPU: Garfield, Harrison, & Son was established in 1819 at East Lee, MA. The company employed 20 and made 1,000 pounds a day. The "and son" was Henry Garfield. See: Forest Mill

JPV: Garlock, O. & H., was established in 1880 at Chittenango, NY. The company employed 4 and made 1,500 pounds a day. See: Garlock Mill

JPW: Garner, J.N., had offices in Chatham, NY. The company made 3,500 pounds a day. See: New York Mill

JPX: Garrett, C.S, & Son was established in 1851 at 12 Decatur St., Philadelphia. The company employed 51 and made 14,000 pounds a day. See: Union Mill, Beaver Dam Mill, Keystone Mill

JPY: Garrett, E.T., had offices in Darby, PA. The company made 1,000 pounds a day. See: Darby Mill

JPZ: Garrett, Harvey S., was established in 1850 at Cheney, PA. The company employed 6 and made 8,000 pounds a day. See: Williston Mill

JQA: Garrett, John, & Sons was established in 1789 at Red Clay Creek, DE. The "and son" was Horatio Gates. See: Garrett Mill

JQB: Gaskell, James, had offices in South Windsor, CT. The company made 2 tons a day. See: West Dudley Mill, Valley Mill

JQC: Gaunt & Derrickson was established in 1843 at South Orange, NJ. See: Phoenix Mill

JQD: George, Moore & Co. was established in 1850 at Catskill, NY. See: Nonatum Mill

JQE: George, S., had offices in Wellsburgh, WV. See: George Mill

JQF: Gevel, Lewis, was established in 1781 at Downington, PA. See: Gevel Mill

JQG: Ghent Paper Co. was established in 1874 at Ghent, NY, and made 4,000 pounds a day. See: Ghent Mill

JQH: Gibbons, James M., was established in 1795 at E. Fallowfield Township, PA. See: Buck Run Mill, Old

JQI: Gilbert & Whiting was established in 1882 at Menasha, WI. The company made 9,000 pounds a day, and later became the Gilbert Paper Co. See: Menasha Mill

JQJ: Gilbert, Beach & Co. was established in 1838 at Seymour, CT. See: Seymour Mill

JQK: Gilbert, Frank, was established in 1872 at Waterford, NY. The company employed 45 and made 8,000 pounds a day. See: Mohawk Mill, Hudson Mill

JQL: Gilman & Sibley was established in 1818 at Rochester, NY. Principals were Harvey Gilman and Derick Silby. See: Rochester Mill

JQM: Gilpin, J., & Co. was established in 1787 in Wilmington, DE, and had offices in Philadelphia at 149 S. Front St. The company employed 100. See: Brandywine Mills

JQN: Gladfelter, P.H., was established in 1864 at Spring Forge, PA. The company employed 100 and made 50,000 pounds a day. See: Spring Grove Mill

JQO: Glass-Edsell Paper Co. was established in 1885 at Delaware, OH. The company made 5 tons a day. See: Delaware Mills (2)

JQP: Glen Manufacturing Co. was established in 1883 at Berlin, NH. See: Glen Mill

JQQ: Glen Mills Paper Co. had offices in Glen Mills, PA. Principal was Joseph M. Dohan. See: Glen Mill

JQR: Glendale Paper Mill Co. had offices in Atlanta, GA. The company made 4,000 pounds a day. See: Glendale Mill

JQS: Glengary Mill was established in 1882 at Lee, MA. The company made 7,000 pounds a day. See: Glengary Mill

JQT: Glens Falls Paper Mill Co. was established in 1864 at Glens Falls, NY. The company made 12 tons a day. See: Glens Falls Pulp Co., Glens Falls Mill

JQU: Glessner & Gilbert had offices in Zanesville, OH. The company made 1,300 pounds a day. See: Cox Mill

JQV: Glessner, Jacob, was established in 1868 at Zanesville, OH. The company employed 20 and made 1,300 pounds a day. See: Cox Mill

JQW: Globe Tissue Paper Co. was established in 1883 at Elkhart, IN. The company employed 20 and made 2,000 pounds a day. See: Globe Tissue Mill

JQX: Godfrey & Clark was established in 1860 at Pittsburgh, PA. The company employed 50 and made 10,000 pounds a day. See: Tarrentown Mill, Elk Horn Mill

JQY: Goodman, Campbell & Co. was established in 1842 at Lafayette, IN.

JQZ: Goodwin, George, & Co. was established in 1778 at Manchester, CT. The "and company" was Sarah Ledyard and Hannah Watson. See: Five Miles Mill

JRA: Gookin & Co. was established in 1818 at Rutland, VT. See: Rutland Mill, Meads Falls Mill

JRB: Gore, H.M., was established in 1880 at Freeland, MD. The company employed 2 and made 2,000 pounds a day. See: Andover Mill

JRC: Gore, John W., was established in 1833 at Bentley Springs, MD. The company employed 4 and made 2,000 pounds a day. See: Valley Mill

JRD: Goss, Samuel, was established in 1821 at Berlin, VT. See: White River Mill

JRE: Graham & Co. was established in 1865 at Rockford, IL. The company employed 15 and made 4 tons a day. See: Graham Mill

JRF: Graham Brothers was established in 1825 at Cincinnati, OH. The company employed 72 people. See: Phoenix Mill, Graham Mill, Decker Creek Mill

JRG: Graham, E.A., had offices in Sauquiot, NY. The company made 3 tons a day. See: Fox Mill

JRH: Grant, Moses, & Son was established in 1809 at Needham, MA. See: Grant Mill

JRI: Grant, Warren & Co. was established in 1854 at Gardiner ME. The company made 11,000 pounds a day. See: Cumberland Mills, Copesecook Mill

JRJ: Graves, Harvey, was established in 1812 at Nottingham, PA. The company employed 2 and made 800 pounds a day. See: Northeast Mill

JRK: Graves, Harvey, & Son was established in 1890 at East Nottingham, PA. See: Silvermere Mill

JRL: Great Bend Paper Co. was established in 1868 at Great Bend, NY. The company employed 26 and made 6,000 pounds a day. See: Great Bend Paper Co.

JRM: Green, Hon, was established in 1845 at Lancaster, OH. See: Green Mill

JRN: Greene, Joseph A., & Co. was established in 1818 at Good Hope Township, OH. See: Pine Grove Mill

JRO: Greenleaf & Taylor was established in 1866 at Springfield, NJ. See: Springfield Mill

JRP: Greenwood, Dewing & Farliss was established in 1835 at Needham, MA. Principals were Joseph Greenwood, Paul Dewing, and Benjamin Farliss (1847). See: Nehoiden Mill

JRQ: Greenwood, John T., was established in 1880 at W. Medway, MA. The company made 1,000 pounds a day. See: West Medway Mill

JRR: Gregory Bros. was established in 1810 at Green River Village, VT. Principals were Samuel and William Gregory. See: Gregory Bros.

JRS: Guie, James, & Sons was established in 1833 at Downington, PA. The company made 4 tons a day. See: Eagle Mill

JRT: Hagey & Sons was established in 1755 at Lower Merion, PA. The company made 1.5 tons a day. Principal was Hans Hagey and son William Hagey (1785–1832). See: Hagey Mills

JRU: Haldeman Paper Co. was established in 1854 at Lockland, OH. The company employed 50 and made 12 tons a day. See: Lower Mill, Athern Upper Mill, Haldeman Mills (2)

JRV: Halderman & Parker was established in 1858 at Lockland, OH. See: Athern Upper Mill

JRW: Haley & Holden was established in 1850 at Dover, NH. See: Salmon Falls Mill

JRX: Hall, Charles H., was established in 1855 at Portland ME. See: Skowhegan Mill, Messalonskee Mill

JRY: Hallett, Charles, was established in 1868 at Riverhead, NY. The company employed 6 and made 2,000 pounds a day. See: Hallett Mill

JRZ: Halstead, John, was established in 1872 at Blossvale, NY. The company employed 10 and made 4,000 pounds a day. See: Blossvale Mill

JSA: Hamilton & Wright was established in 1830 at Chatham Center, NY. See: Columbia Mill

JSB: Hamilton Straw Lumber Co. had offices in Lawrence, KS, but was out of business after 1880. See: Hamilton Straw Lumber Mill

JSC: Hamilton, Samuel P., had offices in Savannah, GA. The company made 10,000 pounds a day. See: Savannah Mill

JSD: Hamilton, W.C., & Sons was established in 1857 at 100 Chestnut St., Philadelphia, PA. The company employed 100 and made 12 tons a day. See: Riverside Mill

JSE: Hammond, W.A., Paper Co. was established in 1882 at Jackson, MI. The company employed 10 and made 2 tons a day. See: Hammond Mill

JSF: Hampden Glazed Paper Co. was established in 1881 at South Hadley, MA. See: Hampden Glazed Paper Co.

JSG: Hampden Paper Co. was established in 1862 at Holyoke, MA. The company employed 110. Principals were James H., John C., Daniel H., and Moses Newton. See: Hampden Mill

JSH: Hampshire Paper Co. was established in 1866 at S. Hadley Falls, MA. The company employed 175 and made 2 tons a day. See: Hampshire Mills

JSI: Hand, Allen F., was established in 1872 at 51 Beekman St., New York. The company employed 11 and made 1 ton a day. See: Putnam Mill

JSJ: Haner, M.L., & Son had offices in Chatham Center, NY. The company employed 24 and made 10,000 pounds a day. See: Columbia Mill

JSK: Hanmer & Forbes Co. was established in 1864 at Burnside, CT. The company employed 32 and made 4 tons a day. See: Burnside Mill

JSL: Hannah & Clemens, Agents, had offices in Wheeling, WV. The company made 1 ton a day. See: Fulton Mill

JSM: Harbottle, Thomas, & Son had offices in Brooklyn, NY. See: Harbottle Mill

JSN: Harder, Herman, had offices in Albany, NY. The company made 5,000 pounds a day. See: Albany City Mills

JSO: Harding Paper Co. was established in 1872 at Franklin, OH. The company employed 270 and made 5 tons a day. See: Franklin Mill, Excello Mill

JSP: Harding, W.W., was established in 1865 at Philadelphia, PA. The company made 8,000 pounds a day. See: Philadelphia Inquirer Mill

JSQ: Hare, Geo. H., had offices in Grave Run Mills, MD. See: Grave Run Mill

JSR: Harknes & Allison was established in 1892 at Atglen, PA. See: Lyndonette Mill

JSS: Harlan & Bro. was established in 1854 at Elkton, MD. The company employed 12 and made 3,000 pounds a day. See: New Leeds Mill, Government Mill

JST: Harned, S.L., Agt. was established in 1866 at South Avon, NY. The company employed

8 and made 3,000 pounds a day. See: Genesee Mill

JSU: Harper, James, had offices in Westville, CT. The company made 2,000 pounds a day. See: Pond Lilly Mill

JSV: Harper, L.K., had offices in Westville, CT. The company made 2,000 pounds a day. See: Pond Lilly Mill

JSW: Harris & Cox was established in 1813 at North Yarmouth, ME. See: Harris & Cox Mill

JSX: Harris, Richard C., was established in 1849 at Green Bank, NJ. The company made 5,000 pounds a day. See: Harris Mill

JSY: Harris, W.O., was established in 1815 at Nashville, TN. See: Whites Creek Mill

JSZ: Hartford Manila Co. was established in 1882 at E. Hartford, CT. The company employed 60 and made 6 tons a day. See: Woodland Mills

JTA: Hartford Paper Co. was established in 1872 at Poquonock, CT. The company employed 95 and made 9,000 pounds a day. See: Rainbow Mills

JTB: Hartje, August, had offices in Steubenville, OH. The company employed 37 and made 6,000 pounds a day. See: Clinton Mill

JTC: Hartland Paper Co. was established in 1884 at Middleport, NY. The company employed 19 and made 5,000 pounds a day. See: Hartland Mill

JTD: Harvey & Bro. was established in 1853 at Wellsburgh, WV. The company employed 16 and made 3 tons a day. See: Harvey Mill

JTE: Harwood Mfg. Co. was established in 1884 at N. Leominster, MA. The company employed 40 and made 6,000 pounds a day. See: Harwood Mill (rebuilt 1884)

JTF: Harwood, J.A. & N., was established in 1868 at N. Leominster, MA. The company employed 40 and made 6,000 pounds a day. See: Harwood Mill (1884)

JTG: Hasting Paper Co. had offices in Springfield, OH. The company made 7 tons a day. See: Hasting Mill

JTH: Havens, J.S. & C.S., was established in 1874 at Patchogue, NY. The company employed 5 and made 1 ton a day. See: Patchogue Mill

JTI: Haverhill Paper Co. was established in 1884 at Bradford, MA. The company employed 60 and made 20,000 pounds a day. See: Haverhill Mill

JTJ: Hawes C.L. & Co. was established in 1885 at Dayton, OH. The company employed 150 and made 35,000 pounds a day. See: Aqueduct Mills (3)

STA: Hawes, R.L., was established in 1845 at Boston as a stationer.

JTK: Hawkhurst & Ray had offices in Indiana, PA. The company employed 7 and made 1200 pounds a day. See: Indiana Mill

JTL: Hawkins, E., had offices in Islip, NY. The company made 1,000 pounds a day. See: Hawkins Mill

JTM: Hawley & Co. was established in 1808 at Moreau, NY. See: Hawley Mill

JTN: Hazen, Gideon, was established in 1840 at Knoxville, TN. See: Middle Brook Creek Mill

JTO: Heath, Orton was established in 1849 at East Lee, MA. See: Greenwater Mill

JTP: Heilman, John Adam, was established in 1793 at Swatara, PA. See: Heilman Mill

JTQ: Helmbold, George Christopher, was established in 1788 at West Manayank, PA. See: Schutz Mills, Jones Mill

JTR: Henchman, Daniel, was established in 1728 at Cornhill, MA. See: Neponset Mills

JTS: Henderson, William S., was established in 1864 at Millburn, NJ. The company employed 25 and made 12,000 pounds a day. See: Fandango Mill

JTT: Henshaw & Eaton was established in 1800 at Weybridge Upper Falls, VT. The company employed 9. Principals were Joshna Henshaw and Joel Eaton. See: Henshaw Mill

JTU: Herkimer Paper Co. had offices in Herkimer, NY. The company made 10,000 pounds a day. See: Herkimer Mill

JTV: Heyser, William, was established about 1883 at Chambersburg, PA. The company employed 16 and made 2 tons a day. See: Hollywell Straw Mill

JTW: Hezekaiah Paper had offices in Millburn, NJ. See: Calico Mill

STA: Hilger, Louis, was established in 1819 at Philadelphia, PA, as a paper merchant.

JTX: Hill Paper Co. was established in 1860 at Washington, D.C. The principal, George Hill, was briefly captured by Confederates in 1863. See: Hill Mill

JTY: Hill, Albert S., was established in 1852 at Northville, CT. The company employed 4 and made 1 ton a day. See: Northville Mill

JTZ: Hill, E., & Co. had offices in Charles River

Village, MA. The company employed 8 and made 4,000 pounds a day. See: Dover Mill

JUA: Hill, George, Jr., had offices in Washington, DC. The company made 5,000 pounds a day. See: Squannacook Mill, Potomac Mills, Hill Mill

JUB: Hillabrandt, Joseph, was established in 1850 at Sammonsville, NY. The company employed 6 and made 3,000 pounds a day. See: Cayudulta Mill

JUC: Hinsdill & Gibbs had offices in North Bennington, VT. See: Hinsdill Mill

JUD: Hinsdill & Walbridge was established in 1812 at North Bennington, VT. See: Hinsdill Mill

JUE: Hodge, W.C., was established in 1860 at Rainbow, CT. The company employed 14 and made 8,000 pounds a day. See: Rainbow Tissue Mill

JUF: Hodges, Alexander, was established in 1840 at Taunton, MA. See: Park Mill

JUG: Hoes, John, was established in 1840 at Stuyvesant Falls, NY. See: Stuyvesant Falls Mill

JUH: Hoffman, Wm. H., & Sons was established in 1775 at Ashland, MD. The company made 3 tons a day. See: Rockdale Mill, Clipper Mills, Eagle Mills, Gunpowder Mill, Hoffman Mill, Marblevale Mill

JUI: Hoggs & Jackson was established in 1781 at Needham, MA. Principals were Edward Jackson and William Hoggs. See: Nehoiden Mill, Island Mills

JUJ: Hoggs, William, & Son was established in 1790 at Needham, MA. See: Island Mills

JUK: Hoglen, C.H., had offices in Medway, OH. The company made 6,000 pounds a day. See: Medway Mill

JUL: Holbrook & Pierce was established in 1853 at East Nottingham Township, PA. Principals were Benjamin Holbrook and Robert Pierce. See: Spring Lawn Mill

JUM: Holdship, Henry and George W., was established in 1824 at Pittsburgh, PA. The company employed 88. See: Clinton Steam Mill, Anchor Steam Mill

JUN: Holebrook & Fessenden was established in 1815 at Brattleboro, VT. Principals were Patty Fessenden and John Holbrook. See: Whetstone Creek Mill, Black River Mill

JUO: Hollingsworth & Vose had offices in Boston, MA. The company made 12,000 pounds a day. See: West Groton Mill, East Walpole Mill

JUP: Hollingsworth & Whitney Co. was established in 1840 at 36 Federal St., Boston, ME. The company made 14 tons a day. See: Pequosette Mill, Aroostook Mill, Cobbossee Mill, New, Monattaquot Mill

JUQ: Hollingsworth, Lyman, had offices in Boston, MA. The company made 4,000 pounds a day. See: Bridgewater Mill

JUR: Holt, J.S., was established in 1865 at Foxboro, MA. The company employed 8 and made 2,000 pounds a day. See: Leland Mill, Blackwater Mill

JUS: Holyoke Paper Mfg. Co. was established in 1860 at Holyoke, MA. The company employed 250 and made 15K a day. See: Holyoke Mill

JUT: Hoogland, Daniel, and Abraham Coles was established in 1801 at Roslyn, NY. See: Onderdonk Mill

JUU: Horton, Jere, had offices in York, PA. The company made 1.5 tons a day. See: Codorus Mill

JUV: House & Co. had offices in Rainbow, CT. The company employed 7 and made 1,000 pounds a day. See: House Mill

JUW: House, F.A., was established in 1883 at Haddam, CT. The company made 2,000 pounds a day. See: Haddam Neck Mill

JUX: Howard & Lathrup was established in 1830 at Canal Village, MA. See: Howard & Lathrup Mill

JUY: Howard, Charles, was established in 1861 at Salt Lake City, UT. See: Sugar House Mill

JUZ: Howell, Frank, was established in 1814 at Lockport, OH. See: Howell Mill

JVA: Howland & Co. was established in 1873 at Sandy Hill, NY. The company employed 60 and made 10 tons a day. See: Howland Mills

JVB: Howland & Palmer was established in 1864 at East Chatham, NY. The company employed 12 and made 20,000 pounds a day. Successor to the firm was E.G. Palmer. See: East Chatham Mill

JVC: Howland, John, and Thomas was established in 1831 at Troy, NY. Peleg Howland became partner in 1849. See: Howland Mill

JVD: Hoyt, John, & Co. was established in 1814 at Manchester, NH. The company employed 60 and made 5 tons a day. See: Amoskeag Mill

JVD.1: Hubbard A.H. & Co. was established in 1818 at Norwich, CT. & Co. Principal was Russell Hubbard and the company made 3,000 pounds a day. The firm dissolved in 1857.

JVE: Hudson & Goodwin was established in

1779 at N. Manchester, CT. See: East Hartford Mill, Silver Stream Mill

JVF: Hudson River Pulp & Paper Co. was established in 1869 at Tribune Bldg., New York. The company employed 170 and made 14 tons a day. See: Palmer Falls Mill, Cornith Mill

JVG: Hudson, Henry & Son was established in 1802 at N. Manchester, CT. The "and son" was Melancthon and grandsons; William and Phillip W. See: Oakland Mill

JVH: Hudson-Cheney Paper Co. was established in 1864 at N. Manchester, CT. See: Oakland Mill

JVI: Humphrey, General David, was established in 1801 at Seymour, CT. See: Rimmon Falls Mill

JVJ: Hunter & Shiland had offices in Essex Junction, VT. The company employed 22 and made 2.5 tons a day. See: Essex Mill

JVK: Hunter, John, was established in 1828 at Ashland, MD. See: Marblevale Mill

JVL: Huntington & Bushnell was established in 1790 at Norwich, CT. See: Norwich Mill

JVM: Hurd & Crehore was established in 1831 at Needham, MA. Principals were William Hurd and Lemore Crehore. See: Rice & Garfield Mill, Ware Mill

JVN: Hurd, William, was established in 1815 at Needham, MA. See: Ware Mill

JVO: Hurlbut Paper Co. was established in 1862 at South Lee, MA. The company employed 200 and made 8,000 pounds a day. Principals were Thomas Otis and Henry Clay Hurlbut. See: Oakland Mill, Hurlbut Mill

JVP: Idler, William, was established in 1803 at Alsace Township, PA. See: Idler Mill

JVQ: Illinois Valley Paper Co. was established in 1884 at Marseilles, IL. The company employed 20 and made 9,000 pounds a day. See: Illinois Valley Mill

JVR: Imgard, Albert, had offices in 2 Barclay St., New York. The company employed 35 and made 12,000 pounds a day. See: Eclipse Straw Board Mill

JVS: Indiana Paper Co. had offices in Indianapolis, IN. The company employed 25 and made 7 tons a day. See: State Mill, Meridian Mill

JVT: Ingalls & Co. was established in 1873 at S. Manchester, CT. The company employed 15 and made 6,000 pounds a day.

JVU: Ingersoll & Benton was established in 1831 at East Lee, MA. Principals were Jared Ingersoll and Caleb Benton. See: Old Forest Mill

JVV: Ingersoll, Jared, was established in 1833 at East Lee, MA. See: Ingersoll Mill

JVW: Ingersoll, Milton, was established in 1833 at Tyringham, MA.

JVX: Ingersoll, R.L., & Co. had offices in Pulaski, NY. The company made 3,000 pounds a day. See: Ingersoll Mill

JVY: Ingham, Hezekai, and Isaiah was established in 1835 at Chillicothe, PA. See: Ingham Mill

JVZ: Ingham, Jonathan and Samuel, was established in 1800 at Solebury Township, PA. See: Great Spring Mill

JWA: Ingham, Mills & Co. had offices in Chillicothe, OH. The company made 15,000 pounds a day. Principals were Hezekiah and Isaiah Ingham. See: Hydraulic Mill, Crouse Mill, Great Spring Mill

JWB: International Paper Co. was established in 1898 at Corinth, NY. See: Corinth Mill

JWC: Isenburg Bros. was established in 1878 at Louisville, KY. The company made 3,000 pounds a day. See: Kentucky Mill

JWD: Ivanhoe Mfg. Co. was established in 1850 at Paterson, NJ. The company made 13,000 lbs. a day. See: Ivanhoe Mill

STA: Ives, Sturges & Co. was established in 1835 at Lee, MA, as stationers.

JWE: J.J. Henderson had offices in Millburn, NJ. See: Fandango Mill

JWF: J.W. Butler Paper Co. was established in 1850 at St. Charles. The company employed 80 and made 8,000 pounds a day. See: West Side Mill, Devitt Mill

JWG: reserved. Jackson & Sharpless was established in 1796 at Brownsville, PA. Principals were Samuel Jackson and Jonathan Sharpless. See: Redstone Mill

JWH: Jackson & Sharpless was established in 1791 at Brownsville, PA. See: Redstone Mill

JWI: Jackson Paper Co. was established in 1883 at Jackson, MI. The company made 4 tons a day. See: Jackson Mill

JWJ: Jackson, Colonel Ephraim, was established in 1801 at Needham, MA. See: Jackson Mill

JWK: Jarvis, Russell, was established in 1850 at Claremont, NH. The company employed 20 and made 2,500 pounds a day. See: Jarvis Mill

JWL: Jennings Bros. was established in 1867 at Fairfield, CT. The company employed 10. See: Japanese Paper Ware Mill

JWM: Jersey City Paper Co. had offices in Jersey

City, NJ. The company made 5,000 pounds a day. See: Jersey City Mill

JWN: Jessup & Moore was established in 1843 at Philadelphia, PA. The company made 33 tons a day. See: York Mill, Augustine Mill, Delaware Mills, Rockland Mill, Chester Mill

JWO: Jessup, Alfred DuPont, was established in 1860 at York, PA. See: Codorus Mill

JWP: John Davidson was established in 1811 at Trenton, NJ. See: Davidson Mill

JWQ: John, Fulton (see Fulton, John), was established in 1809 at Oxford, PA. See: Fulton Mill

JWR: Jones, C.W., had offices in Dundee, MI. The company made 4 tons a day. See: Dundee Mill

JWS: Jones, Evan, was established in 1815 at Lower Merion Township, PA. See: McCleaghan Mill

JWT: Jones, Frank, had offices in Ballston Spa, NY. The company employed 25 and made 2,500 pounds a day. See: Fonda Mill, Ballston Spa Mill

JWU: Jones, James, was established in 1880 at Shortsville, NY. The company employed 8 and made 3,000 pounds a day. See: Ontario Mill

JWW: Jones, Paul, was established in 1793 at West Manayank, PA, and had offices in Philadelphia at 147 S. Front St. See: Jones Mill

JWX: Jones, W.R., had offices in Greenville, SC. The company made 1,200 pounds a day. See: Jones Mill

JWY: Jones, William F., was established in 1868 at Leroy, NY. The company employed 6 and made 2,000 pounds a day. See: Genesee County Mill

JWZ: Judd Bros. was established in 1840 at S. Hadley, MA. The company employed 6 and made 1,000 pounds a day. See: Manila Mill

JXA: Judy, Knisely & Co. was established in 1873 at New Philadelphia, OH. The company employed 20 and made 6,000 pounds a day. See: New Philadelphia Mill

JXB: Kalamazoo Paper Co. was established in 1866 at Kalamazoo, MI. The company employed 126 and made 16,000 pounds a day. See: Kalamazoo Mill

JXC: Kankakee Paper Co. was established in 1873 at Kankakee, IL. The company employed 28 and made 5 tons a day. See: Kankakee Mill

JXD: Katz, Henry, was established in 1750 at Whitemarsh Township, PA. See: Katz Mill

JXE: Kearney, John, had offices in Parkton, MD, and was out of business after 1880. See: Kearney Mill

JXF: Keating, John, was established in 1768 at New York. See: Peekskill Mill, Auburn Mill, Keating Mill

JXG: Keck, Joseph, was established in 1869 at Kecks Center, NY. The company employed 8 and made 2 tons a day. See: Keck Mill

JXH: Keeney & Wood Mfg. Co. was established in 1870 at N. Manchester, CT. The company employed 12 and made 2 tons a day. See: Keeney & Wood Mill

JXI: Keeney Brothers had offices in N. Manchester, CT. See: Rockford Mill, Keeney & Wood Mill

JXJ: Keeney, A.W., was established in 1873 at Rockford, IL. The company employed 28 and made 8 tons a day. See: Globe Mills

JXK: Keeney, Boswell Co. was established in 1860 at East Hartford, CT. The company employed 12 and made 2 tons a day.

JXL: Keeney, M.D., had offices in Wilmington, IL. The company employed 25 and made 5 tons a day. See: Enterprise Straw Board Mill

JXM: Keith Paper Co. had offices in Turner Falls, MA. The company made 5 tons a day. See: Keith Mill

JXN: Keller, Abraham, & Son was established in 1790 at Exeter Township, PA. The "and son" was Jacob Keller. See: Keller Mill

JXO: Keller, George W., had offices in Houcksville, MD. The company employed 6 and made 2,000 pounds a day. See: Pine Grove Mill

JXP: Kellogg, J., had offices in Cleveland, OH. See: Kellogg Steam Mill

JXQ: Kellogg, James, was established in 1843 at St. Louis, MO. See: Kellog Mill

JXR: Kelty, Anthony, was established in 1822 at E. Fallowfield Township, PA. See: Great Spring Mill, Buck Run Mill, Old

JXS: Kendall, Amos, was established in 1821 at Elkhorn KY. See: Franklin Mill

JXT: Kendell, Jonas & Sons was established in 1823 at Leominster, MA. See: Kendall Mill

JXU: Kenney, M.D., & Son had offices in Rockton, IL. The company employed 20 and made 3 tons a day. See: Enterprise Mill

JXV: Kent Paper Co. was established in 1857 at Herkimer, NY. See: Herkimer Mill

JXW: Kent, Hermon L., Co. had offices in Westfield, NY. The company employed 12 and made 5,500 pounds a day. See: Westfield Mill

JXX: Kenyon & Dixon had offices in Baldwinsville, NY. The company employed 12 and made 6,000 pounds a day. See: Baldwinsville Mill

JXY: Ketcham, D.P., was an agent for the mill in 1866 at Stanfordville, NY. The company employed 9 and made 4,500 pounds a day. See: Stone Mill

JXZ: Kills Paper Co. was established in 1845 at Sprintfield, OH. See: Kills Mill

JYA: Kilmer, Chauncey, was established in 1853 at Rock City Falls, NY. The company employed 30 and made 8,500 pounds a day. See: Rock Falls Mills

JYB: Kimball, Daniel, was established in 1817 at Plainfield, NH. See: Billings Mill

JYC: Kimberly & Clark Co. was established in 1881 at Neenah, WI. The company employed 134 and made 4 tons a day. Principals were J.A. Kimberly, H. Babcock, F.C. Shattuck, and C.B. Clark. See: Vulcan Mill, Little Badger Mill, Globe Mill, Neenah Mill, Tioga Mill

JYD: King, David, was established in 1866 at Clifton, OH. See: Clifton Mill

JYE: King, Pendery & Athern was established in 1840 at Lockland, OH. See: Athern Lower Mill, Lockland Mill

JYF: King, Phillip J., & Co. was established in 1798 at Cordorus Creek, PA. The "and company" was George King. See: York Mill, Congress Mill

JYG: Kingsbury, Maj. Elisha, was established in 1793 at Bellows Falls, VT. See: Rockingham Mill, Cold River Mill

JYH: Kingsland, J. & R., was established in 1836 at North Belview, NJ. The company employed 60 and made 4,000 pounds a day. See: Passic Mill

JYI: Kingsley Brothers was established in 1879 at Salisbury, VT.

JYJ: Kingston, William, was established in 1872 at Little Falls, NY. The company employed 11 and made 3 tons a day. See: Rock Mill

JYK: Kirk, Josiah and Lewis was established in 1814 at East Nottingham, PA. See: Kirk Mill

JYL: Kirk, Timothy, was established in 1821 at East Nottingham, PA. See: Kirk Mill

JYM: Knerr, Lewis, Jr., was established in 1885 at Ruckawa, OH. The company employed 20 and made 6,000 pounds a day. See: Ruckawa Mill

JYN: Knowlton & Rice was established in 1824 at Watertown, NY. The principals were George Knowlton and Clark Rice. See: Auburn Mill

JYO: Knowlton Bros. was established in 1861 at Watertown, NY. The company employed 52 and made 6,000 pounds a day. See: Knowlton Bros. Mill, Auburn Mill

JYP: Knowlton, Amos & F.D., was established in 1879 at Dansville, NY. The company employed 6 and made 3,000 pounds a day. See: Eagle Mill

JYQ: Knowlton, F.D., was established in 1880 at Dansville, NY. The company employed 6 and made 3,000 pounds a day. See: Eagle Mill

JYR: Knox & McClure was established in 1819 at New Hope, PA. See: Eagle Mill, Old, Union Mill

JYS: Knox, Lore & Scott was established in 1836 at Connellsville Township, PA. The principals were D.S. Knox, M. Lore, and John Scott. See: Youghiogheny River Mill

JYT: Koons Brothers had offices in Shickshinny, PA. The company made 1 ton a day. See: Huntington Mills

JYU: Kownslar, Conrad, & Son was established in 1804 at Mill Creek, VA. The "and son" was Remington B. Kownslar. See: Kownslar Mill

JYV: Krepps & Carter was established in 1832 at Brownsville, PA. See: Brownsville Mill

JYW: Kroh, Phillip A., was established in 1833 at White Hall, MD. See: Excelsior Mill

JYX: Krum & Williams was established in 1840 at Middleburgh, NY. The company made 5,000 pounds a day. See: Middleburgh Mill

JYY: Krum, Franklin, was established in 1849 at Middleburgh, NY. The company made 5,000 pounds a day. See: Middleburgh Mill

JYZ: Kugler, Matthias & Son was established in 1814 at Sycamore Township, OH. See: Waldschmidt Mill

JZA: La Fayette Paper Co. was established in 1882 at Lafayette, IN. The company employed 30 and made 6 tons a day. See: La Fayette Mill

JZB: Laflin, W.W. & C., was established in 1826 at Lee, MA. The company employed 60. The principals were Walter, Winthrop, and Cutler Laflin. See: Housatonic Mill, Old, Crow Hollow Mill

Directory of Paper Mill Owners 273

JZC: Lake George Pulp & Paper Co. was established in 1878 at Ticonderoga, NY. The company employed 50 and made 8,000 pounds a day. See: Lake George Pulp & Mill

JZD: Lakeside Paper Co. was established in 1883 at Skaneateles, NY. The company employed 16 and made 10,000 pounds a day. See: Lakeside Mill

JZE: Lamborn, Daniel, was established in 1810 at Bel Air, MD. See: Winters Run Mill

JZF: Lamden & Son was established in 1828 at Wheeling, VA. The "and son" was Christopher Lamden. See: Lamden Mill

JZG: Lamden, Christopher, was established in 1835 at Wheeling, WV. See: Virginia Mill

JZH: Lamme, Keiser & Co. was established in 1834 at Rock Bridge, MO. See: Columbia Mill

JZI: Lancaster Mfg. Co. was established in 1864 at Lancaster, NH. The company employed 12 and made 3,000 pounds a day. See: Lancaster Mill

JZJ: Lang, John, was established in 1873 at 24th & Vine Street, Philadelphia, PA. The company employed 25 and made 16,000 pounds a day. See: Park Mill

JZK: Latham & Son was established in 1874 at Suffield, CT. The principals were William B. Latham and son, E.W. Latham. See: Eagle Mill, Old

JZL: Lawless, D.T., was established in 1881 at Penfield, NY. The company employed 11 and made 1 ton a day. See: Lawless Mill

JZM: Lawless, M.J., & Co. had offices in Marcellus Falls, NY. The company made 4 tons a day. See: Phoenix Mill, Eagle Mill

JZN: Lawrence Paper Co. was established in 1883 at Lawrence, KA. The company employed 15 and made 10,000 pounds a day. See: Lawrence Mill

JZN.1: Laws, I., was established by 1823 at Claremont, NH. See: Laws Mills

JZO: LeBourgois, E.C., was established in 1884 at Convent, LA. The company employed 12 and made 5,000 pounds a day. See: LeBourgois Mill

JZP: Lee & Guiles was established in 1873 at Manilus, NY. The company employed 8 and made 2,500 pounds a day. See: Lee Mill

JZQ: Lee, Charles F., was established in 1883 at Manilus, NY. The company employed 8 and made 2,500 pounds a day. See: Lee Mill

JZR: Leeds, Jones, & Bissell was established in 1831 at Richmond, IN. See: Richmond Mill

JZS: Lefever, Newton, was established in 1881 at Fallsburg, NY. The company employed 10 and made 2 tons a day. See: Fallsburgh Mill

JZT: Leffingwell, Christopher, was established in 1767 at Norwich, CT. See: Leffingwell Mill

JZU: Lehman, A.S., was established in 1856 at Chambersburg, PA. The company employed 16 and made 2 tons a day. See: Willow Grove Straw Board Mill

JZV: Leidy, T.H., had offices in Pennsburg, PA. The company made 1,200 pounds a day. See: Douglass Mill

JZW: Levan, Abraham, was established in 1798 at Kutztown, PA. See: Kutztown Mill

JZX: Levering & Co. was established in 1810 at Gwynns Falls, PA. The "and company" was Henry Payson, and John, Nathan, and Aaron Levering. See: Franklin Mill

JZY: Levis & Lewis was established in 1807 at Middletown Township, PA. The principals were Seth Levis and Edward Lewis. See: Spring Hill Mill

JZZ: Levis, Issac, was established in 1790 at Middletown Township, PA. See: Spring Hill Mill

KAA: Levis, Samuel (Sr.), was established in 1778 at Upper Darby Township, PA. See: Tuscarora Mills, Glenwood Mills, Lamb Mill

KAB: Levis, Samuel F., was established in 1793 at Upper Darby Township, PA. See: Glenwood Mills

KAC: Levis, Samuel G., William, and Osborn, was established in 1813 at Upper Darby Township, PA. See: Glenwood Mills, Clifton Mill

KAD: Levis, Thomas, Sr., & Co. was established in 1790 at Springfield Township, PA. The "and company" was John Levis. See: Springfield Mill

KAE: Levis, William, & Son was established in 1793 at Upper Darby Township, PA. The "and son" was John Levis. See: Tuscarora Mills, Lamb Mill, Ridley Creek Mill

KAF: Lewis, Elizabeth, was established in 1833 at Nether Providence Township, PA. See: Franklin Mill

KAG: Lewis, F.R., was established in 1879 at Flint, MI. The company employed 20 and made 7,000 pounds a day. See: Lewis Straw Board Mill

KAI: Lewis, John Howard, was established in

1859 at Swarthmore, PA. The company employed 33 and made 5 tons a day. See: Franklin Mill, Beehive Mill

KAJ: Lewis, John, Jr., was established in 1779 at Springfield Township, PA. See: Crum Creek Mill

KAK: Lewis, W.P., had offices in Dayton, OH. The company made 3,000 pounds a day See: Valley City Mills

KAL: Lewis, H.N., & Co. was established in 1881 at Walpole, MA. The company employed 17 and made 3 tons a day. See: Walpole Mill

KAM: Lick Mills Paper Mfg. Co. was established in 1880 at 100 California St., San Francisco, CA. The company employed 35 and made 7,000 pounds a day. See: Lick Mill

KAN: Lima Paper Mills was established in 1870 at Lima, OH. The company employed 75 and made 14 tons a day. See: Lima Mills

KAO: Lincoln, L., & Co. was established in 1850 at North Dighton, MA. The company made 6,000 pounds a day. The principals were Caleb and Lorenzo Lincoln. See: Lincoln Mill

KAP: Livingston Paper Co. was established in 1880 at Linlithgo, NY. The company employed 10 and made 2,000 pounds a day. See: Livingston Mills

KAQ: Lockport Mfg. Co. was established in 1884 at Lockport, NY. The company employed 20 and made 12,000 pounds a day. See: Lockport Mill

KAR: Lockport Paper Co. was established in 1872 at Lockport, IL. The company made 30 tons a day. See: Lockport Mills

KAS: Logansport Paper Co. was established in 1864 at Logansport, IN. The company employed 10 and made 8,000 pounds a day. See: Logansport Mill

KAT: Long, Minard, had offices in Gargoa, NY. The company made 1,500 pounds a day. See: Long Mill

KAU: Longstreth (a.k.a. Langstroth), Thomas & John, was established in 1792 at Moreland Township, PA, on Pennypack Creek. The firm was listed in 1823 Philadelphia directory as "Paper Hanging Manufacturer." See: Pennypack Creek Mill

KAV: Loomis & Norton was established in 1871 at Suffield, CT. See: Loomis & Norton Mill

KAW: Lord, Albert E., was established in 1883 at Brownville, NY. The company employed 18 and made 6 tons a day. See: Black River Mill

KAX: Los Angeles Paper Co. was established in 1884 at Compton, CA. The company made 4,000 pounds a day.

KAY: Loud, Caleb, was established in 1846 at Loudville, MA. The company employed 6 and made 1,000 pounds a day. See: Manhan Mill

KAZ: Lungren, John, was established in 1785 at Upper Providence Township, PA, around Ridley Creek. Survived by sons William and Charles Lungren. See: Ridley Creek Mill, Duckett Mill

KBA: Lunman, Phiney, & Co. was established in 1850 at East Lee, MA. The "and company" was Harrison Garfield, Caleb Benton, and Joseph Kroh. See: Washington Mill

KBB: Lydall & Foulds Paper Co. was established in 1870 at N. Manchester, CT. The company made 8,000 pounds a day. See: Lydall & Foulds Mill

KBC: Lydig & Mesier was established in 1808 at Hemlock Grove, NY. The principals were David Lydig and Peter A. Mesier. See: Lydig Mill

KBD: Lydig, David, was established in 1800 at Bronx, NY. See: Bronx Mill

KBE: Lynn, George W., & Co. was established in 1859 at East Lee, MA, as a leasing company. See: National Mill, Greenwater Mill

KBF: Lyon Brothers was established in 1824 at Needham, MA. The principals were Peter, Amos, and Jessie Lyon. See: Lyon Mill

KBG: Lyon, Amos, was established in 1822 at Needham, MA. See: Jackson Mill

KBH: Lyon, B.F., & Co. was established in 1872 at Kalamazoo, MI. The company employed 30 and made 4,000 pounds a day. See: Commonwealth Mill

KBI: Lyon, Col. Matthew, was established in 1794 at Fair Haven, VT. See: Fair Haven Mill

KBK: Lyon, Peter, was established in 1810 at Needham, MA. See: Rice & Garfield Mill, Island Mills, Lyon Mill

KBL: Lyon, N.C., was established in 1882 at Wellsville, CT. The company employed 10 and made 1 ton a day. See: Eagle Mill

KBM: Lyons Paper Co. was established in 1873 at Lyons, IA. The company employed 50 and made 6 tons a day. See: Lyons Mill

KBN: Lysle, Wilson, was established in 1825 at Chesterville, PA. The company employed 3 and made 500 pounds a day. See: Franklin Mill

KBO: Mack, Audrus & Woodruff was established in 1838 at Ithaca, NY. See: Falls Creek Mill

KBP: Manchester Paper Co. had offices in Richmond, VA. The company made 1 ton a day. See: Manchester Mill

KBQ: Manlius Paper Co. had offices in Manlius, NY. The company employed 6 and made 3,000 pounds a day. See: Manlius Mill

KBR: Manning & Paine was established in 1866 at Troy, NY. The company employed 16 and made 6,000 pounds a day. See: Orrs' Mill

KBS: Manning & Peckham was established in 1840 at Troy, NY. The company employed 20 and made 5,000 pounds a day. See: Mt. Ida Mill

KBT: Manning, John A., was established in 1882 at Troy, NY. The company employed 18 and made 7,000 pounds a day. See: Crystal Palace Mill

KBU: Mansfield Paper Co. had offices in Mansfield, OH. The company employed 13 and made 2 tons a day. See: Mansfield Mill

KBV: Marietta Paper Mfg. Co. was established in 1858 at Marietta, GA. The company employed 20 and made 5,000 pounds a day. See: Marietta Mill

KBW: Marinette & Menominee Paper Co. was established in 1878 at Marinette, WI. The company employed 125 and made 18,000 pounds a day. See: Menominee Mill, Marinette Mill

KBX: Markle & Drum was established in 1812 at West Newton, PA. See: Mill Grove Mill

KBY: Markle Brothers had offices in West Newton, PA. The principals were S.B. Markle and Cyrus P. Markle. See: West Newton Mill, Old, Mill Grove Mill, South Huntington Mill

KBZ: Markle, Sheppard R., had offices in West Newton, PA. The company made 5,000 pounds a day. See: Mill Grove Mill

KCA: Marshall, Thomas S., & Sons was established in 1856 at Yorklyn, DE. The company employed 8 and made 2,000 pounds a day. See: Marshall Mill

KCB: Martin Manufacturing Co. had offices in Manchester, NH. See: Amoskeag Mill

KCC: Martin, General Walter, was established in 1807 at Martinsburg, NY. See: Martin Mill

KCD: Martin, William, was established in 1825 at Upper Providence Township, PA. See: Ridley Creek Mill

KCE: Marylandville Paper Co. had offices in Philadelphia, PA. See: Marylandville Mill

KCG: Mason, Perkins & Co. was established in 1871 at Bristol, NH. The company employed 40 and made 5 tons a day. See: Bristol Mills

KCH: Massasoit Paper Mfg. Co. was established in 1872 at Holyoke, MA. The company employed 100 and made 4 tons a day. See: Massasoit Mill

KCI: Massillon Paper Co. was established in 1867 at Massillon, OH. The company employed 40 and made 3 tons a day. See: Massillon Mill

KCJ: Mathews, Edward, was established in 1864 at Zanesville, OH. The company employed 40 and made 4,000 pounds a day. See: Novelty Mill

KCK: Matthews, John, & Son was established in 1799 at 393 Market St., Philadelphia. The "and son" was Thomas Matthews. See: Beehive Mill, Union Mill

KCL: Mattson, Aaron, was established in 1790 at Aston Township, PA. See: Mattson Mill

KCM: May, E. & S., was established in 1837 at Lee, MA. The company employed 52 and made 2,000 pounds a day. See: Middle Mill, Ingersoll Mill, Mahaiwe Mill

KCN: McAllister, C.A., was established in 1883 at St. Johnsville, NY. The company employed 6 and made 2,000 pounds a day. See: McAllister Mill

KCO: McAlpin Bros. & Co. was established in 1884 at East Lee, MA. The principals were Robert, Alex, and Charles McAlpin. See: Mahaiwe Mill

KCP: McArthur Bros. was established in 1868 at Danbury, CT. The company employed 10 and made 2 tons a day. See: Beaver Brook Mill

KCQ: McClenachan, George, was established in 1799 at Lower Merion Township, PA. See: McCleaghan Mill

KCR: McCready Bag & Paper Co. was established in 1885 at Lambertville, NJ. The company employed 70 and made 8 tons a day. See: Perseverance Mill No.2, Perseverance Mill No.1

KCS: McDonald, George, had offices in Philadelphia PA, around 1840. The company employed 80 and made 6,000 pounds a day. See: Schuylkill Mill

KCT: McDonald, J.C. was established in 1863 at Hickory Hill, PA. The company employed 2 and made 1,000 pounds a day. See: Elk Grove Mill

KCU: McDowell, Joseph, had offices in Philadelphia, PA. See: Pennypack Creek Mill

KCV: McEwan & Son was established in 1873 at Caldwell, NJ. The company employed 10 and made 4,000 pounds a day. See: Caldwell Mill

KCW: McHenry Paper Co. was established in 1872 at Mount Vernon, PA. See: Stone Mill

KCX: McHenry, M.E., had offices in Russellville, PA. The company employed 4 and made 4,000 pounds a day. See: Collomer Mill

KCY: McHenry, W.H., was established in 1872 at Mount Vernon, PA. The company employed 2 and made 1,000 pounds a day. See: Mt. Vernon Mill

KCZ: McInnes, H., was established in 1881 at Bridgeport, PA. The company employed 25 and made 5.5 pounds a day. See: Rebecca Mill

KDA: McKnight, B.N., was established in 1870 at Conyers, GA. The company employed 15 and made 4,000 pounds a day.

KDB: Mead Paper Co. was established in 1846 at Dayton, OH. The company employed 130 and made 9.5 tons a day. Principal was Daniel Mead. See: Dayton Mills (3), Chillicothe Pulp and Paper Mill

KDC: Meeker, J. Edgar, was established in 1866 at Springfield, NJ. The company employed 8 and made 2 tons a day. See: Mill Town Mill

KDE: Meeter, Thomas, & Sons was established in 1789 at Newark, DE. The company employed 48. The "and sons" were Samuel and William. See: Milford Mill, Meeter Mill

KDF: Megargee Bros. was established in 1840 at 20 S. 6th, Philadelphia, PA. The company made 3,000 pounds a day. See: Spring Lawn Mill, Dove Mill

KDG: Mercer, John, was established in 1860 at Atglen, PA. See: Sadsbury Mill

KDH: Merrill, L.S., had offices in Conway, NH. The company made 3,000 pounds a day. See: Merrill Mill

KDI: Merrimac Paper Co. had offices in Lawrence, MA. The company made 8 tons a day. See: Merrimac Mill

KDJ: Mesick Paper Co. was established in 1868 at Chatham, NY. The company employed 28 and made 6 tons a day. See: Empire Mill

KDK: Messalonskee Mfg. was established in 1848 at Skowhegan, ME. See: Skowhegan Mill

KDL: Metzel, H.D., had offices in Ellicott City, MD. The company made 2,000 pounds a day. See: Lanvale Mill

KDM: Miami Valley Paper Co. had offices in Miamisburg, OH. The company made 2.5 tons a day. See: Miami Valley Mill

KDN: Michigan Paper Co. was established in 1870 at Gedds, MI. See: Michigan Paper Mills

KDO: Mickle, Phillip, had offices in Chatham, NY. The company made 2,500 pounds a day. See: Brown Leaf Mill

KDP: Middleport Paper Mill Co. was established in 1872 at Middleport, NY. The company employed 19 and made 5,000 pounds a day. See: Hartland Mill

KDQ: Middletown Paper Co. had offices in Cleveland, OH. The company made 5 tons a day. See: Valley Mill

KDR: Milan Paper Co. was established in 1884 at Milan, IL. See: Milan Mill

KDS: Miller's River Paper Co. was established in 1883 at Equitable Bldg., Boston, MA. The company made 2 tons a day. See: Miller's River Mill

KDT: Miller, Frank P., was established in 1881 at Downington, PA. The company employed 8 and made 2500 pounds a day. See: Solitude Mill

KDU: Miller, Irvin, was established in 1882 at St. Johnsville, NY. The company employed 9 and made 3,000 pounds a day. See: Miller Mill

KDV: Miller, Warner, was established in 1868 at Herkimer, NY. See: Herkimer Mill

KDW: Milton & Co. was established in 1867 at Medusa, NY. The company employed 7 and made 3,000 pounds a day. See: Medusa Mill

KDX: Minneapolis Straw Board Paper Co. was established in 1882 at Minneapolis, MN. The company employed 25 and made 4 tons a day. See: Minneapolis Straw Board Mill, McKnight Mill

KDY: Minot & Phelps was established in 1884 at Marshall, MI. The company employed 11 and made 6,000 pounds a day. See: Marshall Mill

KDZ: Minot, W.H.H., was established in 1884 at Marshall, MI. The company employed 11 and made 6,000 pounds a day. See: Marshall Mill

KEA: Mitchell & Son was established in 1873 at Palmyra, MI. The company employed 10 and made 2,000 pounds a day. See: Monroe Mill

KEB: Mitchell, Andrew, was established in 1836 at Monroe, MI. The company employed 10 and made 2,000 pounds a day. See: Palmyra Mill, Monroe Mill

KEC: Mode, Alexander & William, was established in 1824 at Chester Co., PA. See: Buck Run Mill, Old

KED: Moline Paper Co. was established in 1852 at Moline, IL. See: Moline Mill

KEE: Monroe Mfg. Co. was established in 1883 at Monroe, MI. The company employed 25 and made 8,000 pounds a day. See: Monroe Mfg. Mill

KEF: Monroe Paper Co. was established in 1883 at Monroe, MI. See: Monroe Straw Mill

KEG: Montague Paper Co. was established in 1872 at Turner Falls, MA. The company employed 250 and made 20 tons a day. See: Montague Mill

KEH: Mooney, Isaac, was established in 1826 at Madison, IN. See: Big Creek Mill

KEI: Moore & Wilson was established in 1883 at Waterford, NY. The company employed 18 and made 10,000 pounds a day. See: Monarch Mill

KEJ: Moore, Arms & Thompson had offices in Bellows Falls, VT. The company employed 75 and made 10,000 pounds a day. See: Windham Mill

KEK: Moore, Brown & Co. was established in 1869 at Sammonsville, NY. The company employed 6 and made 1 ton a day. See: Onondaga Mill

KEL: Moore, John T., had offices in Bellows Falls, VT. The company made 1,800 pounds a day. See: Moore Mill

KEM: Moorehouse Brothers was established in 1879 at 13th & Buttonwood St., Philadelphia, PA. The company made 11,000 pounds a day. See: Moorehouse Brothers Mill

KEN: Morgan & Lowell was established in 1882 at Battle Creek, MI. The company employed 25 and made 18,000 pounds a day. See: Morgan Mill

KEO: Morgan, Ephraim, was established in 1802 at Troy, NY. See: Troy City Mill

KEP: Morgan, John C., was established in 1882 at Battle Creek, MI. The company employed 25 and made 18,000 pounds a day. See: Morgan Mill

KEQ: Morris & Boise had offices in Chatham, NY. The company employed 10 and made 5,000 pounds a day. See: Diamond Mills

KER: Morrison Mill was established in 1859 at Petersborough, NH. The company made 3,500 pounds a day. See: Morrison Mill

KES: Morrison, Bare & Cass was established in 1866 at Pittsburgh, PA. The company employed 75 and made 15,000 pounds a day. See: Tyrone Mill, Roaring Springs Mill

KET: Mosher, Judd & Co. had offices in Stillwater, NY. The company employed 14 and made 3,000 pounds a day. See: Mosher Mill

KEU: Moshier, John S., was established in 1831 in Connecticut. See: Moshier Mill

KEV: Mousam Mfg. Co. was established in 1876 at East Poland, ME. The company employed 20 and made 9,000 pounds a day. See: Mousam Mills, Little Androscoggin Mill

KEW: Mt. Holly Paper Co. was established in 1858 at Mount Holly Springs, PA. The company employed 170 and made 6,000 pounds a day. See: Mt. Holly Mills

KEX: Muir, Charles A., was established in 1880 at Morristown, NJ. The company employed 15 and made 2,000 pounds a day. See: Muir's Mill

KEY: Muir, James A., was established in 1884 at Morristown, NJ. The company made 4,000 pounds a day. See: Morristown Mill

KEZ: Mulford, I.W., & Son was established in 1883 at Bridgeton, NJ. The company employed 16 and made 3,000 pounds a day. See: Shirley Mill

KFA: Mullin, W.A. & A.F., had offices in Mount Holly Springs, PA. The company made 4,500 pounds a day. See: Union Mill, Mullin Mill

KFB: Munroe Felt and Paper Co. was established in 1881 at 179 Devonshire St., Boston, MA. The company employed 50 and made 6 tons a day. See: Munroe Mill

KFC: Napanee Paper Co. was established in 1884 at Napanoch, NY. The company made 5,000 pounds a day. See: Napanoch Mill

KFD: National Wall Paper Co. was established in 1890. Bought by National Wall Paper Co. See: York Wall Paper Mill

KFE: Neal Bros. & Brooks was established in 1873 at Lockport, NY. The company employed 15 and made 6,000 pounds a day. See: Niagara Mill

KFF: Neenah Paper Co. was established in 1874

at Neenah, WI. The company employed 50 and made 3 tons a day. See: Neenah Mill

KFG: Nestell & Bauder had offices in Fort Plain, NY. The company made 3,000 pounds a day. See: Nestell Mill

KFH: Nestell, D.E., had offices in Fort Plain, NY. The company made 3,000 pounds a day. See: Nestell Mill

KFI: New Castle Paper Co. was established in 1882 at New Castle, PA. The company made 3 tons a day. See: New Castle Mill

KFJ: New Hampshire Pulp & Paper Co. had offices in Bristol NH, around 1870. The company employed 29 and made 4,500 pounds a day. See: Grafton Mill

KFK: New Orleans Bulletin was established in 1847 at New Orleans, LA. See: Bulletin Mill

KFL: New York & Felt Mills Co. had offices in South Orange, NJ, around 1875. The company made 8,000 pounds a day. See: Phoenix Mill, Armour Bros. Mill

KFM: New York Paper Co. had offices at 28-34 Read St., New York. The company made 3 tons a day. See: Antelope Mill

KFN: Newark Paper Co. was established in 1882 at Newark, OH. The company employed 20 and made 7,000 pounds a day. See: Newark Paper Co.

KFO: Newberry Bros. was established in 1866 at Vernon, MI. The company employed 20 and made 7,000 pounds a day. See: Shiawassee Mill

KFP: Newcomb & Buddington was established in 1875 at Stone Bridge, NY. The company employed 10 and made 2,500 pounds a day. See: Stone Ridge Mill

KFQ: Newcomb, A.S., & Co. was established in 1883 at High Falls, NY. The company made 6 tons a day. See: Newcomb Mill

KFR: Newhouse & Franklin was established in 1746 at Philadelphia, PA. The principals were Anthony Newhouse and Benjamin Franklin. See: Trout Creek Mill

KFS: Newman, G.H., had offices in Loudville, MA. The company made 2,000 pounds a day. See: Easthampton Mill

KFT: Newton Paper Co. was established in 1876 at Holyoke, MA. The company employed 67 and made 15 tons a day. The principals were Moses Newton, James Rampage, and George A. Clark. See: Newton Mill

KFU: Newton, O.I., was established in 1871 at Sparta, WI. The company employed 19 and made 9,500 pounds a day. See: Newton Mills

KFV: Niagara Falls Paper Mfg. Co. was established in 1862 at Niagara Falls, NY. The company employed 32 and made 6 tons a day. See: Niagara Falls Mill

KFW: Niagara Wood Paper Co. was established in 1884 at Niagara Falls, NY. The company employed 30 and made 5 tons a day. See: Niagara Wood Mill

KFX: Nichols & Kendall was established in 1796 at Leominster, MA. The principals were William Nichols, with Jonas Kendall and sons. See: Leominster Mills (2)

KFY: Niel Brothers & Brooks was established in 1873 at Honeoye Falls, NY. The company employed 13 and made 600 pounds a day. See: Enterprise Mill

KFZ: Niles Paper Mill Co. was established in 1872 at Niles, MI. The company employed 30 and made 12,000 pounds a day. See: Niles Mill

KGA: Nixon Bros. was established in 1850 at Clifton, OH. See: Clifton Mill

KGB: Nixon, Martin & W.H., was established in 1844 at 515 Commerce St., Philadelphia, PA. The company employed 327 and made 43,000 pounds a day. See: Flat Rock Mills, City Mill

KGC: Nixon, Thomas, was established in 1839 at Richmond, IN. The company employed 25 and made 1,000 pounds a day. See: Nixon Mill

KGD: Nonotuck Paper Co. was established in 1880 at Holyoke, MA. The company employed 175 and made 20,000 pounds a day. See: Nonotuck Mill

KGE: Northrup & Eldridge was established in 1855 at East Lee, MA. See: Northrup Mill

KGF: Norwich Savings Bank was established in 1884 at N. Westchester, CT. The company made 2,000 pounds a day. See: Strong's Mill

KGG: O'Neil Bros. & Co. was established in 1880 at Soquel, CA. The company employed 16 and made 5,000 pounds a day. See: South Coast Mill

KGH: Oakland Paper Co. was established in 1830 at N. Manchester, CT. The company employed 50 and made 6,000 pounds a day. See: Oakland Mill

KGI: Oglesby, Moore & Co. was established in 1853 at Middletown, OH. The company employed 65 and made 14,000 pounds a day. See: Oglesby-Moore Mills

KGJ: Ohio Mfg. Co. was established in 1883 at Niles, MI. The company employed 80 and made 6 tons a day. See: Ohio Mills

KGK: Ohio Paper Co. was established in 1880 in Ohio. The company employed 52 and made 2 tons a day. See: Ohio Mills

KGL: Ohio Straw Board Co. was established in 1885 at Upper Sandusky, OH. The company employed 30 and made 20,000 pounds a day. See: Upper Sandusky Mill

KGM: Ondawa Paper Co. was established in 1882 at Middle Falls, NY. The company employed 30 and made 4 tons a day. See: Ondawa Mill

KGN: Onderdonk, Andrew & Co. was established in 1773 at Roslyn, Hempstead Township, NY. The "and company" was Hugh Gaine, publisher, and Henry Remsen. See: Onderdonk Mill

KGO: Orr, Spencer & Wall was established in 1875 at Piqua, OH. The company employed 20 and made 7,000 pounds a day. See: Hydraulic Mill

KGP: Orrs & Co. was established in 1835 at Troy, NY. The company employed 110 and made 24,000 pounds a day. See: Troy Mill, Mt. Vernon Mill

KGQ: Otis Falls Pulp and Paper Co. was established in 1881 at Jay, ME. Principal was Hugh J. Chisholm. See: Otis Falls Mill, Jay Bridge Mill, Falmouth Mill

KGR: Oswego Paper Works had offices in Cooperstown, NY. The company made 1,200 pounds a day. See: Oswego Mill

KGS: Outterson, F.E., & Co. was established in 1840 at Pulaski, NY. The company employed 6 and made 1 ton a day. See: Pulaski Straw Board Mill

KGT: Owen & Hurlbut was established in 1822 at South Lee, MA. Principals were Charles Owen and Thomas Hurlbut. Firm dissolved in 1862. See: South Lee Mill, Irving Mill, Phoenix Mill, Potter Mill

KGU: Owen Paper Co. was established in 1862 at Housatonic, MA. The company made 2,000 pounds a day. Principals were Charles Owen, Edward Owen, and Henry D. Cone ('73). See: Owen Mill, Old

KGV: Paddack, S.D., was established in 1865 at Elbridge, NY. The company employed 12 and made 4,000 pounds a day. See: Paddack Mill (1865), Paddack Mill (1868)

KGW: Page & Dains was established in 1856 at E. Litchfield, CT. The company employed 6 and made 1,000 pounds a day. See: Litchfield Mills

KGX: Page Paper Co. had offices in Furnace, MA. The company employed 40 and made 8,000 pounds a day. See: Hardwick Mill

KGY: Palmer, Lowell M., had offices in Brooklyn, NY. The company made 6,000 pounds a day. See: Palmer Mill

KGZ: Pancost, John, was established in 1826 at Nether Providence Township, PA. See: Franklin Mill

KHA: Pansler, Ludwig, was established in 1780 at Alsace Township, PA. See: Pansler Mill

STA: Park & Watson was established in 1845 in New York as a stationer.

KHB: Park, Lincoln, & Park was established in 1833 at Taunton, MA. See: Park Mill

KHC: Park, Richard, & Co. was established in 1809 at Taunton, MA. See: Park Mill

KHD: Parker & Bassett was established in 1840 at Lowell, MA. The company employed 12 and made 3,000 pounds a day. See: Beaver Brook Mill

KHE: Parker & Servis was established in 1856 at Lockland, OH. See: Athern Lower Mill

KHF: Parker, H.A., & Co. was established in 1834 at E. Pepperell, MA. The company employed 8 and made 6,000 pounds a day. See: Nissitissitu Mills

KHG: Parker, Joseph, & Son was established in 1840 at New Haven, CT. The company employed 40 and made 2 tons a day. See: West Rock Mill

KHH: Parkhurst Bros. was established in 1883 at Westfield, NJ. The company made 2,000 pounds a day. See: Branch Mill

KHI: Parsons Paper Co. was established in 1846 at Holyoke, MA. The company employed 325 and made 12 tons a day. Principals were Joseph C. Parsons, Col. Aaron Bragg, and Whiting Street. See: Parsons Mill

KHJ: Parsons, Joseph C., was established in 1846 at Suffield, CT. Later principal of Parsons Paper Co. See: Eagle Mill, Old

KHK: Patten Paper Co. had offices in Neenah, WI, around 1870. The company employed 50 and made 3 tons a day. See: Neenah Mill

KHL: Patten, A.W., was established in 1880 at

Appleton, WI. The company made 4.5 tons a day. See: Valley Pulp & Paper Mill

STA: Patterson & Lambdin was established in 1818 at Pittsburgh PA, as a merchant.

KHM: Patterson, Edgar, was established in 1805 on P Street in Washington, DC. See: Rock Creek Mill

KHN: Patterson, J., & Co. was established in 1824 at Northern Liberties, PA. See: Pittsburgh Steam Mill

KHO: Patten Paper Co. was established in 1882 at Appleton, WI. The company employed 75 and made 7 tons a day. See: Patten Mill

KHP: Paul, Charles H., was established in 1883 at Clinton, CT. The company employed 7 and made 1,000 pounds a day. See: Clinton Mill

KHQ: Payn, Louis F., had offices in Chatham, NY. The company made 10,000 pounds a day. See: Stony Brook Mill

KHR: Pearl Paper Co. was established in 1883 at S. Hadley, MA. The company employed 7 and made 2,000 pounds a day. See: Pearl Mill

KHS: Peaslee, George H., had offices in Ancram, NY. The company made 5.5 tons a day. See: Ancram Mill

KHT: Peaslee, J.N., was established in 1885 at Malden Bridge, NY. The company employed 26 and made 3 tons a day. See: Malden Bridge Mill

KHU: Peaslee, W.W., was established in 1845 at Malden Bridge, NY. The company employed 26 and made 3 tons a day. See: Malden Mill

KHV: Peck & Streeter was established in 1882 at Allegan, MI. The company employed 20 and made 8,000 pounds a day. See: Allegan Mills

KHW: Peck, Everard & Co. was established in 1820 at Rochester, NY. See: Rochester Mill

KHX: Peninsular Paper Co. was established in 1872 at Ypsilanti, MI. The company employed 90 and made 11,000 pounds a day. See: North Mill, South Mill

KHY: Pennsylvania Pulp & Paper Co. had offices in Philadelphia, PA. The company made 12 tons a day. See: Pennsylvania Pulp & Paper (2)

KHZ: Perrine Paper Co. was established in 1880 at Franklin, OH. The company made 2.5 tons a day. See: Perrine Mill

KIA: Persee & Brooks was established in 1854 at Windsor Locks, CT. See: Hibernian Mill

KIB: Peters, James, & Co. had offices in Latrobe, PA. The company made 8 tons a day. See: Loyalhana Mills

KIC: Pettebone Paper Co. was established in 1883 at Niagara Falls, NY. The company employed 31 and made 6 tons a day. See: Pettebone Mill

KID: Phelps & Field was established in 1836 at Lee, MA. Principals were George H. Phelps and Marshall Field. See: Columbia Mill

KIE: Phoenix Mfg. Co. was established in 1872 at Battenville, NY. The company employed 15 and made 6,500 pounds a day. See: Phoenix Mill

KIF: Phoenix Paper Mfg. Co. was established in 1884 at Phoenix, NY. The company made 2,000 pounds a day. See: Phoenix Mill

KIG: Philadelphia Paper Co. was established in 1884 at 30th and Hamilton St. in Philadelphia, PA. The company employed 12 and made 4 tons a day. See: Union Pacific Mills, Philadelphia Mill

KIH: Phillips & Speer was established in 1834 at Brookville, IN. See: Brookville Mill

KII: Phillips, Samuel, was established in 1789 at Lawrence, MA. See: Shawshin River Mill

KIJ: Phillips, Spear & Co. was established in 1835 at Cincinnati, OH. See: Cincinnati Steam Mill

KIK: Phillips, Thomas & Co. was established in 1872 at Akron, OH. The company employed 63 and made 4,500 pounds a day. See: Akron Mill

KIL: Philmont Paper Co. had offices in Philmont, NY. The company made 8,000 pounds a day. See: Philmont Mill

KIM: Philp, H.A., & Co. had offices in Brooklyn, NY. The company made 4 tons a day. See: Brooklin Steam Mill

KIN: Phinneys & Todd was established in 1806 at Toddsville, NY. Principals were Jeheil Todd and Elihu Phinney. See: Cooperstown Mill

KIO: Pickering, Joseph, was established in 1827 at N. Windham, CT. See: Pickering Mill

KIP: Pickerson, Judge W.P., was established in 1868 at Nashville, TN.

KIQ: Pierce Bros. was established in 1885 at St. Johnsbury, VT. The company employed 11 and made 1.5 tons a day. See: Valley Falls Mill

KIR: Pioneer Mfg. Co. was established in 1852 at Athens, GA. The company employed 26 and made 4,000 pounds a day. See: Pioneer Mill

KIS: Piqua Straw Board & Paper Co. had offices in Piqua, OH. The company made 30,000 pounds a day. See: Piqua Mill

KIT: Pitkin, Elisah, was established in 1813 at Stuyvesant Falls, NY. See: Hope Mill

KIU: Plain City Paper Co. was established in 1883 at Plain City, OH. The company employed 30 and made 10,000 pounds a day. See: Darby Mill

KIV: Platner & Ingersoll was established in 1833 at East Lee, MA. Principals were George Platner and Jared Ingersoll. See: Turkey Mill

KIW: Platner & Porter Mfg. Co. was established in 1847 at Unionville, CT. The company made 14,000 pounds of book paper and 4,000 pounds of stationery a day. See: Washington Mill, Tunxis Mill, Union Village Mill

KIX: Platner & Smith was established in 1839 at Lee, MA. Principals were George Platner and Elizur Smith. See: Union Mill, Baker Mill, Defiance Mill, Eagle Mill, Housatonic Mill, Laurel Mill, Pleasant Valley Mill

KIY: Platner Brothers was established in 1842 at Ancram, NY. See: Ancram Mill

KIZ: Pleasant Mills Paper Co. was established in 1861 at 609 Chestnut St., Philadelphia, NJ. The company employed 18 and made 6,000 pounds a day. See: Pleasant Mills

KJA: Plover Paper Co. was established in 1892 at Whiting, WI. The company employed 175 and made 15 tons a day. See: Plover Mill

KJB: Portage Straw Board Co. was established in 1883 at Akron, OH. The company employed 350 and made 66 tons a day. See: New Portage Mill, Circleville Mill

KJC: Porter & Clark was established in 1825 at Bath Island, NY. Principals were Henry W. Clark and Albert H. Porter. See: Bath Island Mill

KJD: Powers & Brown was established in 1867 at Dalton, MA. See: Berkshire Mill

KJE: Powers Paper Co. was established in 1889 at Holyoke, MA. See: Powers Paper Mill

KJF: Prather & Jacob was established in 1816 at Louisville, KY, as merchants. Principals were Thomas Prather and John Jacob.

KJG: Pratt, L., was established by 1823 in Claremont, NH. See: Suncook Falls Mills

KJH: Prentiss, Bemis & Co. was established in 1834 at Alsted, VT. See: Cold River Mill

KJI: Priest, Thurston, was established in 1861 at Newton, MA. See: Nehoiden Mill

KJJ: Puget Sound Pulp and Paper Co. was established in 1892 at Everett, WA. See: Puget Sound P&P Mill

KJK: Pultz & Walkley was established in 1871 at Plantsville, CT. The company employed 13 and made 4,000 pounds a day. See: Glen Mill

KJL: Pusey & Jones was established in 1848 at Newark, DE. See: Seymour Mill, Nonatum Mill

KJM: Pusey Paper Co. had offices in Windsor Locks, CT, around 1810. See: Hibernian Mill

KJN: Putnam, G., had offices in Jordan, NY. The company made 2,500 pounds a day. See: Jordan Mill

KJP: Queechy Paper Co. was established in 1877 at Canaan Four Corners, NY. The company employed 16 and made 2 tons a day. See: Queechy Mill

KJQ: Quincy Paper Co. had offices in Quincy, IL. The company made 21 tons a day. See: Gem City Mill, Atlas Mill (1880)

KJR: Quinnesee Water Power Co. was established in 1892 at Marinette, WI. See: Quinnesee Mill

KJS: Rahn, Adam, & Co. was established in 1809 at Maidencreek Township, PA. The "and company" was Conrad Herbst. See: Rahn Mill

KJT: Ramsdell, H. Powell, was established in 1864 at Salisbury Mills, NY. The company employed 44 and made 8,500 pounds a day. See: Arlington Mill

KJU: Randall, W.T., was established in 1880 at Beloit, WI. The company employed 20 and made 3 tons a day. See: Enterprise Mill

KJV: Randolph, Phillip, was established in 1870 at 8 Decatur St., Philadelphia. The company employed 15 and made 4,000 pounds a day. See: Pleasant Garden Mill

KJW: Ravine Mill had offices in Coeymans, NY. The company made 5,000 pounds a day. See: Ravine Mill

KJX: Rayle & McKinney was established in 1835 at Knoxville, TN. The company employed 15 and made 2,000 pounds a day. See: Knoxville Mill

KJY: Reckless & Kennedy had offices in Council Bluffs, IA. The company made 8,000 pounds a day. See: Council Bluffs Mill

KJZ: Record, Alvin, was established in 1844 at Livermore Falls, ME. The company employed 30 and made 3,000 pounds a day. See: Record

Leather Board Mills, Jay Bridge Mill, Messalonskee Mill, Falmouth Mill

KKA: Red River Paper Co. was established in 1883 at Fergus Falls, MN. The company employed 25 and made 12,000 pounds a day. See: Red River Mill

KKB: Reid, J. & W., had offices in Norwich, CT. The company made 8,000 pounds a day. See: Fort Ned Mills

KKC: Remington Paper Co. was established in 1881 at Watertown, NY, and made 9 tons a day. See: Remington Mill

KKD: Remington, A.D., was established in 1878 at San Francisco, CA. Principal was Alfred Denison Remington. See: Remington Mill, Lick Mill

KKE: Remington, C.R., & Son was established in 1881 at Watertown, NY. The company employed 30 and made 10,000 pounds a day. See: Remington Mill

KKF: Reynolds, E.S., was established in 1869 at South Bend, IN. The company employed 15 and made 3,000 pounds a day. See: South Bend Mill

KKG: Reynolds, James L. was established in 1883 at Hoboken, NJ. The company employed 12 and made 6,000 pounds a day. See: Hoboken Mill

KKH: Rhodes, Utter & Co. was established in 1880 at Rockford, IL. See: Rockford Mill

KKI: Rice & Crane was established in 1831 at Needham, MA. Principals were Thomas Rice, Sr., and Luther Crane. See: Thatcher Mill, Longfellow Mill

KKJ: Rice & Garfield was established in 1850 at Needham, MA. Principals were Thomas Rice, Jr., and Moses Garfield (sole owner after 1863). See: Ware Mill, Rice and Garfield Mill

KKK: Rice, F.W., Estate had offices at 552 Devonshire St., Boston. The company made 2 tons a day.

KKL: Rice, Frederick, was established in 1873 at Needham, MA. See: Jackson Mill

KKM: Rice, General Charles, was established in 1818 at Newton, MA. See: Nehoiden Mill

KKN: Rice, Thomas, Jr., Paper Co. was established in 1852 at Needham, MA. See: Thomas Rice Mill, Foster Mill, Jackson Mill, Rice & Garfield Mill

KKO: Rice, Thomas, Sr., was established in 1835 at Needham, MA. See: Thomas Rice Mill

KKP: Rich, C.W., & Co. was established in 1882 at Ausable Chasm, NY. The company employed 6 and made 4,000 pounds a day. See: Ausable Chasm Mill

KKQ: Richards Paper Co. was established in 1884 at Gardiner, ME, and made 5 tons a day. See: Gardiner Mills

KKR: Richardson Paper Co. was established in 1883 at Monroe, MI. The company employed 24 and made 12,000 pounds a day. See: Richardson Mill

KKS: Richmond Mfg. Co. was established in 1883 at Providence, RI. The company made 30,000 pounds a day. See: Richmond Mill

KKT: Richmond Paper Co. was established in 1834 at Richmond, VA. The company made 5,000 pounds a day. See: James River Mill

KKU: Richmond, G.N. & Bro. was established in 1860 at Appleton, WI. The company made 5,000 pounds a day. See: Richmond Mill

KKV: Richtine, George, & Co. was established in 1857 at Zanesville, OH. See: Cox Mill

KKW: Riege, John L., & Son was established in 1866 at Riegelsville, NJ. The company employed 21 and made 4,000 pounds a day. See: Riege Mill

KKX: Rimmon Paper Co. was established in 1840 at Seymour, CT. See: Rimmon Falls Mill

KKY: Ripley Mfg. Co. was established in 1882 at Unionville, CT. The company employed 50 and made 12,000 pounds a day. See: Sunny Side Mill

KKZ: Rittenhouse, William, & Sons was established in 1690 at Philadelphia, PA. See: Rittenhouse Mill

KLA: Riverside Paper Co. was established in 1867 at Holyoke, MA. The company employed 200 and made 8,000 pounds a day. Principal was Charles O. Chapin. See: Riverside Mill

KLB: Roaring Brook Paper Mfg. Co. was established in 1872 at East Glastonbury, CT. The company employed 7 and made 2 tons a day. See: Roaring Brook Mill

KLC: Roberts & Co. was established in 1874 at North Amherst, MA. The company employed 4 and made 1,000 pounds a day. See: Oak Mill

KLD: Roberts, John, & Son was established in 1883 at New London, CT. The company employed 14 and made 2 tons a day. See: Roberts Mill, New London Mills

KLE: Roberts, H.J., was established in 1882 at

Huntington, CT. The company employed 7 and made 1,000 pounds a day. See: Oronoque Mills

KLF: Robertson Brothers was established in 1882 at Holyoke, MA. The company employed 12 and made 2,200 pounds a day. See: Robertson Brothers Mill

KLG: Robertson, C.M., was established in 1851 at Montville, CT. The company employed 20 and made 5,000 pounds a day. See: Montville Mill, Rockland Mill

KLH: Robertson, G. & G.A., & Co. was established in 1842 at Hinsdale, NH. The company employed 18 and made 1,500 pounds a day. See: Hindsdale Mills

KLI: Robertson, John, & Son was established in 1870 at Bellows Falls, VT. The company employed 16 and made 3,000 pounds a day. See: Robertson Mill

KLJ: Robertson, William, & Sons was established in 1830 at Putney, VT. The company employed 16 and made 2,500 pounds a day. See: Owl Mills

KLK: Robinson & Black was established in 1875 at Holyoke, MA. The company made 2,500 pounds a day. See: Holyoke Manila Mill

KLL: Robinson, G.A., was established in 1880 at West Mansfield, MA. The company made 1,000 pounds a day. See: Wading River Mill, Springdale Mill

KLM: Rochester Paper Co. was established in 1864 at Rochester, NY. The company employed 90 and made 25,000 pounds a day. See: Genesee Mill

KLN: Rock River Paper Mills had offices in Beloit, WI. The company made 48,000 pounds a day. See: Rock River Mills

KLO: Rockland Mfg. Co. was established in 1825 at Rockland, DE. See: Delaware Mills

KLP: Rogers & Walker was established in 1812 at Connellsville Township, PA. Principals were Daniel and Joseph Rogers and Zodiac Walker. See: Youghiogheny River Mill

KLQ: Rogers, H.E., was established in 1830 at S. Manchester, CT. The company employed 20 and made 4,000 pounds a day. See: Atlantic Mill, Adriatic Mill

KLR: Rogers, Harper W., was established in 1850 at Hudson, NY. The company employed 10 and made 600 pounds a day. See: Rogers Mill

KLS: Rogers, Samuel S., was established in 1854 at East Lee, MA. See: Mahaiwe Mill

KLT: Rommel, G.B., had offices in Pittston, PA. The company made 4,000 pounds a day. See: Wyoming Valley Mill

KLU: Rossman, J.W., & Son was established in 1872 at Stockport, NY. The company employed 45 and made 14,000 pounds a day. See: Stockport Mills

KLV: Roudolph, Frank M., was established in 1895. The company made 2 tons a day. See: York Wall Paper Mill

KLW: Rudolphs, S.A., & Sons was established in 1881 at 506 Market St., Philadelphia. The company employed 75 and made 1,000 pounds a day. See: Ashland Mill

KLX: Rufner, S., was established in 1848 at Lockland, OH. See: Athern Lower Mill

KLY: Rumford Falls Paper Co. was established in 1897 at Rumford Falls, ME. See: Rumford Falls Mill

KLZ: Russell & Andrews was established in 1882 at Penn Yan, NY. The company employed 35 and made 10,000 pounds a day. See: Milo Mill

KMA: Russell & Haswell was established in 1783 at North Bennington, VT. See: Vermont Gazette Mill

KMB: Russell Paper Co. was established in 1864 at 53 Devonshire St., Boston. The company employed 300 and made 10 tons a day. Principal was William A. Russell. See: Russell Mill, Old, Russell Mill

KMC: Russell Willard & Co. had offices in Bellows Falls, VT. The company employed 25 and made 5,000 pounds a day. See: Russell Willard & Co. Mills

KMD: Salisbury & Vinton Paper Co. was established in 1869 at Indianapolis, IN. The company employed 50 and made 7,000 pounds a day. Firm incorporated in 1878. See: Central Mill

KME: Sampson, Edwin H., & Son was established in 1868 at Boston, MA. The company made 6,000 pounds a day. See: Sampson Mill

KMF: Sandusky Paper Co. was established in 1880 at Sandusky, OH. The company employed 11 and made 1 ton a day. See: Sandusky Mill

KMG: Sandusky Straw Board Co. was established in 1882 at Upper Sandusky, OH. The company employed 90 and made 26,000 pounds a day. See: Upper Sandusky Mill

KMH: Satterly, John, was established in 1850 at Little Falls, NY. See: Satterly Mill

KMI: Saunderson, Issac, was established in 1801 at Milton, MA. See: Saunderson Mill

KMJ: Savannah Palm Paper Co. had offices in Savannah, GA. The company made 10,000 pounds a day. See: Savannah Mill

KMK: Savannah Palm Paper Mfg. Co. had offices in Valley Forge, PA. The company made 2.5 tons a day. See: Africa Mill

KML: Savels, John, was established in 1811 at Gardiner ME. See: Cobbossee River Mill

STA: Sawn & Co. was established in 1800 at Philadelphia, PA, as a merchant.

KMM: Schauck, Peter, was established in 1824 at Bentley Springs, MD. See: Schauck Mill

KMN: Schmidt & Ault Paper Co. was established in 1897. The company made 200 tons a day. See: Schmidt & Ault Mill

KMO: Schutz, Conrad and Francis, was established in 1790 at Lower Merion Township, PA. See: Schutz Mills

KMP: Schutz, Frederick, was established in 1788 at Lower Merion Township, PA. See: Dove Mills

KMQ: Scioto Straw Board Co. was established in 1882 at Kenton, OH. The company employed 50 and made 10 tons a day. See: Scioto Mill

KMR: Scott & Bayless had offices in Steubenville, OH. See: Clinton Steam Mill

KMS: Scott, Sam, had offices in Nashville, TN, around 1850. See: Nashville Mills

KMT: Scully, Thomas F., was established in 1883 at Atlanta, GA. The company employed 11 and made 2,000 pounds a day. See: Fulton Mill

KMU: Seaman, James, had offices in Ridgewood, NY. The company made 2,000 pounds a day. See: Seaman Mill

KMV: Sedwick & Sabin was established in 1844 at Lenox, MA. Principals were Thomas Sedwick and George Sabin. See: Pleasant Valley Mill

KMW: Seeley, E.A., was established in 1853 at Scotch Plains, NJ. The company employed 25 and made 5,000 pounds a day. See: Falls Mills

KMX: Sensenich, John H., was established in 1860 at Compassville, PA.

KMY: Seymour Paper Co. was established in 1854 at 45 John St., New York. The company employed 200 and made 25,000 pounds a day. See: Pacific Mill

KMZ: Sharpless, Haskins & Wallace was established in 1810 at Brownsville, PA. See: Redstone Mill

KNA: Shaver, Abraham, was established in 1852 at Freeland, MD. The company employed 3 and made 2,000 pounds a day. See: Glenmount Mill

KNB: Sheeder, J. Fred, was established in 1871 at Kimberton, PA. The company employed 10 and made 6,000 pounds a day. See: Kimberton Mill

KNC: Sheets, John, was established in 1827 at Madison, IN. See: Indian Kentuck Creek Mill

KND: Sheffield, J.B., & Son was established in 1828 at Saugerties, NY. The company employed 215 and made 8 tons a day. See: Sheffield Mills, Saugerties Mill

KNE: Sherer, Robert, was established in 1829 at East Nottingham Township, PA. See: Spring Lawn Mill

KNF: Sherman & Co. was established in 1870 at Belfast ME. The company employed 5 and made 1,000 pounds a day. See: Sherman Mill

KNG: Sherman Bros. had offices in Marcellus Falls, NY. The company employed 11 people. See: Marcellus Mills

KNH: Shober, Gottlieb, was established in 1806 at North Salem, NC. See: Shober Mill

KNI: Shober, John A., was established in 1866 at Lancaster, PA. The company employed 30 and made 2 tons a day. See: Slackwater Mill

KNJ: Shue & Apel was established in 1872 at Lyons Falls, NY. The company employed 4 and made 2,000 pounds a day. See: Lewis Mill

KNK: Shue Bros. & Co. was established in 1845 at Lyons Falls, NY. The company employed 4 and made 2,000 pounds a day. See: Lewis Mill

KNL: Shufelt, John D. was established in 1883 at Brainard, NY. The company employed 16 and made 4 tons a day. See: Nassau Mill

KNM: Shyrock & Johns was established in 1790 at Chambersburg, PA. Principals were John Shyrock and Thomas Johns. See: Hollywell Mill

KNN: Shyrock Brothers was established in 1865 at 914 Filbert, Philadelphia. The company employed 25 and made 5,000 pounds a day. See: St. Papyrus Mill

KNO: Shyrock, George A., & Co. was established in 1827 at Chambersburg, PA. The company made 2,400 pounds a day. See: Papyrus Mill, Hollywell Straw Mill, Mammoth Mill

KNP: Simons, H.E., had offices in Valparaiso, IN. The company made 8,000 pounds a day. See: Valpariso Mill

KNQ: Simmons, William, was established around 1823 in Leominster, Worcester Co., MA. See: Leominster Mills

KNR: Singerly, William M. was established in 1881 at Fair Hill, MD. The company employed 50 and made 10,000 pounds a day. See: Providence Mill

KNS: Skaneateles Paper Co. was established in 1875 at Skaneateles, NY. The company employed 35 and made 6,000 pounds a day. See: Skaneateles Mill

KNT: Skelly, F.J., was established in 1885 at Otis, IN. The company employed 3 and made 1,000 pounds a day. See: Otis Board Mill

KNU: Skidmore, T.J., was established in 1854 at Leona, NY. The company employed 6 and made 1,200 pounds a day. See: Willow Dale Mill

KNV: Skinner, J.C., & Co. was established in 1848 at Hamilton, OH. The company employed 15 and made 1,500 pounds a day. See: Hamilton Tissue Mill

KNW: Small, Gould & Co. was established in 1875 at Baldwinsville, MA. The company employed 15 and made 6,000 pounds a day. See: Baldwinsville Mill

KNX: Smart, A.J., was established in 1850 at Sand Lake, NY. The company employed 15 and made 6,000 pounds a day. See: Palm Leaf Mill

KNY: Smart, Joseph W. was established in 1853 at Troy, NY. The company employed 34 and made 11,000 pounds a day. Principals were Joseph Smart followed by Joseph and Andrew ('58) and Robert T. Smart ('75). See: Troy City Mill, Gold Leaf Mill

KNZ: Smith & Bassett was established in 1834 at Seymour, CT. Principals were Sylvester Smith and Samuel Bassett. See: Smith & Bassett

KOA: Smith & Field was established in 1840 at Russell, MA. Principals were J.R. Smith and Cyrus W. Field. See: Russell Mill

KOB: Smith Paper Co. was established in 1864 at Lee, MA. The company employed 300 and made 50,000 pounds a day. See: Valley Mill, Columbia Mill, Eagle Mill, Housatonic Mill, Laurel Mill, Niagara Mill

KOC: Smith, Elizur, was established in 1834 at Lee, MA. See: Turkey Mill

KOD: Smith, G.F., was established in 1858 at Moriches, NY. The company made 1,000 pounds a day. See: Moriches Mill

KOE: Smith, Harrison, & Co. was established in 1852 at East Lee, MA. The "and company" was David S. May. See: Thatcher Mill

KOF: Smith, Henry, was established in 1868 at Tecumseh, MI. The company employed 14 and made 5,000 pounds a day. See: Tecumseh Mill

KOG: Smith, J.H., & Co. was established in 1868 at Smithton, PA. The company employed 15 and made 4,800 pounds a day. See: Smithton Mill

KOH: Smith, Jeremiah, was established in 1741 at Milton, MA. See: Neponset Mills

KOI: Smith, Leonard, was established in 1867 at Mellenville, NY. The company employed 6 and made 1,600 pounds a day. See: Squampaumick Valley Mill

KOJ: Smith, Lincoln & Charles was established in 1863 at Merrick, NY. The company employed 4 and made 1,000 pounds a day. See: Union Mill

KOK: Smith, S.J., was established in 1865 at Conneaut, OH. The company employed 16 and made 3,500 pounds a day. See: Conneaut River Mill

KOL: Smith, S.I., had offices in Elkhart IN around 1869 and closed in 1882. See: Elkhart Combination Board Mill

KOM: Smith, Valentine, had offices in Freeport, NY. The company made 1,500 pounds a day. See: Smith Mill

KON: Smith, William, & James W. Roll was established in 1852 at Millburn, NJ. See: Smith Brothers Mill

KOO: Smiths, W.W., was established in 1831 at Seymour, CT. The company employed 6 and made 2,000 pounds a day. See: Smiths Mill

KOP: Snider, Louis, & Sons was established in 1855 at Hamilton, OH. The company employed 136 and made 19,500 pounds a day. See: Franklin Mill, Fair Grove Mill, Fordham Mill

KOQ: Snow Paper Co. was established in 1884 at South Fitchburg, MA. The company employed 20 and made 6 tons a day. See: Snow Mill

KOR: Snyder, John, was established in 1819 at Selinsgrove, PA. See: Middle Creek Mill

KOS: Sonn, J.H., & Co. was established in 1851 at Medina, NY. The company employed 20 and made 4 tons a day. See: Shelby Mill

KOT: South Lee Mfg. Co. was established in 1857 at South Lee, MA. Principal was Edward Owen. See: South Lee Mill, Irving Mill

KOU: South Sudbury Mfg. Co. was established in 1876 at West Cummington, MA. See: Sudbury Mill

KOV: Southford Mfg. Co. was established in 1881 at Southford, CT. The company employed 36 and made 5 tons a day. See: Southford Mill

KOW: Southwestern Paper Co. had offices in Lawrence, KS, and closed after 1880. See: Southwestern Mill

KOX: Southworth Mfg. Co. was established in 1839 at Mittineague, MA. The company employed 100 and made 4,000 pounds a day. See: Southworth Mill

KOY: Spaulding Brothers was established in 1877 at Townsend Harbor, MA. The company employed 7 and made 4,000 pounds a day. See: West Townsend Leather Board Mill, Spaulding Brothers Mill

KOZ: Spelman, J.H., was established in 1875 at Chatham, NY. The company employed 6 and made 2 tons a day. See: Stone Mill

KPA: Sprague, W.N., was established in 1869 at Middle Falls, NY. The company employed 33 and made 7,000 pounds a day. See: Battenkill Mill

KPB: Springdale Paper Co. was established in 1835 at Westfield, MA. This was the Jessups family mill. See: Jessups Mill

KPC: Springfield Paper Co. was established in 1864 at Rainbow, CT. The company employed 30 and made 5,000 pounds a day. See: Springfield Mill

KPD: St. Joseph Paper & Pail Co. was established in 1881 at St. Joseph, MI. The company employed 30 and made 18,000 pounds a day. See: St. Joseph Mills

KPE: Stahl & Martin had offices in Rockwood, NY. The company employed 7 and made 3,000 pounds a day. See: Rockwood Mill

KPF: Stahl & Spencer was established in 1877 at Rockwood, NY. The company employed 7 and made 3,000 pounds a day. See: Rockwood Mill

KPG: Star Paper Co. had offices in Shortsville, NY. The company made 13 tons a day. See: Star Mill, Diamond Mill

KPH: Starin, Henry W., was established in 1810 at Esperance, NY. See: Starin Mill

KPI: Stark Paper Co. was established in 1850 at North Bennington, VT. The company employed 40 and made 7,500 pounds a day. See: State Line Mill, Paran Creek Mill

KPJ: Starr, William & Kirk was established in 1848 at Montoursville, PA. The company employed 10 and made 2,000 pounds a day. See: Fairfield Mill

KPK: Steadman, E.H., was established in 1828 at Elkhorn, KY. See: Franklin Mill

KPL: Steel, General John, was established in 1788 at Steelville, PA. James Steel (brother) joined the firm in 1800. See: Steelville Mill

KPM: Steel, James, was established in 1808 at Steelville, PA. See: Steelville Mill

KPN: Steever, J.B., was established in 1864 at Brodalbin, NY. The company employed 6 and made 3,000 pounds a day. See: Keneetee Mill

KPO: Stelwagons, Joseph & Son was established in 1853 at 325 Commerce St., Philadelphia, PA. The company employed 25 and made 12,000 pounds a day. See: Severance Mill

KPP: Sterling Paper Co. had offices in Sterling, IL. The company made 12,000 pounds a day. See: Sterling Mill

KPQ: Stevens & Son was established by 1823 in Claremont NH. Principal was Josiah Stevens.

KPR: Stevens & Thompson had offices in North Hoosick, NY. The company made 7 tons a day. See: North Hoosick Mill

KPS: Stewart & Carmichael was established in 1867 at Amsterdam, NY. The company employed 50 and made 5 tons a day. See: Forest Mill

KPT: Stimpson, Green & Fairbanks was established in 1819 at Centre Rutland, VT. See: Owl Mills

KPT.1: Stone Fort Paper Co. was established in 1879 at Manchester, TN. The company employed 35 and made 17,000 pounds a day. See: Stone Fort Mill

KPT.2: Stovekin, J., & Co. was established in 1872 at Kaukauna, WI. Principals were John Stoveken and Henry Hewitt. See: Eagle Mill

KPU: Stowe, J.D., & Sons was established in 1848 at Scitico, CT. The company employed 10 and made 3,000 pounds a day. See: Suffield Mill

KPV: Strong, C.W., was established in 1870 at North Westchester, CT. The company made 2,000 pounds a day. See: Strong's Mill

KPW: Stubbs, Joseph I., was established in 1838 at East Nottingham, PA. See: Stubbs Mill

KPX: Sturges & Costar was established in 1843 at East Lee, MA. Principals were Samuel D. Sturges and William Costar. See: Greenwater Mill

KPY: Sugar Creek Mill had offices in Atlanta, GA. The company made 1,000 pounds a day. See: Sugar Creek Mill

KPZ: Sugar River Paper Mill Co. was established in 1866 at Claremont, NH. The company employed 75 and made 14,000 pounds a day. See: Sugar River Mill

KQA: Sumington, Joseph, & Co. was established in 1883 at Lancaster, PA. The company employed 22 and made 6,000 pounds a day. See: Beltonford Mill

KQB: Summer, William, had offices in Milton, MA. See: Eagle Mill

KQC: Susquehanna Water Power & Paper Co. was established in 1883 at Conowingo, MD. The company made 12 tons a day. See: Susquehanna Water Power & Mill

KQD: Sutphin & Wrenn had offices in Middletown, OH. The company made 6,000 pounds a day. See: Sutphin-Wrenn Mill

KQE: Sutton, John W. & Bro. was established in 1853 at Indiana, PA. The company employed 13 and made 5,500 pounds a day. See: Indiana Straw Board Mill

KQF: Swan, Walter, & George Reid was established in 1815 at Poughkeepsie, NY. See: Walsh Mill, Slee's Mill

KQG: Syms & Dudley Paper Co. was established in 1881 at Holyoke, MA. The company employed 150 and made 8 tons a day. See: Syms & Dudley Mill

STA: Syracuse Pulp & Paper Company had offices in New York.

KQH: Taggart & Brown was established in 1880 at Pulaski, NY. The company employed 5 and made 3,000 pounds a day. See: Salmon River Board Mill

KQI: Taggart Brothers was established in 1865 at Watertown, NY. The company employed 50 and made 3 tons a day. See: Taggart Brothers Mill

KQJ: Tait & Son was established in 1848 at Trumbull, CT. The company employed 9 and made 4,000 pounds a day. See: Oak Grove Mill

KQK: Talbot County Paper Mills was established in 1880 at Easton, MD. The company employed 45 and made 6 tons a day. See: Talbot County Mills

KQL: Talbot County Paper Mills Co. had offices in Vernon Depot, CT, and closed after 1880. See: Oak Grove Mills, Granite Mill

KQM: Tama Paper Co. had offices in Tama City, IA. The company made 4,000 pounds a day. See: Tama Mill

KQN: Tangeman Paper Co. was established in 1863 at Lockland, OH. See: Friend & Tangeman Mill

KQO: Tangeman, George P., & Co. had offices in Hamilton, OH. The company made 10 tons a day. See: Rockdale Mills

KQP: Tangeman, J.H., & Co. was established in 1851 at Lockland, OH. The company made 9,000 pounds a day. See: Star Mill

KQQ: Tanner & Faxon was established in 1881 at East Lee, MA. The company employed 30 and made 7,000 pounds a day. See: Greenwood Mill

KQR: Taylor, Bradford L., had offices in Westfield, NY. The company employed 12 and made 5,500 pounds a day. See: Westfield Mill

KQS: Taylor, Frank, & Co. was established in 1872 at South Hadley, MA. The company employed 6 and made 1,000 pounds a day. See: Manila Mill

KQT: Taylor, S.P., & Co. was established in 1856 at Taylorville, CA. The company employed 50 and made 5,000 pounds a day. See: Pioneer Mill, San Geronimo Mill

KQT.1: Tempest, Francis, was established in 1815 at Birmingham Township, PA. The company made 1,000 pounds a day. See: Sunnydale Mill

KQT.2: Tennessee Paper Co. had offices in Nashville, TN. See: Tennessee Mill

KQT.3: Terre Haute Paper Co. was established in 1883 at Terre Haute, IN. The company employed 50 and made 3 tons a day. See: Ellsworth Mills

KQU: Thatcher & Ingersoll was established in 1840 at East Lee, MA. The company made 1,000 pounds a day. Principals were Stephen Thatcher and Jared Ingersoll. See: Waverley Mill, Thatcher Mill

KQV: Thatcher, Stephen, was established in 1824 at East Lee, MA. See: Thatcher Mill

KQW: Thilmany Pulp Co. was established in 1883 at Kaukauna, WI. Principal was Oscar Thilmany. See: Thilmany Mill

KQX: Thomas & Belaney was established in 1820 at Millburn, NJ. See: Rahway River Mill

KQY: Thomas Sedwick & Co. was established in 1835 at Lenox, MA. See: Pleasant Valley Mill

KQZ: Tompkins & Davis was established in 1846 at Nassau, NY. Principals were John B. Davis and Peter C. Tompkins. See: Rensselaer Mill

KRA: Thompson, David, & Sons was established in 1885 at Ballston Spa, NY. The company employed 25 and made 2,500 pounds a day. See: Fonda Mill, Ballston Spa Mill

KRB: Thompson, John, was established in 1829 at Westville, CT. The company employed 8 and made 2,000 pounds a day. See: Thompson Mill

KRC: Thorp, E., was established by 1823 at Leominster, Worcester Co., MA. See: Leominster Mills (2)

KRD: Thurber, Samuel, Jr., & Co. was established in 1780 at Providence, RI. See: Thurber Mill

KRE: Ticonderoga Pulp & Paper Co. was established in 1870 at Ticonderoga, NY. The company employed 100 and made 12,000 pounds a day. See: Ticonderoga Pulp & Mill

KRF: Tileston & Hollingsworth was established in 1801 at Groton Center, MA. The company employed 40 and made 31,000 pounds a day. See: Tileston & Hollingsworth Mill, Eagle Mill, Fuller Mill, Gillespie Mill, Groton Center Mill, Mattapan Mill, Neponset Mills, Boies Family Mill

KRG: Tilton, Fredrick G., & Co. was established in 1879 at Fort Edward, NY. The company made 4 tons a day. See: Fort Edward Mill

KRH: Tipp Paper Co. was established in 1882 at Tippecanoe City, OH. The company employed 24 and made 14,000 pounds a day. See: Tipp Mill

KRI: Tippecanoe Paper Co. was established in 1880 at Monticello, IN. The company employed 20 and made 5,000 pounds a day. See: Tippecanoe Mills

KRJ: Tompkins, John D., was established in 1870 at Brainard, NY. The company employed 16 and made 8,000 pounds a day. See: XX Mill

KRK: Tompkins, Peter & Staats, was established in 1838 at East Chatham, NY. See: East Chatham Mill, Stuyvesant Falls Mill

KRL: Tompkins, Smith P., was established in 1850 at Stanford, NY. The company employed 10 and made 1,000 pounds a day. See: Tompkins Mill

KRM: Tondes, Daniel, was established in 1842 at Lafayette, IN. See: Tondes Mill

KRN: Townsend, James A., had offices in Newburgh, NY. The company employed 100 and made 3,000 pounds a day. See: Highland Mill

KRO: Tracy, H.P., was established in 1867 at Elmwood, IL. The company employed 24 and made 8,000 pounds a day. See: Elmwood Mill

KRP: Tracy, John W., was established in 1882 at Lineboro, MD. See: Tracy's Mill

KRQ: Trendley, Fredrick, was established in 1841 at Cleveland, OH. The company made 4,000 pounds a day. See: Newton Falls Mill

KRR: Trentmen & Dinnen was established in 1883 at Ft. Wayne, IN. The company employed 20 and made 8,000 pounds a day. See: Fort Wayne Mill

KRS: Trimble, William, was established in 1799 at Chester Creek, PA. See: Trimble Mill

KRT: Truman, Morris, had offices in Upper Darby Township, PA, near Darby Creek. See: Darby Creek Mill

KRU: Trumbull, E.S., was established in 1850 at Gargoa, NY. The company employed 7 and made 3,000 pounds a day. See: Trumbull Mill

KRV: Turnbull, B., was established in 1846 at Clifton, OH. See: Clifton Mill

KRW: Turner's Falls Paper Co. was established in 1879 at Turner Falls, MA. The company employed 60 and made 5 tons a day. See: Turner's Falls Mill

KRX: Tytus Paper Co. was established in 1873 at Middletown, OH. The company made 22,000 pounds a day. See: Tytus Mill

KRY: Uhler, Sidney L., had offices in Uhlersville, PA. The company made 2.5 tons a day. See: Uhler Mill

KRZ: Underhill, R.L., was established in 1820 at Urbana, NY. See: Underhill Mill

KSA: Union Mills Paper Mfg. Co. was established in 1880 at New Hope, PA. The company employed 60 and made 6 tons a day. See: Union Mill

KSB: Union Paper Mfg. Co. was established in 1870 at Holyoke, MA. The company employed 147 and made 4 tons a day. Principals were Henry and Edwin Dickinson, and J.E. Taylor. See: Union Mill

KSC: United Wall Paper Co. was established in 1889. This was the former York Card and Paper Co. See: York Wall Paper Mill

KSD: Unkley, George, was established in 1802 at Lower Saucon Township, PA. See: Unkley Mill

KSE: Updegraff & Walker was established in 1802 at Mt. Pleasant Township, OH. Principals were Nathan Updegraff and Lewis Walker. See: Short Creek Mill

KSF: Urlick, Jacob, was established in 1820 at Upper Oxford Township, PA.

KSG: Valentine, John, was established in 1853 at Chicopee, MA. See: Cox Mill

KSH: Valentine, Myers, had offices in Roslyn, NY. The company employed 6 and made 1 ton a day. See: Valentine Mill

KSI: Valentine, Phillip, & Co. was established in 1829 at Suffield, CT. See: Eagle Mill, Old

KSJ: Valentine, William, & Son had offices in Roslyn, NY. The "and son" was William, Jr. See: Valentine Mill

KSK: Valley Falls Paper Mfg. Co. was established in 1873 at Valley Falls, NY. The company employed 17 and made 6,000 pounds a day. See: Valley Falls Mill

KSL: Valley Paper Co. was established in 1864 at Holyoke, MA. The company employed 250 and made 6 tons a day. See: Valley Mills

KSM: Valley Paper Co. was established in 1876 at Hulton, PA. See: Valley Paper Mill

KSN: Valley Pulp and Paper Co. had offices in Appleton WI, about 1875. The company made 4.5 tons a day. See: Valley Pulp & Paper Mill

KSO: Van Alstyone & Ryan was established in 1883 at Shepardston, WV. The company employed 26 and made 5 tons a day. See: New Dominion Mill

KSP: Van Courtland, General Pierre, was established in 1819 at Annsville, NY. See: Van Courtland Mill

KSQ: Van de Carr, H.S., was established in 1860 at Stockport, NY. The company employed 25 and made 4.5 tons a day. See: Eurica Mill

KSR: Van Nortwick Paper Co. was established in 1869 at Batavia, IL. The company employed 70 and made 24,000 pounds a day. See: Van Nortwick Mills

KSS: Van Reed, Charles L., was established in 1826 at Reading, PA. The company made 2,500 pounds a day. See: Van Reed Mill

KST: Van Reed, John, was established in 1812 at Lower Heidelberg Township, PA. See: Van Reed Mill

KSU: Van Weghten & Son was established in 1808 at Schenectady, NY. See: Schenectady Mill

KSV: Vandalia Paper Mill Co. was established in 1884 at Vandalia, IL. The company employed 31 and made 8 tons a day. See: Vandalia Mill

KSW: Vandusen, S.G., was established in 1880 at Dansville, NY. The company employed 6 and made 1,500 pounds a day. See: California Mill

KSX: Vernon Bros. & Co. was established in 1850 at 63 Duane St., New York. The company made 2 tons a day. See: Russell Mill, Ithaca Falls Mills, Northampton Mill

KSY: Verran, John, was established in 1881 at Lee, MA. The company employed 35 and made 2,000 pounds a day. See: New England Mills

KSZ: Victoria Paper Mills Co. was established in 1880 at Fulton, NY. The company employed 24 and made 5,000 pounds a day. See: Victoria Mill

KTA: Vinton, Timothy, had offices in Brattleboro, VT. The company made 3,000 pounds a day. See: Vinton Mill

KTB: Virginia Paper Co. had offices in Richmond, VA. The company employed 15 and made 3,000 pounds a day. See: Dominion Mill, Old

KTC: Voss, Daniel, was established in 1769 at Milton, MA. See: Neponset Mills

KTD: Wade, W.W. was established in 1835 at Millburn, NJ. See: Wade Mill

KTE: Wade, Izries & Condit was established in 1810 at Millburn, NJ. See: Thylers Paper Mill

KTF: Wade, Jonas, was established in 1810 at Millburn, NJ. See: Rahway River Mill

KTG: Wade, U.N., was established in 1835 at Millburn, NJ. The company made 3,000 pounds a day. See: Wade Mill

KTH: Wagener, D. was established in 1843 at Wheeling, WV. The company employed 15 and made 4,000 pounds a day. See: Buckeye Mill

KTI: Wagman, Thorpe & Co. was established in 1867 at Fort Miller, NY. The company employed 25 and made 7,200 pounds a day. See: Fort Miller Mill

KTJ: Wait, N.W., Son & Co. was established in 1850 at Sandy Hill, NY. The company made 24,000 pounds a day. See: Wait Mills

KTK: Waite, E.B., & Co. was established in 1858 at Little Falls, NY. The company employed 20 and made 4,000 pounds a day. See: Waite Mill

KTL: Wakeman, David, was established in 1870 at Southport, CT, and made 3,000 pounds a day. See: Sasco Creek Mill

KTM: Walbridge, Ebenezer, & Co. was established in 1786 at Bennington, VT. The "and company" were Joseph Hinsdill, Stebins Walbridge, and Gustavus Walbridge (1810). See: Walbridge Mill

KTN: Waldron Paper Co. was established in 1872 at Waldron, IL. See: Waldron Mill

KTO: Waldschmidt, Christian, was established in 1810 at Cincinnati, OH. See: Waldschmidt Mill

KTP: Wales & Mills was established in 1843 at Needham, MA. Principals were Nathaniel Wales and William Mills. See: Jackson Mill

KTQ: Walker, F.R., & Son had offices in Montgomery, NY. The company employed 120 and made 19,000 pounds a day. See: Eagle Mill, Montgomery Mill

KTR: Wall & Davidson was established in 1884 at City Mills, MA. The company made 4,000 pounds a day. See: Elliot Mill

KTS: Wall & Shepherd was established in 1884 at Piqua, OH. The company employed 20 and made 7,000 pounds a day. See: Hydraulic Mill

KTT: Walloomsac Paper Co. was established in 1874 at North Hoosick, NY. The company employed 130 and made 10 tons a day. See: Walloomsac Mill

KTU: Walover, Peter, was established in 1800 at Lower Merion Township, PA. See: Walbridge Mill, Jones Mill, McCleaghan Mill

KTV: Walsh & Craig was established in 1792 at Newburgh, NY. Principals were Hugh Walsh and James Craig. See: Walsh Mill

KTW: Walsh, J. DeWitt, was established in 1859 at Newburgh, NY. The company made 2,500 pounds a day. See: Walsh Mill

KTX: Walsh, John H., was established in 1796 at Newburgh, NY. The company made 2,500 pounds a day. See: Walsh Mill

KTY: Wampler, Lewis, was established in 1818 at Baltimore, MD. See: Wampler Mill

KTZ: Ward Bros. was established in 1878 at Riverton, CT. The company employed 12 and made 3,000 pounds a day. See: Ward Mill (rebuilt 1883)

KUA: Wardlow, Thomas & Co. had offices in Middletown, OH. The company made 12,000 pounds a day. See: Niagara Mill

KUB: Wardwell & Clark was established in 1869 at Newton Upper Falls, MA. The company employed 25 and made 7,000 pounds a day. See: Newton Mills

KUC: Ware, John, was established in 1790 at Newton, MA. See: Ware Mill

KUD: Warner, Newman & Warner was established in 1868 at North Anthony Falls, MN. The company employed 45 and made 12,000 pounds a day. See: Minneapolis Mill

KUE: Warren Mfg. Co. was established in 1873 at Riegelsville, NJ. The company employed 60 and made 14,000 pounds a day. See: Warren Mill

KUF: Warren, S.D., & Co. was established in 1867 at Boston, MA, and closed after 1880. See: Cumberland Mills, Copesecook Mill

KUG: Warren, S.D., was established in 1875 at Gardiner ME. The company made 3.5 tons a day. See: Warren Mill

KUH: Warren, William J., was established in 1881 at W. Dudley, MA. The company employed 12 and made 3,000 pounds a day. See: West Dudley Mill

KUI: Waterbury, L., & Co. had offices at 139 Front St., New York. The company made 10 tons a day. See: Waterbury Mill

KUJ: Watertown Paper Co. had offices in Watertown, NY. The company made 2.5 tons a day. See: Watertown Mill

KUK: Watson & Chamberlin was established in 1858 at Loudville, MA. The company employed 12 and made 1,000 pounds a day. See: Watson Mill

KUL: Watson & Smith had offices in Canaan Four Corners, NY. The company made 1 ton a day. See: Watson Mill

KUM: Watson, Alex. B., had offices in Canaan Four Corners, NY. The company made 1 ton a day. See: Watson Mill

KUN: Watson, H.F., was established in 1881 at

Erie, Erie Co., PA. The company made 40 tons a day. See: Erie Mill

KUO: Watson, John V., was established in 1884 at Loudville, MA. The company employed 12 and made 1,000 pounds a day. See: Watson Mill, Baltimore Mill

KUP: Watts & Barber was established in 1864 at Madison, IN. The company employed 11 and made 3,000 pounds a day. See: Madison Mill

KUQ: Waugh, William, & Bro. was established in 1870 at Fulton, NY. The company made 4,500 pounds a day. See: Oswego River Mill

KUR: Wauregan Paper Co. was established in 1879 at Holyoke, MA. The company employed 130 and made 6 tons a day. Principals were James H. Newton and Edward T. Newton. See: Wauregan Mill

KUS: Waverly Paper Mill had offices in Waverly, NY. The company made 2,800 pounds a day. See: Waverly Mill

KUT: Wead Paper Co. was established in 1881 at Malone, NY. The company employed 100 and made 15,000 pounds a day. See: Malone Mill

KUU: Webb, Joseph, was established in 1799 at Kennett Square, PA. See: Webb Mill

KUV: Webster, Ensign & Seymore was established in 1793 at Troy, NY. See: New York Mill

KUW: Weeden, J.C., & Co. was established in 1863 at Lambertville, NJ. The company employed 21 and made 2,000 pounds a day. See: Mountain Spring Mill

KUX: Weeks, F.G., had offices in Skaneateles Falls, NY. The company employed 60 and made 5 tons a day. See: Dhacutt Mill, Brick Mill

KUY: Weils, R.C., was established in 1867 at Golden, CO. The company made 4,000 pounds a day. See: Golden Mill

KUZ: Wells, Buchanan & Shartle was established in 1860 at Philadelphia, PA. See: Valley Mill (1810), Valley Mill (1833)

KVA: Wells, Charles, & Co. was established in 1861 in Philadelphia. The company employed 10 and made 2,500 pounds a day. See: Valley Mill, Rising Sun Mill (1810), Valley Mill (1833)

KVB: Wemple & Quackenbush was established in 1881 at Sammonsville, NY. The company employed 11 and made 5,000 pounds a day. See: Sammonsville Mill

KVC: West Jersey Paper Mfg. Co. had offices in Camden, NJ. The company made 5,000 pounds a day. See: West Jersey Mill

KVD: West Newton Paper Co. had offices in West Newton, PA. The company made 16 tons a day. See: West Newton Mills

KVE: West Point Mfg. Co. was established in 1875 at West Point NE. The company made 6,000 pounds a day. See: West Point Mill

KVF: West Ware Paper Co. was established in 1884 at W. Ware, MA. The company made 4 tons a day. See: West Ware Mill

KVG: West, George W., was established in 1863 at Ballston Spa, NY. The company employed 350 and made 20 tons a day. See: West Mill, Baker Mill, Eagle Mill, Empire Mill, Excelsior Mill, Fibre Mill, Glen Union Mill, Island Mill, Pioneer Mill, Silver Lake Mill

KVH: West, John, was established in 1809 at Taunton, MA. See: Taunton Mill

KVI: Weston, Byron, was established in 1863 at Dalton, MA. The company employed 200 and made 4,000 pounds a day. See: Defiance Mill, Centennial Mill

KVJ: Weston, Franklin, was established in 1863 at Dalton, MA. See: Defiance Mill

KVK: Weston, J.D., was established in 1840 at Saugerties, NY. See: Weston Mill

KVL: Westover & Foster was established in 1867 at Richmondville, NY. The company employed 10 and made 3,500 pounds a day. See: Empire State Mill

KVM: Westville Mfg. Co. was established in 1837 at Taunton, MA. Principal was Eldridge Clark. See: Park Mill

KVN: Wetherell & Carpenter was established in 1871 at Newington, CT. The company made 1 ton a day. See: Rockdale Mill

KVO: Weymouth Paper Mills was established in 1860 at Box 375 New York, NJ. The company employed 50 and made 8,000 pounds a day. See: Weymouth Mill, Atlantic Mill

KVP: Wheeler, C., was established in 1846 at, CT. See: Wheeler Mill

KVQ: Wheeler, F.A., & Co. had offices in Montclair, NJ. The company made 10,000 pounds a day. See: Watehung Mill

KVR: Wheeler, Thomas & Backus was established in 1835 at Chatham, NY. See: Wheeler Mill

KVS: Wheelwright, George W., Paper Co. was established in 1858 at Boston, MA. The company made 10,000 pounds a day. See: Wheelwright Mill, Rollstone Mill, Rollstone Mill, Old

KVT: White & Co. was established in 1840 at Hohokus, NJ. The company employed 12 and made 1,200 pounds a day. See: Union Mill

KVU: White & Gale was established in 1817 at Newbury, VT, with Ira White as stationer and bookseller. See: Newbury Mill

KVV: White & Keeney had offices in N. Manchester, CT, and made 1 ton a day. See: Hartford Mill

KVW: White Mills Paper Co. had offices in Chatham, NY. The company made 3.5 tons a day. See: White Mills

KVX: White Paper Co. was established in 1866 at Suffield, CT. The company employed 25 and made 4,000 pounds a day. Principal was Daniel White. See: Suffield Mill, White Mill

KVY: White, James & Son was established in 1862 at Bloomingdale, NJ. The company employed 9 and made 1,000 pounds a day. See: Rahway River Mill, Bloomingdale Paper Works

KVZ: Whiteman, William S., was established in 1806 at Knoxville, TN. See: Middle Brook Creek Mill

KWA: Whiteman, W.S. II, & Co. was established in 1838 at Nashville, TN. See: Cumberland River Mill

KWB: Whiteman, William S. II, was established in 1850 at Nashville, TN. See: Whites Creek Mill

KWC: Whiteman, William S. III, was established in 1861 at Nashville, TN. See: Stone Fort Mill, Old, Nashville Mills

KWD: Whiteside, J.H. & A., was established in 1870 at Champlain, NY. The company employed 21 and made 9,500 pounds a day. See: Champlain Straw Board Mill

KWE: Whitewater Paper Mill Co. had offices in Whitewater, WI. The company made 1.5 tons a day. See: Whitewater Mill

KWF: Whiting Paper Co. was established in 1865 at Holyoke, MA. The company employed 420 and made 30,000 pounds a day. Principals were William Whiting and son, William L. Whiting. See: Whiting Mill No. 2, Whiting Mill No. 1

KWG: Whitman & Porter was established in 1851 at Agawam, MA. See: Agawam Mill

KWH: Whitmore Paper Co. was established in 1881 at Holyoke, MA. See: Whitmore Mill

KWI: Whiton, James, & Son was established in 1831 at East Lee, MA. See: Forest Mill, Old

KWJ: Whittlesey Paper Co. had offices in Windsor Locks, CT, about 1810. See: Whittlesey's Tissue Mill

KWK: Whittlesey, Frank R., was established in 1872 at Windsor Locks, CT. See: Whittlesey's Tissue Mill

KWL: Wilcox, James M., & Co. was established in 1729 at Glen Mills, PA. The company employed 75 and made 75,000 pounds a day. See: Ivy Mill, Glen Mill

KWM: Wilder & Co. was established in 1881 at Boston. The company made 10 tons a day. See: Wilder Mills

KWN: Wilkes Barre Paper Mfg. Co. was established in 1883 at Wilkes Barre, PA. The company employed 2 and made 6,000 pounds a day. See: Wilkes Barre Mill

KWO: Wiley & Button had offices in Schaghticoke, NY. The company employed 21 and made 3.5 tons a day. See: Empire Mills

KWP: Wiley, Clarence A., had offices in Portsmouth, OH. The company made 2,000 pounds a day, and was out of business by 1885. See: Wiley Mill

KWQ: Wilkinson & Phelps was established in 1884 at Banning, GA. The company employed 14 and made 3,000 pounds a day. See: Wilkinson Mill

KWR: Wilkinson Bros. & Co. was established in 1872 at 72-74 Duane St., New York. The company employed 80 and made 10 tons a day. See: Derby Mills

KWS: Wilkinson, J.R., & Co. was established in 1870 at Banning, GA. The company employed 14 and made 3,000 pounds a day. See: Wilkinson Mill

KWT: Willamette River Pulp and Paper Co. was established in 1889 at Oregon City, OR. See: Willamette Mill, Columbia Mill

KWU: Willcox Paper Co. was established in 1851 at Ivy Mills, PA. See: Ivy Mill, Glen Mill

KWV: Willcox, John, & Son was established in 1765 at Upper Providence Township, PA. The "and son" was Mark Willcox. See: Ridley Creek Mill, Ivy Mill

KWW: Willcox, Mark, was established in 1779 at Ivy Mills, PA. See: Ivy Mill

KWX: Willcox, Thomas M., was established in 1729 at Ivy Mills, PA. See: Ivy Mill

KWY: Williams & Co. was established in 1876

at Dayton, IL, but closed after 1880. See: Dayton Mill

KWZ: Williams, C.K., was established by 1823 at Claremont, NH. See: Williams Mill

KXA: Williams, B.F., was established in 1860 at Schuylkill, PA. The company employed 2 and made 1,000 pounds a day. See: Williams Mill

KXB: Williams, S.W., had offices in Streator, IL, and closed after 1880. See: Pembroke Mill, Streator Mill

KXC: Wilson & Bird was established in 1828 at Troy, NY. See: Troy City Mill

KXD: Wilson, Jabez, had offices in Ercildoun, PA. The company made 1 ton a day. See: Wilson Mill

KXE: Wilson, Osborne, & Gibbs was established in 1847 at Pittsfield, MA. See: Crane Bros. Mill, Colt Mill

KXG: Winchester Paper Co. was established in 1870 at Winchester, VA. The company employed 50 and made 7 tons a day. See: Winchester Mill

KXH: Windham Paper Co. was established in 1876 at Chaplin, CT. See: Windham Paper Mill

KXI: Wingate & Cummings was established in 1880 at Mount Vernon, PA. The company employed 2 and made 1,000 pounds a day. See: Mt. Vernon Mill

KXJ: Winnebago Paper Mills was established in 1874 at Neenah, WI. The company made 14,000 pounds a day. See: Winnebago Mills

KXK: Winnipiseogee Paper Co. was established in 1870 at 53 Devonshire St., Boston. The company employed 250 and made 20 tons a day. See: Winnipiseogee Mill

KXL: Winona Paper Co. was established in 1880 at Holyoke, MA. The company employed 200 and made 24,000 pounds a day. See: Winona Mill

KXM: Winton, C.E., was established in 1883 at Hotchkissville, CT. The company made 4,000 pounds a day. See: Waterbury Mill

KXN: Wise, William, had offices in White Hall, MD. The company made 2,000 pounds a day. See: Whitehall Mill

KXN.1: Wiswall, A.C., was established in 1860 at Lincolntown, NC. See: Lincolnton Mill

KXO: Wiswall, A.C., & Son was established in 1869 at Needham, MA. The company employed 12 and made 4,000 pounds a day. See: Wiswall's Mill

KXP: Wiswall, Thomas, & Co. was established in 1823 at Northumberland, NH. See: Northumberland Mill

KXQ: Witmer, Abraham, had offices in Carlisle, PA. The company made 4,000 pounds a day. See: Middlesex Mill

KXR: Witmer, R.S., had offices in Carlisle, PA. The company made 4,000 pounds a day. See: Middlesex Mill

KXS: Wolverine Car Roofing & Mfg. Co. was established in 1881 at Detroit, MI. The company made 8,000 pounds a day. See: Wolverine Mill

KXT: Wood & Reddington was established in 1809 at Schoharie Bridge, NY. See: Great Western Mill

KXU: Woodruff Paper Co. had offices in Dansville, NY. The company employed 40 and made 2 tons a day. See: Straw Pulp Mill

KXV: Woodward, Joseph A., was established in 1885 at New Hope, PA. The company employed 4 and made 1 ton a day. See: Clear Spring Mill

KXW: Woodworth, Oliver, was established in 1832 at New London, CT. The company employed 14 and made 4,000 pounds a day. See: Waterford Mill, Old

KXX: Worth, Davis R., was established in 1870 at Mount Vernon, PA. See: Stone Mill

KXY: Worthington, R. & I., was established in 1872 at Cooperstown, NY, as bankers. See: Otsego County Paper Works

KXZ: Worthy Paper Co. was established in 1872 at Mittineague, MA. The company employed 75 and made 6,000 pounds a day. See: Worthy Mill

KYA: Wright, John, was established in 1873 at Kennett Square, PA. The company employed 4 and made 4,000 pounds a day. See: Willow Brook Mill

KYB: Wyman, Franklin had offices in Westminster, MA. The company made 10,000 pounds a day. See: Wyman, Franklin Mills

KYC: Xenia Paper Co. was established in 1882 at Xenia, OH. The company employed 25 and made 8,000 pounds a day. See: Xenia Mill

KYD: Yantic Paper Co. was established in 1873 at Greenville, CT. The company employed 15 and made 3,000 pounds a day. See: Yantic Mill

KYE: Yarnall, Issac, was established in 1800 at Great Crossing, KY. See: Yarnall Mills

KYF: Yoran, Levi, was established in 1884 at Fort Plain, NY. The company employed 7 and made 5,000 pounds a day. See: Yoran Mill

KYG: York Card and Paper Co. was established in 1888. This was originally the A.A. Yerks Wall Paper Co. See: York Wall Paper Mill

KYH: York Haven Paper Mill was established in 1885 at 429 Walnut St., Philadelphia, PA. The company made 10 tons a day. See: York Haven Mill

KYI: Young & Humphrey had offices in Napanoch, NY. The company made 5,000 pounds a day. See: Ulster Mill

KYJ: Young, D. & J.B., was established in 1881 at Bentley Springs, MD. The company made 2,000 pounds a day. See: Young Mill

KYL: Young, D.S., & Sons was established in 1884 at Ercildoun, PA. The company employed 4 and made 2,000 pounds a day. See: Buck Run Mill

KYM: Young, G.F., was established in 1880 at Atglen, PA. The company employed 2 and made 1,000 pounds a day. See: Lyndonette Mill, Old

KYN: Young, George F., was established in 1870 at Atglen, PA. See: Fern Rock Mill

KYO: Young, William, was established in 1793 at Rockland, DE. See: Delaware Mills

KYP: Ypsilanti Paper Co. had offices in Ypsilanti, MI. The company made 9 tons a day. See: Ypsilanti Mill, Superior Mill

KYQ: Zellerbach, A., & Sons was established in 1899 at San Francisco, CA. See: Zellerbach Mill

Chapter Notes

Chapter 1

1. Dard Hunter, *Papermaking: The History and Technique of an Ancient Craft*, 2nd ed., 1978, 241–274.
2. Ibid.
3. Joel Munsell, *Chronology of Paper and Paper Making*, 1876, 81–84.
4. Owen and Hurlbut (hereafter cited as O&H), correspondence, 37.01.11.
5. Lyman Horace Weeks, *A History of Paper Manufacture in the U.S., 1690–1916*, 176.
6. Hunter, *Papermaking*, 309.
7. Hunter, *Papermaking*, 520
8. Munsell, *Chronology of Paper*, 54.
9. O&H, 37.09.13; 43.12.04.
10. This and the preceding quote are all from O&H, 27.08.04.
11. O&H, 41.09.10; 39.12.09; 41.12.25; 42.11.02; 38.09.06.
12. O&H, 36.06.23; 42.02.12; 38.02.26; 31.06.22; 36.06.23.
13. O&H, 46.07.01; 44.08.27.
14. O&H, 37.11.14.
15. Tileston and Hollingsworth (hereafter cited as T&H), correspondence, 46.01.24; 47.02.13.
16. O&H, 36.09.03; 37.12.08.
17. T&H, 50.09.24.
18. Munsell, *Chronology of Paper*, 65–170.
19. O&H, 45.12.25; 52.04.15.

Chapter 2

1. Harrison Elliot, "Benjamin Franklin, Paper Mill Promoter and Patron," *The Paper Maker*, Vol. 20, 1951, 35–40.
2. Weeks, *A History of Paper Manufacture*, 92.
3. William Bond Wheelwright, "The Gilpins of Delaware," *The Paper Maker*, Vol. 10, 1941, 6–11.
4. Sidney M. Edlestein, "Papermaker Joshua Gilpin Introduces the Chemical Approach to Papermaking in the United States," *The Paper Maker*, Vol. 30, 1961, 3–12.
5. Ibid.
6. Wheelwright, "The Gilpins of Delaware."
7. Robert H. Clapperton, *The Papermaking Machine*, 331–334.
8. Joan Evans, *The Endless Web, 1804–1954*, 45–57.
9. Wheelwright, "The Gilpins of Delaware."
10. A.J. Ward, "John Dickinson and the Brandywine," from *History of Herfordshire* Web site, www.hertfordshiregenealogy.co.uk, March 2006.
11. Wheelwright, "The Gilpins of Delaware."
12. Weeks, *A History of Paper Manufacture*, 115–126.
13. Ibid.
14. Ibid.
15. John W. Maxson, Jr., "Coleman Sellers: Machine Maker to American's First Mechanized Paper Mills," *The Paper Maker*, Vol. 30, 1961, 13–27.
16. Ibid.
17. McGaw, *Most Wonderful Machine*, 162–163.
18. O&H, 31.01.05.
19. Harold B. Hancock and Norman B. Wilkinson, "The Gilpins and Their Endless Papermaking Machine," *The Pennsylvania Magazine of History and Biography*, Vol. 5, 1957, 391–405.
20. Weeks, *A History of Paper Manufacture*, 176–177.

Chapter 3

1. Munsell, *Chronology of Paper*, 54–55.
2. McGaw, *Most Wonderful Machine*, 71–73.
3. C.M. Hyde and Alexander Hyde, *The Centennial Celebration and Centennial History of the Town of Lee, Mass.*, 1878, 304–312.
4. O&H, 22.04.30.
5. McGaw, *Most Wonderful Machine*, 61–62.
6. Ibid.
7. O&H, 27.12.09.
8. O&H, 31.10.22; 36.08.11; 37.11.12; 38.09.11; 39.01.22; 39.02.20; 39.04.15.
9. O&H, 37.12.05.
10. McGaw, *Most Wonderful Machine*, 39–40.
11. Leo L. Lincoln and Lee C. Drickamer, *Postal History of Berkshire County, Massachusetts, 1790–1981*, 165–170.
12. O&H, 41.01.10.
13. O&H, 41.04.10; 42.08.05; 43.10.02.
14. O&H, 45.02.25; 46.05.05; 45.04.21; 45.04.29.
15. O&H, 44.10.14.
16. O&H, 46.07.14.
17. O&H, 50.05.29.
18. Herman Melville, "The Paradise of Bachelors and the Tartarus of Maids," *Herman Melville: Selected Tales and Poems*, ed. Richard Chase.

Chapter 4

1. Elliot, "Benjamin Franklin, Paper Mill Promoter."
2. Hunter, *Papermaking*, 128.

3. Maxson, "Coleman Sellers: Machine Maker."
4. O&H, 29.07.20.
5. Arthur E. James, "The Paper Mills of Chester County, Pennsylvania, 1779–1967," *The Paper Maker*, Vol. 39, 1970, 3–18.
6. Ibid.
7. O&H, 45.02.12; 52.01.21.
8. Maxson, "Coleman Sellers: Machine Maker."
9. Weeks, *A History of Paper Manufacture*, 139–140.
10. Ibid.
11. T&H, 46.08.31; 44.11.30.
12. T&H, 45.04.07; 47.08.04.
13. This and the several paragraphs that follow are from Clarence A. Wiswall, *One Hundred Years of Paper Making: A History of the Industry on the Charles River at Newton Lower Falls Massachusetts.*
14. O&H, 52.04.27; 52.08.02.
15. This and the several paragraphs that follow are from John W. Maxson, Jr., "American Papermakers in the Great Tariff Debate," *The Paper Maker*, Vol. 31, 1962, 14–34.
16. O&H, 42.05.10; 55.11.12; 56.11.24.
17. Weeks, *A History of Paper Manufacture*, 130–131.
18. B.G. Watson, "Bryan Donkin: Pioneer Paper-Machine Manufacturer," *The Paper Maker*, Vol. 35, 1966, 21–32.
19. Smith, Winchester and Co. (hereafter cited as S&W), correspondence, 52.01.03. O&H, 51.10.30.
20. O&H, 52.05.27; 52.09.08.
21. O&H, 32.02.20; 31.03.10.
22. O&H, 40.05.19.
23. O&H, 45.12.24.
24. O&H, 47.03.21.
25. O&H, 53.11.01.
26. O&H, 54.01.18.

Chapter 5

1. Letter of J. Bingley Blake to Kilmer and Ashton, dated March 31, 1848.
2. Munsell, *Chronology of Paper*, 116–118 and 150–156.
3. Munsell, *Chronology of Paper*, 120–164.
4. Weeks, *A History of Paper Manufacture*, 196.
5. Ibid.
6. This and succeeding paragraphs are from Munsell, *Chronology of Paper*, 127–133.
7. O&H, 52.05.24; 48.01.28.
8. O&H, 30.06.04; 37.05.13; 37.05.26.
9. O&H, 30.07.14.
10. O&H, 43.07.23.
11. O&H, 31.01.20.
12. This Knowlton and Rice letter along with several more appearing in this section are from Smith, *History of Papermaking in the United States*, 32–37.
13. R.M. Snell, *The Story of Papermaking in the United States*, 112.
14. Ibid.
15. Receipts in the author's collection.
16. O&H, 32.06.13; 37.04.05.
17. O&H, 38.09.06.

18. O&H, 38.06.06.
19. O&H, 36.02.16.
20. Munsell, *Chronology of Paper*, 139–164.
21. O&H, 58.07.02; 59.02.19.
22. O&H, 45.05.15; 48.02.19, and the 1846 letter is printed in Ralph Snell, ed., *Hurlbut's Papermaker Gentleman* (4 vols.) January 1933–September 1935.
23. Cyrus Field, letter, 50.09.03 (hereafter cited as Field).
24. Field, 54.05.27.
25. This missive from the author's collection.

Chapter 6

1. This and much of the information in the rest of this section are from McGaw, *Most Wonderful Machine.*
2. Hyde, *Lee: The Centennial Celebration*, 236–280.
3. O&H, 34.05.27; 31.03.28.
4. O&H, 31.03.21.
5. O&H, 31.06.29.
6. O&H, 31.01.05; 33.08.01.
7. O&H, 37.02.15; 37.01.07.
8. O&H, 37.02.15; 37.02.27.
9. O&H, 37.02.27.
10. O&H, 37.02.27.
11. O&H 37.05.03.
12. McGaw, *Most Wonderful Machine*, 73–74.
13. O&H, 37.01.11.
14. John Butler, letter, 38.07.27.
15. O&H, 37.05.03.
16. O&H, 38.02.27; 38.08.13.
17. O&H, 39.04.11; 39.08.05.
18. O&H, 39.02.09.

Chapter 7

1. This and much that follows are from R.H. Clapperton, "The Invention and Development of the Endless Wire, or Fourdrinier, Paper Machine," *The Paper Maker*, Vol. 23, 1954, 1–17.
2. B.G. Watson, "Bryan Donkin: Pioneer Paper-Machine Manufacturer," *The Paper Maker*, Vol. 35, 1966, 21–32.
3. Ibid.
4. Evans, *The Endless Web*, 42.
5. Edwards, "Connecticut Paper Mills."
6. O&H, 28.05.05.
7. This and several paragraphs to follow are from Clapperton, *The Papermaking Machine.*
8. Munsell, *Chronology of Paper*, 108.
9. O&H, 47.07.09.
10. Smith, Winchester and Co. (hereafter cited as S-W), correspondence, 51.09.03.
11. S-W, 51.04.17.
12. O&H, 43.11.27.
13. This and several paragraphs to follow are from Clapperton, *The Papermaking Machine.*
14. Frances Edwards, "Connecticut Paper Mills: The Franklin Mill in Suffield," *The Paper Maker*, Vol. 35, 1966, 11–16.

15. S-W, 52.09.02.
16. This and the rest of this section are from McGaw, *Most Wonderful Machine*.

Chapter 8

1. Hunter, Papermaking, 243–244.
2. O&H, 51.01.21.
3. Norris F. Schneider, "Pioneer Zanesville Paper Mill Once City's Leading Industry," *Zanesville Sunday Times Signal*, April 27, 1958.
4. Weeks, *A History of Paper Manufacture*, 205.
5. This and much of the rest of the section are from R.M. Snell, *The Story of Papermaking in the United States*.
6. Ibid.

Chapter 9

1. Hunter, *Papermaking*, 266
2. O&H, 49.02.02.
3. O&H, 57.03.27.
4. "Rags to Riches: The Story of Virginia Papermaking," *Arts in Virginia*, Vol. 4, No. 3, Spring 1964, 1–26.
5. T&H, 45.08.09.
6. Smith, *History of Papermaking in the United States*, 107–108.
7. Munsell, *Chronology of Paper*, 171–195.
8. This and much of this section are from Wheelwright, "A Strange Chapter in Southern Papermaking."
9. "Rags to Riches: The Story of Virginia Papermaking."
10. Weeks, *A History of Paper Manufacture*, 181–184.

Chapter 10

1. This and much of the early history of the Willcoxes are from Hunter, *Papermaking*, 250–297.
2. This and much of the early history of the Cranes are from McGaw, *Most Wonderful Machine*.
3. McGaw, *Most Wonderful Machine*, 177.
4. Pierce, *The First 175 Years of Crane Papermaking*.
5. R.R. Bowker, "Great American Industries, VI: A Sheet of Paper," *Harper's*, June 1887.
6. Crane and Co. circular dated Nov. 1, 1845, and postmarked Dalton, Mass., May 20, with the note "$5 of the Hudson River Bank under date of July 1, 1843."
7. Winthrop S. Boggs, "Ten Decades Ago: 1840–1850," *American Philatelic Society*, 1949.
8. Wells and Boston, *The Annual of Scientific Discovery or Year-book of Facts in Science and Art for 1856*.
9. Crane, correspondence of Crane Museum, Dalton, Mass. (hereafter cited as C&C), 52.05.03.
10. C&C, 58.07.19.
11. O&H, 52.06.10.
12. C&C, 61.07.09.
13. This and much of the rest on Crane and Co. are from Wadsworth R. Pierce, *The First 175 Years of Crane Papermaking*, 1977.
14. Ralph Snell, ed., "Founders of American Paper Industry: Rittenhouse, DeWees, Brown, Willcox, Ivy and Glen Mills," *Superior Facts*, Vol. 2, No. 6, December 1928.
15. R.R. Bowker, "Great American Industries VI: A Sheet of Paper," *Harpers*, June 1887.
16. Information on the Oakland Mill courtesy of Manchester Historical Society. The treasury department's paper mill from Hunter, *Papermaking*, 565–66.
17. McGaw, *Most Wonderful Machine*, 181–183.
18. Pierce, *The First 175 Years of Crane Papermaking*, 33–34.
19. Willcox, "Paper Making in the United States."

Chapter 11

1. Much of the information in this chapter is courtesy of Holyoke Historical Society, www.holyokemass.com.
2. O&H, 85.10.28.

Chapter 12

1. This and much of the information on Smith Paper in the rest of this chapter are from McGaw, *Most Wonderful Machine*.
2. O&H, 48.05.10, 47.04.24, 49.02.07.
3. O&H, 38.03.22.
4. Munsell, *Chronology of Paper*, 127–175.
5. This and much information on pioneering mills are from Hunter, *Papermaking*, 490–491.
6. Hunter, *Papermaking*, 310.
7. This and much of the information on Vermont are from Marcus A. McCorison, "Vermont Papermaking 1784–1820," *The Paper Maker*, Vol. 33, 1964, 19–28, 23–31.
8. www.franklinnh.org.

Chapter 13

1. This and much of the information on the Palmers Falls mills are from www.hudsonrivermillproject.org, presented by the Corinth Museum and Hudson River Mill Historical Society.
2. McGaw, *Most Wonderful Machine*, 214.
3. O&H, 42.07.07.
4. O&H, 42.09.19.
5. O&H, 43.09.08.
6. O&H, 43.12.04.
7. O&H, 38.09.07.
8. O&H, 29.07.20; 30.08.19. Also see Maxson, "Coleman Sellers: Machine Maker."
9. O&H, 45.05.15, 52.01.01.
10. O&H, 53.11.12, 54.02.27, 54.09.19, 56.06.12.
11. O&H, 51.02.14.
12. Charles Thomas Davis, *Manufacture of Paper*, 204–219.
13. O&H, 54.09.19.

14. Seth Hayden, "An Example to Manufacturers (A Visit to a Paper Mill in 1871)," *The Paper Maker*, Vol. 34, 1965, 41–46.

Chapter 14

1. The information about Fox River Valley derived from source material published by *Fox Cities Online*, www.focol.org.
2. O&H, 43.10.23.
3. William B. Beatty, "Early Papermaking in Utah," *The Paper Maker*, Vol. 28, 1959, 9–20.
4. Ibid.
5. Florence Donelly, "Pioneer Paper Mill of the West," *The Paper Maker*, Vol. 18, 1949, 15–19.
6. Ibid.
7. This and the rest of this section from Florence Donelly, "The Beautiful Mill," *The Paper Maker*, Vol. 19, 1950, 23–31.
8. Florence Donelly, "Camas Paper Mill: First in Washington," *The Paper Maker*, Vol. 29, 1960, 23–32.
9. Florence Donelly, "A Fortune from Straw," *The Paper Maker*, Vol. 19, 1950, 11–17.

Bibliography

Books

Bryan, Clark W. *The Paper Mill Directory of the World*, Springfield, MA: C.W. Bryan, 1885.
Clapperton, Robert H. *The Papermaking Machine*, New York: Pergamon, 1967.
Davis, Charles Thomas. *Manufacture of Paper*, Philadelphia: H.C. Baird, 1886.
Evans, Joan. *The Endless Web 1804–1954*, Westport, CT: Greenwood, 1955.
Gravell, Thomas L., George Miller, and Elizabeth Walsh. *American Watermarks 1690–1835*, New Castle, DE: Oak Knoll Press, 2002.
Harper, Wyatt E. *The Story of Holyoke*, Holyoke, MA: Centennial Committee, 1959.
Herring, Richard. *Paper and Paper Making, Ancient and Modern*, London: Longmans, Green, Longman, Robert & Green, 1863.
Hofmann, Carl. *Manufacture of Paper in All Its Branches*, Philadelphia: H.C. Baird, 1873.
Hunter, Dard. *Papermaking: The History and Technique of an Ancient Craft*, 2nd ed., New York: Dover Publications, 1978.
Hyde, C.M., and Alexander Hyde. *The Centennial Celebration and Centennial History of the Town of Lee, Mass.*, Springfield, MA: C.W Bryan, 1878.
Lincoln, Leo L., and Lee C. Drickamer. *Postal History of Berkshire County, Massachusetts, 1790–1981*, Williamstown, MA: L.C. Drickamer, 1982.
Macdonald, Geo. S., ed. *250 Years of Paper Making in America*, East Stroudsburgh, PA: Lockwood Trade Journal, 1940.
Mandl, G.T. *Three Hundred Yeas in Paper*, London: G.T. Mandl, 1985.
McGaw, Judith A. *Most Wonderful Machine: Mechanization and Social Change in Berkshire Paper Making, 1801–1885*, Princeton, NJ: Princeton University Press, 1987.
Munsell, Joel. *Chronology of Paper and Paper Making*, Albany, NY: J. Munsell, 1876.
Pierce, Wadsworth R. *The First 175 Years of Crane Papermaking*, North Adams, MA: Excelsior, 1977.
Post, L.D. *Post's Paper Mill Directory 1905–1906*, New York: L.D. Post, 1905.
Seitz, May A. *The History of the Hoffman Paper Mills in Maryland*, Baltimore: Holiday Press, 1946.
Scull, Penrose. *Papermaking in the Berkshires: The Story of the Hurlbut Paper Company*, South Lee, MA: Hurlbut Paper Co., 1956.
Smith, David C. *History of Papermaking in the United States (1691–1969)*, New York: Lockwood, 1970.
Snell, R.M. *The Story of Papermaking in the United States*, Holyoke, MA: Paper Makers Chemical Corp., 1929.
Snell, Ralph, ed. *Hurlbut's Papermaker Gentleman* (4 vols.) South Lee, MA: Hurlbut Paper Co., January 1933–September 1935.
Snell, Ralph, ed. *Superior Facts* (5 vols.), South Lee, MA: Hurlbut Paper Co., 1928–1933.
Stephenson, Louis T. *Economics of the Paper Industry*, Westport, CT: Greenwood Press, 1940.
Tague, William H., and Robert B. Kimball, eds. *Berkshire, The First Three Hundred Years: 1676–1976*, Pittsfield, MA: Eagle Publishing, 1976
Teele, Albert K. *The History of Milton, Mass.*, Boston: Press of Rockwell and Churchill, 1887.
Weeks, Lyman Horace. *A History of Paper Manufacture in the U.S. 1690–1916*, New York: Lockwood, 1916.
Wells, David A., and A.M. Boston, eds. *The Annual of Scientific Discovery or Year-book of Facts in Science and Art for 1856*. Boston: Gould, Kendall and Lincoln, 1856, 84.
Wilhoit, Megan Murray. *From Appleton to Atlanta: The Institute's First 75 Years*. Sarasota, FL: Sun Fung Museum Catalogs and Books, 2008.
Wiswall, Clarence A. *One Hundred Years of Paper Making: A History of the Industry on the Charles River at Newton Lower Falls, Massachusetts*, Reading, MA: Reading Chronicle Press, 1938.
Woodside, Laurence M., ed. *Paper Machine Felts: Their Manufacture and Application for Improved Papermaking*, Albany: Albany Felt, 1967.

Articles

Beatty, William B. "Early Papermaking in Utah." *The Paper Maker*, Vol. 28, 1959, 9–20.
Bowker, R.R. "Great American Industries VI: A Sheet of Paper." *Harper's*, June 1887.
Carny, Charles T. "Manufacture of Banknote Paper." *Annual of Scientific Discovery*. Boston: Gould and Lincoln, 1855.
Clapperton, R.H. "The Invention and Development of the Endless Wire, or Fourdrinier, Paper Machine." *The Paper Maker*, Vol. 23, 1954, 1–17.
Donelly, Florence. "The Beautiful Mill." *The Paper Maker*, Vol. 19, 1950, 23–31.
_____. "Camas Paper Mill: First in Washington." *The Paper Maker*, Vol. 29, 1960, 23–32.
_____. "A Fortune from Straw." *The Paper Maker*, Vol. 19, 1950, 11–17.
_____. "Pioneer Paper Mill of the West." *The Paper Maker*, Vol. 18, 1949, 15–19.

_____. "The San Lorenzo Paper Mill." *The Paper Maker*, Vol. 20, 1953, 23–32.
Edlestein, Sidney M. "Papermaker Joshua Gilpin Introduces the Chemical Approach to Papermaking in the United States." *The Paper Maker*, Vol. 30, 1961, 3–12.
Edwards, Frances. "Connecticut Paper Mills: The Franklin Mill in Suffield." *The Paper Maker*, Vol. 35, 1966, 11–16.
Elliot, Harrison. "Benjamin Franklin, Paper Mill Promoter and Patron." *The Paper Maker*, Vol. 20, 1951, 35–40.
Elliot, Harrison. "The First Paper Mill in New York." *The Paper Maker*, Vol. 22, 1953, 25–30.
Halley, R.A. "Paper-making in Tennessee." *The American Historical Magazine and Tennessee Historical Society Quarterly*, Vol. IX, 1904, 213–216.
Hancock, Harold B., and Norman B. Wilkinson. "The Gilpins and Their Endless Papermaking Machine." *The Pennsylvania Magazine of History and Biography*, Vol. 5, 1957, 391–405.
Hancock, Harold B., and Norman B. Wilkinson. "Thomas and Joshua Gilpin Papermakers." *The Paper Maker*, Vol. 27, 1958, 1–11.
Hayden, Seth. "An Example to Manufacturers (A Visit to a Paper Mill in 1871)." *The Paper Maker*, Vol. 34, 1965, 41–46.
Hayden, Seth. "Papermaking in Chambersburg, Pennsylvania," *The Paper Maker*, Vol. 37, 1968, 275–290.
James, Arthur E. "The Paper Mills of Chester County Pennsylvania 1779–1967," *The Paper Maker*, Vol. 39, 1970, 3–18.
Kennedy, Joseph S. "Once Upon a Time." *Philadelphia Inquirer*, April 23, 1995.
Maxson, John W., Jr. "American Papermakers in the Great Tariff Debate." *The Paper Maker*, Vol. 31, 1962, 14–34.
_____. "Coleman Sellers: Machine Maker to American's First Mechanized Paper Mills." *The Paper Maker*, Vol. 30, 1961, 13–27.
_____. "Papermaking in America: From Art to Industry, 1690 to 1860." *The Quarterly Journal of the Library of Congress*, Vol. 25, 1968, 116–129.
McCorison, Marcus A. "Vermont Papermaking 1784–1820." *The Paper Maker*, Vol. 33, 1964, 19–28, 23–31.
Melville, Herman. "The Paradise of Bachelors and the Tartarus of Maids." *Herman Melville: Selected Tales and Poems*, ed. Richard Chase, 1966.
"Rags to Riches: The Story of Virginia Papermaking." *Arts in Virginia*, Vol. 4, No. 3, Spring 1964, 1–26.
Schneider, Norris F. "Pioneer Zanesville Paper Mill Once City's Leading Industry." *Zanesville Sunday Times Signal*, April 27, 1958.
Shaw, Merle B., and George W. Bicking. "Research on the Production of Currency Paper in the Bureau of Standards Experimental Paper Mill." *Technologic Papers of the Bureau of Standards*, T329, Washington, DC, June 1926, 94–120.
Snell, Ralph, ed. "Founders of American Paper Industry: Rittenhouse, DeWees, Brown, Willcox, Ivy and Glen Mills," *Superior Facts*, Vol. 2, No. 6, December 1928.
Voorn, Henk. "On the Invention of the Hollander Beater." *The Paper Maker*, Vol. 25, 1956, 1–9.
Watson, B.G. "Bryan Donkin: Pioneer Paper-Machine Manufacturer." *The Paper Maker*, Vol. 35, 1966, 21–32.
Watson, B.G. "John Dickinson and This Paper Machine." *The Paper Maker*, Vol. 36, 1967, 1–7.
Weston, Bryon. "History of Papermaking in Berkshire County, Massachusetts." *Collections of the Berkshire Historical and Scientific Society*, Vol. 11, 1895.
Wheelwright, William Bond. "The Gilpins of Delaware." *The Paper Maker*, Vol. 10, 1941, 6–11.
Wheelwright, William B. "A Strange Chapter in Southern Papermaking." *The Paper Maker*, Vol. 2, 1941, 2–5.
Willcox, James M. "Paper Making in the United States." *Report to the Commissioner of Patents for the Year 1850*, Ex. Doc. No. 32, pp. 405.

Web Sites

The internet has been valuable in identifying a number of paper mills. As web pages come and go the writer updates www.ajvalente.com as a clearinghouse of active links. The following links were active as of this writing:

http://www.holyokemass.com provided a great deal of source information on Holyoke manufacturing.
http://www.franklinnh.org City of Franklin Fire Department provided details of a tragic paper mill fire.
http://www.hudsonrivermillproject.org Hudson River Mill Project funded by a grant from the New York Council for the Humanities did a superb job reporting on early events at this mill.
http://www.focol.org Fox Cities Online local historians posted papermaking related articles directly on-line.

Index

Numbers in ***bold italics*** indicate pages with illustrations.

Acme Felt Company 118
Adams, Peter 8
Adams & Bishop 8
Akron Woolen and Felt Company 118
Albany Argus 158
Albany, New York 64
Albany Register 39
Albion Paper Company 144
Aling & Cory 189
Aling, William 66
Allen, Joseph 100
Allen, Richard N. 123–124
Allen & Co. 100–101
Alston, M.P 106
Alta Californian 191
alum 34; in beater 111
American Philosophical Society 19
American Watchman 19
American Writing Paper Co. 179
Ames, Col. David 20, 121
Ames, David, Jr. 20, 24, 78
Ames, John 20, 24, 83–84, 135
Amies, Thomas 39, 53
Amoskeag Paper Mills Co. 163
aniline 173
Annis, Thomas 48
anti-tariff convention 53
Appleton Woolen Company 118
Apsley Mill 16, 18–19
Athens Mill 114
Auburn Mill 68
Augustine, Richard Smith 70
Austin, William 10
automated cutter 37

Babcock, Dubisson, and Hall 70
Babcock, Duly & Hall 14
Babcock tubular boilers 192
Badger Mill 185
Bailey, Jacob & Francis 40
Bailey, Oliver E. 70
Baker Mill 153
Baldwin, Mr. Henry 52
Baltimore County, Maryland 45
Baltimore Post Office envelopes 57
Bangs, Richards, & Platt 29
bank-note paper 125; with jute 128; with mohair 49
Barclay, Henry 8, 93, ***96***
Barker, Benjamin 159
Barrett, Ebenezer 160
Barstow, George & Edgar 71
Barstow's paper warehouse 71
Barton, George S. 98

Bay State Mill 129
Beach, Hommerken & Kerney 8, 93
Beach, Lewis 158
Beach, Moses Y. 7, 21, 94
Beardsley, E.R.& C. 108
Beaver Creek Mill 103
Beckett, William 107
bed plate 2
Beebe, Jared 137
Beebe & Holbrook 137, 169
Belcher & Burton 65
Bellows Falls canal 165
belt-driven steam dryer 95
Bemis Paper Co. 145
Benedict, William J. 10
Benjamin, Nathan 60
Bentley spelling book 29
Benton, Caleb 100
Benton & Garfield, ***96***, 100
Berkshire County, Massachusetts 8, 26, 29–30
Berkshire Courier 157
Berkshire Life Insurance Co. 35
Berkshire Mill 26, 35, 121–122
Berlin Mills Co. 164
Bermondsey Mill 90
Betts, Edward 99
Bigelow, Jonathan 10
Billings, James and Stephen 165
Bishop Mill 103
Bissell & Pease 93
Blake, Bill 165
Blake, Francis 190
Blake, J.S. 61
Blake, Robbins & Co. 190
Blank, Ephraim F. 72
Blank, Thomas 72
bleaching clay 173
bleaching liquor 16; French 63
Bliss, John 92
Bloxham & Fourdrinier 90
board strainer 67
bogus manila 63; processing 144
Boies, James 45, 160
boiler house construction 78
book paper 16
bookbinding 30
Boston & Portland Railroad 160
Boston, Concord, and Montreal Railroad 163
Boston Congregationalist 47
Boston Weekly Journal 157
Boswell Keene Co. 111
Boyston & Whitcomb 9

Bradford, William 5
Bradley, Stephen 10
Brady, Mathew 118
Brandywine Mill 16, 22, 24
British Parliament 13, 15; ban on machine exports 94
broken paper 149; as wrappings 119
Bronson & Crocker 9
Brooklyn Eagle 158
Brown, Levi L. & Co. 51
Brown, W.W. 164
Brownell & Miller 108
Buchanan, C.S. 64; and rotary bleach boiler 144
Buchanan, James 69
Buck, William 192
Bud, David 26
Buffalo Card & Envelope Co. 178
Buffalo, New York 66
Buffalo Paper Mills 110
Bunking & Van Norstrand's NY warehouse 175
Bunting & Foot 56
Burbank, Abijah 50; in Vermont 165
Burbank, Abijah, Jr. 50; in Vermont 165
Burbank, Caleb 50
Burbank, Elijah 50
Burbank, Isaac 50
Burbank, Silas 165
Burbank & Fales ***96***
Burbank Mill 50, 121
Burgess, Henry 81
Burghardt, Benjamin 158
Burghardt, Erastus 159
Burgharth, Gaston 157
Burke, A.J. 45
Burnap & Babcock 13, 46–47
Burr, Nelson 10
Burrit, Able 12
Butler, Asa 170
Butler, H.V. ***96***
Butler, John 81, 83
Butler, Julius 107
Butler, J.W. Paper Co. 107–108
Butler, Oliver Morris 107
Butler, Simeon & Asa 28, 52, 93
Butler & Hunt 107
Butler & Ward 28
Butterfield, David M. 144
button separation 148
Byrd, James 110

calendering 31, 34; sheet-calender 22, 24, 37; super-calender 138–139, 149; web calender 136, 139
California Farmer 191
Camas Mill (picture) 193
Camp, A.D. & Co. 71
Camp, E. 62
Campbell, J. & S. 70
Campbell, Samuel 70
Canal Village, Massachusetts 21, 133
Candee, Page, & Lester 81
cardboard 72
Carew Mfg Co. 134
Carey, Matthew & Son 19
Carson, David 121
Carson, W. & Sons **96**
Castle Mill 27, 76, 122
Caswell, Gurdon 66
caustic soda 34
census 10, 41
centrifugal pump 97
chain-mould 39, 75, 92
Chapin, Charles O. 144
Charles River 47
Chase, Robert & Co. 163
Chase & Linton 110, 114
Cheboygan Paper Mill 188
Chelty, C. Yaneat 75
Chemical Paper Co. 171
Cheney, F.W. 129
Chester County, Pennsylvania 40–41
Chicago & Northwestern Railway 183
chilled rolls 65
Chisholm, Hugh J. 161
Church, Joseph 76
Church, Leonard 76
Church, Luman 76, 100
Church, Samuel 26, 76
Church & Baptiste & Co. 12
cigarette paper 131
Cincinnati Commercial Gazette 107
Cincinnati Daily Gazette 159
Clapperton, Robert H. 17–18
Clark, Anson H. 34
Clark, George A. 146
Clark, James 70
Clark, William 47, 105
Clay, Henry 53
Cleveland, Grover 162
Clinton Steam Mill 103
Cogswell, William B. 161
Cold River Mill 61, 164
Cole, Thomas 103
Coleman, Ezra 57
Colt Mill 130
Columbia Mill 73, 76, 79, **155**
Columbia River 194
Columbia River Paper Co. 194
Combined Locks Paper and Pulp Mills 188
Commitment River Company 169
Compromise Tariff 53
Cone, Henry D. 175
Congress Mill 39, 100
Congress paper 39
Connecticut 92

Connecticut River 133
Connecticut River Pulp Mill 168
Connecticut River Railroad 136
Continental Congress 15
Cope, Edwin 41
Copesecook Mill 51, 160
copperplate engraving 33
Corliss steam engine 190
cotton fibers and moving-wire machine 91
cotton rags 9–12, 16–17
counting room 37, **162**
Cox, Ezekiel Taylor 104–105
Cox, Horatio J. 105
Cox, Jones L 105
Cox & Co. 104
Craig, Rev. Elijah 109
Craig, Parkers & Co. 109
Crane, Frederick G. 129
Crane, James A. 123
Crane, James Brewer 123
Crane, Lindley Murray 123
Crane, Luther 49, 120–121, 123
Crane, Robert B. 123
Crane, Seymour 123, 129
Crane, Stephen 15, 49, 120
Crane, W. Murray 129–130
Crane, Zenas 81, 120–122
Crane, Zenas & Co. 121
Crane, Zenas, Jr. 129
Crane, Zenas Marshall 123, 125
Crane & Co. **96**, 123, 125–129, 179; stationery contract of, 1853 58–59
Crane Brothers, Inc. 123–124
Crehore, Charles 47
Crehore, Lemuel 47
Crehore & Neal **96**
Crehore Mill 47
Crocker Mill 145
Crocker Mill No. 2 145
Crow Hollow Mill 76
Crown Paper Company 194
cudweed pulp 157
Culver & Mitchell **96**, 97
Cumberland Mill 109
Curtis, A.C. & W. 49, 81, 94, **96**
Curtis, Allen 49, 94
Curtis, Crocker 94
Curtis, George B. 49
Curtis, Solomon 48–49, 94
Curtis, William 49, 94
Curtis & Annis 48
Curtis Mills 49
Curtisville Furnace Co. 157, 159
Curtisville Pulp Mill 157–158
Cutting, Jonah 164
Cuyahoga Falls Mill 104
cylinder moulds 22–**23**
cylinder press 112
cylinder-wire machine 17–18, **25**; Ames 50; homemade 106; operation 36
Cyrus Currier & Sons 111

D & J Ames 7, 24, 78; sell South Hadley Mill 135
dandy roller **23**, 62, 149

Davis, Charles Thomas Manufacture of Paper 174
Davis, John B. 63
Davis, William A 52–53, 63; blue paper 172
Day, Lyon & Co. **96**
DeBuckemyer, Messr. 13
Defiance Mill 122, 153
de-inking recycled paper 112
Delaware County, PA 40
Democratic Press 39
Dennison Mfg. Co. 160
Deseret News 189–190
Devitt, M. 107
Devitt Mill 107
de Warville, Brissot 16
Dewing, Paul 49
Diamond, A. 58
Dickensen & Curtis 9–10
Dickey, Joseph 40
Dickey, Margaret 40
Dickey & Lysle 40
Dickinson, Ann 92
Dickinson, Edwin 145
Dickinson, Henry 145
Dickinson, John 18–19, 91–92; double-cylinder machine 61; envelope machine 56
Dickinson & Clark Mill 146
Didot, Pierre-Francis 89
Didot, St. Ledger 75, 89–90, 92
die cutting machine 56
Diem, F. J. & Co. 105
digester 159; direct-fired 160
Dixon digesters 168
Dohan, Joseph M. 131
Donaghee, James 81
Donkin, Bryan 17, 91–92; rag boiler 55
Donkin, Bryan & Co. 94
Dorchester Mill 45
Dorlan Mill 41
double-cylinder 61
double-dipped handmade paper 171
double-sized paper 130
dryer felting 80
drying cylinder 80
Duff & Keating 58
Duncan, Lewis & Bartow 56
Dwight, Edmund 134

Eagle Mill 28, 41, 45, 135, 153, 183
East Side Mill 183
Eastwood, John 70
Eaton, Arthur W. 179
Eaton, Crane & Pike Paper Co. 179
Eaton, Hurlbut Paper Co. 179
Eaton, Joel 165
Elliott, A.G. & Co. **177**
Elliott, Curtis 48
Elliott, Simon 48
Elliott & Curtis 48
Eminent Philadelphian 19
engine sizing 34, 100
Enterprise Mill 76

envelope cutter 114
envelope machine 149
envelope paper 33, 43, 57–58
Erie Canal 29, 60–61, 69, 103
Erwin, John 107
esparto grass 157–158
Essones Mill 89
Ewing, George 133
Excelsior Mill 45, 139
Exeter Mill 163

Fair Haven Mill 166
Fairbanks, Ebenezer W. 164
Falmouth Mill 161
Farley, Benjamin *96*
Farlowe, David M. 178
Farmington Chronicle 160
Farnham & Blanchard 53
Feeters, R.M. 42
Ferguson, P. & Co. 13
Fermin, Didot & Sons 94
Fessenden, Patty 164
Fessenden, William 164
Field, Cyrus 69, 72, 153 *172*–173
Field, Marshall 73
fire dryer 21
Fisher, Mires 15–16
Flat Rock Mill 41
flax 55; processing 127
Flying Dutchmen paper machine 187–188
foolscap (a.k.a. cap) 43; Congress 54; exchange (writing) 51; super 43
Forest Mill 76, 114
Forest Paper Co. 161
Foster, Joseph 49
Fourdrinier, Henry 17, 18, 90, 115
Fourdrinier, Sealy 17, 18, 90, 115
Fourdrinier machine 20, 99
Fox River Mill *186*
Fox River Paper and Pulp Company 188
Frambach, Henry 184–188
Frambach Paper Co. 185, 190–191
Franklin, Benjamin 15, 38, 119
Franklin Manufacturing Company 110
Franklin Mill Holyoke, MA 137
Franklin Mill in Kentucky 109
Franklin Mill in Richmond, VA 110, *115–117*
Freeman, Hampden 137
Frisbee, T.G. 10
Frogmore Mill 91
Fry, Richard 160
Fuller & Co. 117
Fulton, Henry 28
Fulton, James 41

Gallatin, Albert 12, 53
Gamble, John 90–91
Gardner, Smith 169–171
Garfield, Harrison 100
Garfield & Benton 100
Gaunt & Derrickson *96*
Gavitt, Nelson Co. 189
German States 14

Gibbs, Lucius 166
Gilbert, Daniel 121
Gilbert, William 188
Gilbert & Whiting 187
Gilbert Paper Co. 188
Gill, J.K. 193
Gillespie Mill 45
Gillman, Alonzo 65
Gilman & Silbley 61
Gilpin, Joshua 15–17, 18–19
Gilpin, Thomas 15–17, 19, 24; Baltimore memorial 51
Gilpin Mill *19*
Glen Manufacturing Co. 164
Glen Mill 41, 120, 128
Glen Mills Paper Company 131
Glenmount Mill 45
Globe Mill 184
Goddard, George 190
Goddard, Isaac 80, 98
Goddard, Rice & Co. 98
Goddard, Silas *96*
Goodwin, George, Sr. 81
Goodwin & Co. *96*
Gookin, Richard 164
Gookin, William 164
Goss, Samuel 165
Gould & Banks 52
Government Mill 130, 140
Graham, Benjamin 106
Graham, James 106
Graham, James Brown 106
Graham, Joseph 106
Graham Mill 106
Graham Paper Co. 106
Granite Mill 190
Grant, Dennis, & Co. 51
Grant, Moses & Son 47
Grant, Warren & Co. 51; in Maine 160; refined-rag paper 111
Grant Mill 47
Great Britain 12–13, 15
Greatrake, Laurence 16, 18–19
Greeley, Horace 76
Green, Lawrence H. 164
Green Bay and Mississippi Canal 183
Green Brothers 118
Greenleaf & Taylor 145
Greenwood, Joseph 49
Gregory, Samuel 164
Gregory, William 164
Grifin, Martin L. 168
Grover, Ann 18
Guilford Mill 164
Guilford Smith 96
guitar cases 123
Gunpowder Mill 45

Hadley Falls canal 21, 92–93, 133
Hadley Falls Company 133
Hagy, Jacob 38
Hale, David 69
Haley & Holden *162*, 163
Hall, Charles H. 160
Hall, John H. 70, 90
Hamilton, Saul 66
Hamilton & Wright 63

Hamilton Mill 107
Hamilton Tissue Mill 107
Hampden Glazed Paper Co. 171
Hampden Mill 137
Hampden Paper Company 137
Hampshire Mill *134*–135
Hann & Beebe paper warehouse 69
hard water 8
Harding, Jasper *96*
hardware paper 110
Harper, James 98
Harper Fourdrinier *98*
Hartford Mill 111
Hasford, J. 65
Haswell, Anthony 166
Hawes, R.L. 57
Hawley, S & A & Co. 60
Hawthorn, Nathaniel 35
Hazard, Roland 161
Heathcote, L. 118
hemp 55, 132
Henchman, Daniel 45
Henry, James 64
Henshaw, Joshua 165
Hereckenrath, Schneider & Co. 173
Herkimer Mill 61, 159
Herman & Mellen 71
Hewitt, Henry 183
Hewitt Water Power Co. 185, 188
Hibernian Mill 93
Hig, H. 34
Hinsdill, Ensign Joseph 164, 166
Hoe & Co. 33; in India 75; Lightning Press 72
Hoes, John 64
Hoffman, Prof. Ferdinand 156
Hoffman, William 55; centrifugal pump 97
Hoffman, William H. & Sons 45
hog box 67
Hogan, Perkins & Co. 56
Hoggs, William & Son 49, 120
Holbrook, John 164
Holbrook & Fessenden 66, 164
Holebrook, G.B. 137
hollander (a.k.a. rag engine or beater) 6, 127, 149
Hollingsworth brothers 63
Hollingsworth, Amour 45
Hollingsworth, John 55
Hollingsworth, Lyman 55
Hollingsworth, Mark 45, 55
Hollywell Mill 43
Holme, John 5
Holmes, Dr. Oliver Windell 35
Holyoke, Elizur 134
Holyoke and Western Railroad 136
Holyoke Freeman 137
Holyoke Machine Co. 137, 140
Holyoke Manila Mill 146
Holyoke Mill *141*
Holyoke Paper Company 141, *142*, 169
Holyoke Water Power Company 136
Hopkins, George F. 69

hot-pressing 125
Houghton & Johnson 56
Housatonic Mill 77; sale to Platner & Smith 154
Housatonic Railroad 83
Housatonic River 121
Howard, Charles 25; debts to Cyrus Field 74
Howard, Thomas 189; pulp regulator **96**, 98
Howard & Lathrop 24–25, 134; mill closure 74
Howe, Henry P. 80, 98
Howe & Goddard 80, 98; tandem dryer 82–83
Howland & Palmer 64
Hoyt, D. 66
Hoyt, John & Co. 163
Hubbard, R. & A. H. **96**
Hubbell, C. R. 105
Hudson, Henry 81; custom machine wires 95
Hudson-Cheney Paper Co. 128–131
Hudson River Pulp & Paper Mill 168
Hudson River Pulp Co. 167
Hulbert, William H. & C.M. Paper Co. 58, 72
Hull, Henry 75
Hunt, Joseph 107
Hunter, Dr. Dard 5, 128
Hurd, William 47
Hurd & Crehore 47–48
Hurlbut, Henry Clay 175
Hurlbut, Thomas 28, 29, 71, 83, 153, 170, 175
Hurlbut Mill **178–181**
Hurlbut Paper Co. 178–179
Hydraulic Mill 105
hydroelectric plant 187

imitation handmade paper 18
imitation laid paper 132
imperial newsprint price 44
india paper substitute 131
Ingersoll, Jared 76, 100
Ingersoll, John & Co. 12
Ingersoll Mill 100
Ingham, Hezekiah 105
Ingham, Isaiah 105
insurance losses 62; on South Lee Mill 78
International Paper Co. 161, 179; move to Cornith 168
Ira White 165
Irving Mill 175
Island Mills 49
Ivanhoe Paper Co. 111
Ivy Mill 38, 119; closure 128

J. & R. Kingsland 111, 116
Jackson, Edward 48–49, 120
Jackson, Ephraim 47
Jackson, Samuel 106
Jackson Mill 47
Japanese Mill 124
Jarvis & French 65

Jay Bridge Mill 161
jeans 109
Jefferson Administration 10–11
Jefferson Mill **62**
Jenks, Daniel 51
Jenks, William 51
Jessup & Brothers **96**
Jessup & Moore 99, 159
Johneiden, C.H. 173
Jolly, J&W 140
Jones, John 99
Jordan, Joseph 111
Jordan, Richard 163
Jordan refiner 111
Judd Brothers 135
jute 55, 93, 128, 192

Kaukauna Times 187
Kaukauna Water Power Company 188
Kay, William **96**
Keating, John 60
Keeting Mill 60
Keller, Friedrich 156
Kellog, George 29
Kellogg, J. 104
Kendall, Amos 53, 109
Kendall, Charles S. 51
Kendall Mill 113
Kent Paper Co. 62
Kilmer & Ashman 61
Kimball, Daniel 165
Kimberly & Clark Co. 184, 186, 188
King, George 39, **96**
King, Phillip J. 39
Kingsbury, Elisha 164, 165
Kingsland refiner 111
Kingsley Brothers 166
Kneeland, J.C. 65
Knight dryers 106–107
Knowlton, George 66–67
Knowlton & Rice 67, **96**
Knowlton Brothers Mill 68
Koch, Louis 72
Krah, Joseph 100
Krepps & Carter 24

Laflin, Addison H. 61
Laflin, Bryon 61
Laflin, Butler & Co. 107
Laflin, Mathew 107
Laflin, W.W. & C. 76, 79
Laflin, Walter 76, 99–100
Laflin, Winthrop 76
laid machine wire 61
Lamden, Christopher 103
Lamden Mill 103
Lamme, Keiser & Co. 109
Lancaster Intelligencer & Weekly Advertiser 40
Lasserden, Kendell & Co. 110
Lathrop, Wells debts to Cyrus Field 74
Laurel Mill 154
Laurie, Adam 107
Lavoisne's *Complete Genealogical, Historical, Chronological, and Geographical Atlas* 20

Law, British North America 66
Lawrence University 183
layboy 65
lead monoxide 172
leather board 72
Leeds, Jones, & Bissell 105
Lefever, David 40
Leffingwell, Christopher 92
Leffingwell Mill 92, 95, 97
letter paper 51, 57
Levop, Thomas C. 109
Lick, James 190
Lick Mill 190
lifting-roller 64
lightning press 33
lignin in wood-pulp 159
Lima Mills 168
lime 16, 78, 148
Lincolnton Mill 114
linen-cotton composite paper 132
linen paper 39, 125
linen rags 9–12, 14
Livorno (Leghorn), Grand Dutchy of Tuscany 14
localized fiber process 128, 130
Lockport Felt Company 118
Lockport Paper Co. 108
London Repertory of Arts, Manufacturers, and Agriculture 19
Longfellow, Henry Wadsworth 35
Longfellow Mill 122
Loomis, Riley 76–77
Louisiana Opelousas Courier 112
Louisville Advertiser 109
Louisville Courier 109
Love, James A. 40
Love, Newton B. 40
Lovejoy, Don 10
Lowe, Henry 159
Lower Merton Mill 39
Lyman, George W. 134
Lyon, Amos 47
Lyon, Col. Matthew 166
Lyon, Peter 47, 49
Lyon, Warren W. 9
Lysle, James 40
Lysle, William 40

machine-made paper 47
Macon Telegraph 112
Madison Mill 162
Magarga, Charles 10
Magarga & Cope **96**
Magasgu & Co. 42–43
Mahaiwe Mill 100
Mammoth Mill 44
Manayunk Mills 41
manila belts 123
manila paper 55
March, J.M. & Co. 51
Martin, Gen. Walter 60
Martin Manufacturing Company 163
Maryland 15
Maryland Wall Paper Co. 171
Massachusetts 50–51
Massasoit Paper Company 145, 169

Index

May, Edward 100–101
May, Sylvester 100–101
McAlpine, Robert 153
McCausland, Gen. J.A. 45
McCormick, J.B. 140
McFee, Hugh 24
McFeeters, R. 58
McGaw, Judith 26
M'Chensney, John 71
McWilliams Paper Warehouse 126
Mead, Daniel 105
Mead Corporation 105
Megargee Brothers 159
Megunticook Woolen Company 118
Melville, Herman 28, 35
Melville, Thomas 28
membrane paper 129
Menasha Mill
Menominee Pulp Mill 187
Merrill, Aliajah 28
Merrill, J.T. 28
Merrill, William 28
Messalonskee Mill 161
Meyerhofer, John 72
Middle Mill 100
Middlebury Mill 166
Miller, Hezekiah R. 48
Miller, T.L. 104
Miller, Warner 159
Mills, William 47
Milwaukee Sentinel & Gazette 183
Missouri Gazette 109
Missouri Intelligencer 109
Mohawk & Albany Railroad 65
Montague Paper Co. 158
Montigue, Charles 31, 33
Moody, L.W. 135
Mooney, Isaac 105
Moore, A.D. (NY paper warehouse) 71
Moore, Plato 63
Moore & Leggett 14, 57
Moosalamoo Mill 166
mortar and pestle 5, 119
Mosphett's Superior Bleaching Powder 173
Most Wonderful Machine, Mechanization and Social Change in Berkshire Paper Making 1801–1885 26
Mt. Vernon Mill 40
Mountain Mill 100
moving-wire machine 90–91, **98**, 100, 149
multi-cylinder dryer 79–80, **90**
mummy wrappings 13
Munsell, Joel 13; *A Chronology of Paper and Paper-Making* 6–7
Muskingum Messenger 104
Myers & Minn 71

Naga, W. & Sons 54
Napier press 33
Nash Mill 18
Nashville Banner 109
National Mill 100, 157
National Wall Paper Assoc. 185

natural white 173
Navarino hats 65, 77
Neenah Mill 184
Nehoiden Mill 49
Neponset Mills 45, 120–121
Nesbitt & Co. 59
New England Association of Banks for the Suppression of Counterfeiting 125
New England Roofing and Manufacturing Co. 51
New London Gazette 92
New Orleans Bulletin 110
New Orleans Picayune 110
N.Y. & N.H. Railroad 182
N.Y. Evening Express 158
N.Y. Evening Post 159
N.Y. Gazette 60
N.Y. Herald 72, 154, 159
N.Y. Journal of Commerce 69, 72; conversion to wood-pulp 159
N.Y. Market Exchange 11–12; alternative fibers 55; fire 71
New York State 60
N.Y. Sun 8, 159
N.Y. Times 158, 194
N.Y. Tribune 76; conversion to wood-pulp 159
N.Y. Weekly Times 158
N.Y. Weekly Tribune 72
N.Y. World 158
New Yorker Staats-Zeitung 158
Newhouse, Anthony 38
Newton, Daniel 168
Newton, Daniel H. 137, 144
Newton, Edward T. 139
Newton, James H. 137, 139
Newton, John C. 137, 139, 144, 171
Newton, Moses 137, 146, 171
Newton Mill 146
Newton Paper Company 146
Niagara Falls Paper Company 112
Nicholas-Louis Robert 89–90
Niles Weekly Register 44, 51
Nonatum Mill 99
Nonotuck Paper Co. 171
North Carolina 15
Northampton Mill 31; paper machine 83
Northern Pulp Co. 187
Northrup Mill 100
Northumberland Mill 163
note envelope 58
note paper 51, 58
Nuremberg Ultramarine Works 173

Oakes, Thomas 16
Oakland Mill 128
Old Forest Mill 100
Old Stone Fort Mill 114
Onderonk, Hendrick 60
O'Neill, Edward 194–195
Ordnance of Nullification 53
Oregon City Mill 193
Osborn, John L. 170
Otagauma Pulp and Paper Mill 188

Otis Falls Mill 161
Owen, Charles 28, 43, 52, 71, 83, 103, 153
Owen, Edward 175
Owen & Hurlbut 28–29, 33, 41–42, 51, 54, 56–57, 66, 73–74, **96**, 104, 109–110, 154, 169–172, 189; boiler problems 79; purchase of Potter Mill 175–**176**; ruling machine 135; tandem dryer 82–83
Owen Mill 176, **177**
Owen Paper Co. 175
Owl Mill 164

Pacific Mills 70
Pagenstecher, Albrecht 156
Pagenstecher & Co. 156–157, 159, 167
pails 123
Palmer, Jeremiah 167
Palmer, A. & W. patent paper cutter 153
Palmer Water Power Company 167
Panic of 1837 25, 40, 54–55, 65, 95
paper: boat 123; bonnet 65, 77, 103; buggy box 124; bullet patch 129; collar 123; construction 21–22; hat 77; pharmaceutical bowls 123; racing scull 123; railroad wheels 123; roofing shingles 51; wash tubs 124
paper city 146
The Paper-Making Machine 17–18
Paper Manufacturers Association of New York 112
Paper Mill Village, New Hampshire 161, 164
paper shavings 21, 41
Paper Trade Journal 173
paper trimmer 22, 83–84, 154
Papermaking: The History and Technique of an Ancient Craft 5
Papertown 100
Paperville 190
Papyrus Mill 45
parchment substitute 130–131
Park & Watson 57
Parker, William 9, 56, 65, 78
Parks, William 38
Parson, W.H. paper warehouse 72
Parsons, Joseph C. 135
Parsons Mill **135**, 136, 141
part-cotton paper 132
Parthington digester 161
paste board 72
Patterson, B.W. 9
Pease, E.H. & Co. 64
Peaslee, Horace W. 61
Peck, J. & E. 9
Pennsylvania 15, 39
Pennsylvania Railroad 41
Penobscot Chemical Fiber Company 162
Perkins, Thomas H. 134
Perkins vertical digesters 186
Persee & Brooks **96**

Index

Persee & Brooks Mill 69–70
Pettee, S.E. 58
Phelps, George H. 65, 73
Phelps, James 93, 95–97
Phelps & Field 73, 79
Phelps & Spafford 94–95; tandem dryer 82
Philadelphia 41–42
Philadelphia Bulletin 44
Philadelphia Committee of Safety 8
Philadelphia County, Pennsylvania 41
Philadelphia Gazette 119
Phillips & Speer 105
Phineas Allen 35
Phiney, Lunman & Co. 100
Phoenix cap 43, 171
Phoenix Mill, Cincinnatti 106–107
Phoenix Mill, So. Lee **27**–28, 85
photographic baths 123
piano cases 124
Pickering, Joseph 93
Pictet-Brelaz process 161
Pine, Edward 65
Pioneer Mill 129, 190–191
Pitkin, Elisah 60
Pittock, H.L. 193
Pittsburgh Almanac 103
Pittsfield, Massachusetts 26, 28
Pittsfield & Stockbridge Railroad 33, 83, 100
Pittsfield Eagle 30–31, 33
Pittsfield Gazette 103
Pittsfield Sun 9
plate paper 46–47, 71
Platner, Adeline 154
Platner, George 100, 153
Platner & Porter **96**, 100
Platner & Smith **96**, 153; ream labels 33; shipping on Erie Canal 66; textile dye 172
Pleasant Valley Mill 31–32
plunge washer 68
Poor Richard's Almanac 38
Portland Oregonian 192
Post, V.B. paper warehouse 190
Potter Wall Paper Mill 171
Poulson's Daily Advertiser 19
Preble, J.E. blank book factory 69
Prentiss, Bemis & Co. 164
Prentiss, Fredrick 164
Prentiss, Thomas 164
Prentiss Wire Mills 140
press felt belt 117
pressboard 48
Priesly, J. & Co. 13
Providence Journal 158
Puffer, M.G. 58
pulp dresser 67
pulp lap 160
pulp regulator 97
Pusey, Jones & Co. 99
Pusey, Joshua 99

Qinnesec Water Power Co. 187
quatro post 53

R. Farnham 54, 172
rag bleaching 95
rag boiler 55, 63
rag cutter 7, 147
rag dresser 80–81
rag duster 81, 147
rag room 35
rag washer 95, **97**, 149
Rampage, James 146
Ransom and Millbourn's sizing section 139
Raynor, Samuel envelope manufacturer 69
Readen, John 119
Reay, George H. 59
Record, Alvin 161
recycled paper 112; de-inking 107
Red Mill 70, 121–122, 128–130
Redstone Mill, Missouri 109
Redstone Mill, Pennsylvania 103, 106
Reed, Stephen 165
reel-up 106
Rees' Encyclopedia 41
refined-rag paper 111
Reilly, Calvin 107
Reilly & Laurie 107
Remington Mill 192
retree 149
Revere, Paul 120
Rhode Island 15
ribbed mould 126
Rice, Charles 114
Rice, Clark 66–67
Rice, George M. 98
Rice, John L. 49, 122
Rice, Thomas, Jr. 49, **96**, 114; mill **48**
Rice, Thomas, Sr. 49
Rice & Crane 49, 123; tandem dryer 82
Rice and Garfield Mill (drawing) 49
Rice & Kendall 51
Rice, Barton and Fales Machine and Iron Company 98
Richard & Hoskins **96**
Richmond, G. & N. 183
Richmond Enquirer 110
Richmond Mill **184**
Riley Mill 161
Rising Paper Company 175
Ritchie, Thomas 110
Rittenhouse, Jacob 6
Rittenhouse, William 5
Rittenhouse Mill 5
Riverside Groundwood Mill 164
Riverside Mill, Holyoke 144
Riverside Mill, New Hampshire 164
Robbins, Charles F. 190
Robertson & Black 146
Robertson, William & Sons 164
Robinson, Conway 110
Rochester, NY 66
Rockingham Mill 165
Rockwell, Norman 35
Roger, Henry J. 187

Root, E. & Co. paper dealers 73
Rose, C.W. 56
rosin 34, 72
Rossman, Stephen 64
rotary bleach boiler 64, **138**–146, 148, 149, 169, 182–183, 192, 194; for de-inking 114; explosion 163, 164; spherical 162
rotary cutter 149
Routledge, Thomas 158
Royce, James H. 158
ruling machine 37, **113**; attachment to winder 114; striker 153
Rumford Falls Woolen Company 118
Russell, David 166
Russell, Henry 190
Russell, William A. 158
Russell Paper Co. 129
Russia 17
Rutland Farmer's Library 166

San Francisco Bulletin 191
San Francisco Call Bulletin 192
San Francisco Chronicle 192
San Francisco Examiner 192
San Geronimo Mill 190
San Jose Mercury 192
Saratoga Flag 63
satin letter paper 71
Satterly, John 62
Saugerties Mill 8
Savage & Moore **96**
Savanna Republic 110
Savels, John 160
Scales & Scallion 54
Scandaga Mill 61
Schaffer, A.C. 110
Schaffer, Jacob Christian 156
Schenectady, New York 65
Schmidt, Johan 104
Schuler & Benninghofen 117
Schuylkill City Mill 41
Schuylkill River 38
Scientific American 64, 176
screw press 22, 122
Seal, Joshua 99
Seaverns, John L. 45
Sedwick & Sabin 31, 33
Sedwick, Thomas & Co. 31
Sellers, Charles Lee 78
Sellers, Coleman 22, 29, 39, 44; on coloring 172; pulp dresser 67; sheet-calenders 77–78
Sellers, James 39
Sellers, Nathan 16, 39, 119
Sergeant, William 69
Sergeant Brothers rag dealers 69
Seymour & Co. 69
Seymour & Son 69
Seymour, Johanthan paper warehouse 69, 122
Seymour Paper Co. 70
Shattuck, Surner 48
Shaver, Abraham 45
Sheets, John 105
Sheffeld, J.B. & Son 8, 61
shipping trunks 123

shrinkage 111, 127
Shyrock, George 44
Shyrock, John 43, 63
Shyrock & Co. 45
Silby, Hiram 35
silk paper 129, 171
Silver Lake Mill 61
sizing 34, 71
sizing soap in beater 111
Skinner, Colin 70
Skinner, J.C. & Co. 107
Smeaton, John 90
Smith, Charles 94, 96
Smith, David **96**
Smith, Dewitt 157
Smith, Elizur 153–154
Smith, Hamilton 69
Smith, J.R. 153
Smith, Jeremiah 45
Smith, Joseph 189
Smith, Wellington 157
Smith & Bassett 93
Smith Paper Co. 157–158
Smith, Winchester & Co. 96, 111
Smith, Wm. J. & Co. 65
smultz 172
soda ash 148, 161
soda-pulp process 159–160
Solvay process 161
South Hadley Mill 7, 21, 24–25, 134
South Lee Manufacturing Co. 85
South Lee Mill 9–12, 14, 24, 26, **27**–31, 33, 39, 76; retrofit 82–83; use of ultramarine 173
Southworth Mfg. Co. 43, 135
Spafford, George 93–95
Spears, Otis 56
spool paper cutter 105
Springfield, Massachusetts 20
Springfield Mill 20
stable fiber method 64
stamping machine 37
Staniar, William 116
Staniar & Laffey 23
starch 34
Starin, Henry W. 60
Starks, E. 10
Steadman, Ebenezer Hiram 104, 109
steam dryer 18, 24, 36
steam engines: for calenders 191; for cylinder-wire machine 18; for moving-wire machines 168; for rag engines 103–104, 106, 177; for wood grinders 167, 190
steam heated calender 65
steam heated dryer 78, 92
Steel, James 40
Steele, Gen. John 40
Steelville Mill 40
Steinway, Theodore 156
Stephens & Thomas 116
Stephens, George, paper warehouse 178
Stidolph, William 16
Stimpson, Solomon 164
Stoddard & Freeman **96**

Stone, Joseph F. 73
Stone Mill, Adams, MA 51
Stone Mill, Dalton, MA 125, 129
Stone Mill, Mt. Vernon, PA 40
Stone Mill, Westfield, MA 123,
Stoveken, John 183, 187
Stowe, Joseph Daniel 93
straw-manila paper 63
straw newsprint 44; wrapping paper 106
structural dome 124
Stuart, R.S. 46
Stuart, S.A. 46
stuff chest 149, 91
Sturge, Ivar & Co. 12
Sugar House Mill 189–190
sulphite process 161, 167
super post 43
superfine bath 71
Swann, John 160
Syracuse Paint and Paper Co. 171
Syracuse Pulp & Paper Company 171

tandem dryer **17**, 82–83; regulated in series 99
tarred sheeting paper 51
Taylor, Frank 54, 76
Taylor, J.E. 145
Taylor, Samuel Penfeld 190
Taylor, S.P.& Co. 190
telegraph paper 105
textile dye 172
Thatcher, Jethro 77
Thatcher, Stephen 77, 81, 101, 122
Thatcher & Ingersoll 100
Thilmany, Oscar 187
Thilmany Pulp Co. 187
Thistle Mill 70
Thomas & Woodcock 65, 67, 107
Thompson, F.A. 123
thrasher 147
Tileson, Edmund P. 45
Tileson & Hollingsworth 11, 45–46, **96**; *New Orleans Picayune* 110; plate paper 71
Tilghman, Benjamin 161–162
Tilton & Co. **58**
tobacco packaging paper 93
Tompkins, Peter C. 63
Tompkins, Staats 63
Tondes, Daniel 108
Toppan Carpenter & Co. 126–127
Tower, Ashley B. 141
Tower, David H. 141
Trendley, Frederick 104
triple-cylinder 61
triple-tandem copper dryers 106
Trout Creek Mill 38, 41
Troy, New York 65
tub sizing 34, 71, 175–**176**
Turkey Mill 100, 153
Two Waters Mill 91

ultramarine 172–173
Umbagog Mill 161
Underhill, R.L & Co. 61
Union Bag and Paper Co. 179

Union Mill, Holyoke, MA **139**
Union Mill, Lee, MA 27, 76
Union Paper Company 145
Union Pulp Company 188
United Box Board and Paper Co. 179
U.S. Post Office 57–58; cards 129; envelopes 58–59; stamps 126–127
United States Spelling-Book 103
U.S. Treasury Department 11, 15; currency 43, 127–130; notes 119, 126; stamps 129
U.S. War Department 105, 188
Upper Forest Mill 100

Valley Gleaner 154, 168
Valley Paper Co. 144
Van Cortland, Gen. Pierre 61
Vancouver Independent 193
Van Rensselaer, Stephen 65
Van Veghten & Son 65
vat tax 51
Vebote 158
Vermont Gazette 166
Vermont Intelligencer 164
Vilas, Samuel 11
Virgil (John Baskerville) 38
SS *Virginia* 46
Voelter, Henrich 156, 158
Voss, Daniel 45, 49
vulcanized rollers 127

wagon box 124
Walbridge, Charles 65
Walbridge, Gen. Ebenezer 164, 166
Waldo, Samuel 160
Waldschmidt, Christian 104
Wales, Nathaniel 47
Wales & Mills 47, **96**
Walker Tariff 54
wall paper 62, 89, 103; tariffs 54; white litharge 170–171
Walsh, J.H. & Son **96**
Walsh, John 60, 81
Walsh & Co. 60
Wardwick, Samuel 29–29
Ware, John 47
Ware, Reuben 47
Ware & Clark 47
Warren, George 166
Warren, Samuel Dennis 160
Warren, S.D. & Co. 160, 178
Warren, William S. 130
Warren Mill 159
Washington, George 60
Water Shops Mill 21
Waterbury & Sons 118
waterleaf 34
waterproof paper 72
Waters, E. & Sons
Watertown Freeman 68–68
Watertown Register 67
waterwheel: in basement 136; brest wheel 6; iron gudgeon (bearing) **115**, 122; overshot 6, 100, **115**; undershot **25**; Van de Water 98
Wauregan Paper Co. 139

Waverly Mill 100
Webster, Ensign, & Seymour 26
Webster, Noah Minerva
Webster's American Spelling Books 50
Webster's New Quarto Dictionary 21
Weiss & Son 118
Wells & Woodruff 73
Wendell, D.S. 10
West, Charles Smith 193
West Groton Mill 55
West Side Mill 107–108
West Stockbridge, Massachusetts depot 9, 30
Westbrook, Thomas 160
Western Budget 65
Western Railroad 34, 51, 123
Western Star 30
wet machine 48
Weymouth, Thomas 58
Wheeler, Backus and Thomas 63
Wheeler Mill 63
Whetstone Creek Mill 164
whisky rye 64
White, Norman 61
White & Corbin 58
White & Gale 165
white lead 170
White Lead Company 170
White Mountain Pulp and Paper Co. 164

White River Mill 165
Whiteman, William S. 109, 114, 172
Whiteman, William, III 114
White's Creek Mill 114
Whiting, William 139, 145
Whiting Mill No. 1 *135*, 139, *147*
Whiting Mill No. 2 145
Whiting Paper Co. 146
Whitmore Manufacturing Co. 171
Whitney, James 55
Whyte, Alex **96**
Wilkins, Rice, & Kendall 51
Willamette River Pulp and Paper Company 194
Willcox, Henry B. 128
Willcox, J.M. & Co. 43
Willcox, James M. 41, 119–120, 132; death 126
Willcox, John 119
Willcox, Joseph 120
Willcox, Mark 119, 120
Willcox, Thomas 38, 119
Willcox Brothers 128
Willcox Mill 15
Willcox Paper Co. 128–131
Wilson & Steel 12
Winchester, Arthur 96
Winchester, Harvey 96
Winchester Arms Co. 129
Winnipesaukee Mill 163
Winslow, Rev. M. 75

Wisconsin Papermakers Association 188
Wiswall, A. C. & Co. 110
Wiswall, Augustus Curtis 110, 114, 163
Wiswall, Thomas & Co. 163
Wiswell, Henry 121
Wolf Pit Springs 123
Wood & Reddington 60
wood bark 160–161
wood grinder 156, **185**
wood-pulp 156; cheeses 158
Woodbridge Geography & Atlas 29
wove marks 99
wove mould 15–16, 38
wrapping paper 45, 171
Wright, Eleazer 121
Writing Paper Manufacturers of America 111–112, 154
Wurtzbach, Fredrich 158
Wyatt, J. 19
Wyatt, S. 110

Yankee paper machine 188
Yarmouth Mill 161
Yarnall, Isaac 109
Yeatman, H.J. 110
Young, Brigham 189

Zanesville City Times 105
Zellerbach, A. & Sons 195

www.ingramcontent.com/pod-product-compliance
Ingram Content Group UK Ltd.
Pitfield, Milton Keynes, MK11 3LW, UK
UKHW050542150426
5217IPUK00026B/2035